MONOGRAPHS ON NUMERICAL ANALYSIS

General Editors

E. T. GOODWIN, L. FOX

D1806155

MONOGRAPHS ON NUMERICAL ANALYSIS

Already published

THE NUMERICAL SOLUTION OF TWO-POINT
BOUNDARY PROBLEMS IN ORDINARY DIFFERENTIAL EQUATIONS

By L. FOX   1957

THE ALGEBRAIC EIGENVALUE PROBLEM

By J. H. WILKINSON   1964

AN INTRODUCTION TO
NUMERICAL
LINEAR ALGEBRA

BY

L. FOX, M.A., D.Sc.

DIRECTOR, UNIVERSITY COMPUTING LABORATORY, AND
PROFESSOR OF NUMERICAL ANALYSIS,
OXFORD

CLARENDON PRESS · OXFORD

Oxford University Press, Ely House, London W.1

GLASGOW   NEW YORK   TORONTO   MELBOURNE   WELLINGTON
CAPE TOWN   IBADAN   NAIROBI   DAR ES SALAAM   LUSAKA   ADDIS ABABA
DELHI   BOMBAY   CALCUTTA   MADRAS   KARACHI   LAHORE   DACCA
KUALA LUMPUR   SINGAPORE   HONG KONG   TOKYO

Casebound: ISBN 0 19 853402 7
Paperback: ISBN 0 19 853407 8

© *Oxford University Press*, 1964

FIRST PUBLISHED 1964
REPRINTED LITHOGRAPHICALLY IN GREAT BRITAIN
FROM CORRECTED SHEETS OF THE FIRST EDITION
BY WILLIAM CLOWES & SONS, LIMITED
LONDON, BECCLES AND COLCHESTER
1967, 1973

Preface

This series of Monographs on Numerical Analysis owes its existence to the late Professor D. R. Hartree who, defying the walrus, thought that the time had come to talk of one thing at a time, at least in this field. Indeed the various areas of Numerical Analysis have expanded so rapidly that it is now virtually impossible to write a single book which gives more than a very elementary introduction in all fields. We even need a variety of books on each single topic, as in other branches of science and mathematics, to meet the various requirements of the undergraduate, the research student, and those who spend their working life in solving numerical problems in specific contexts.

Numerical analysis was introduced in 1959 into the Oxford undergraduate mathematical syllabus, and it seemed to me preferable to talk about numerical linear algebra in the first place and to leave for subsequent courses the theory and practice of approximation and its applications to the solution of differential and integral equations in which, of course, linear algebra plays a large part. The material of this book is therefore based on this first set of lectures, and I generally cover some two-thirds of it in about 28 lectures, treating less thoroughly Chapters 4, 5 and 7 and the later parts of Chapters 8, 10 and 11.

I had considerable difficulty with Chapter 2. Instead of introducing linear equations via vector spaces, linear transformations and matrices I started with linear equations and tried to show how the algebra of matrix manipulation 'hangs together' and simplifies not only our notation but the proofs of our numerical operations. Though this does not give a beautiful mathematical theory I made the deliberate choice for three reasons. First, the Oxford undergraduates learn the mathematical theory from other lecturers. Second, I think that with that theory they do not easily acquire the facility with matrix manipulation which they will need in numerical work and even for the study of more advanced theoretical texts. Third, the Director of a Computing Laboratory must also consider the engineers and scientists who use the computer to solve numerical problems, and in my experience these workers have some antipathy to and even fear of words like 'space', 'rank' and even 'matrix' which, they feel, represent strange and impractical mathematical abstractions. And yet the use and manipulation of matrices, in the elementary form given here and which is sufficient for many practical purposes, is really very easy! This is true also of 'norms', introduced in an elementary way in this Chapter and which are so valuable for measuring the convergence of series and iterative processes.

And of course mathematical rank and matrix singularity have less importance in practical work. Here the data is rarely exact, and instead of a matrix A we have to consider the matrix $A+\delta A$, where we may know only upper and lower bounds to the elements of δA. Even if A is exact (in a 'mathematical' problem) our numerical methods involve arithmetic which is rarely exact. We must face the fact, and numerical analysts do not apologise for it, that the question of error analysis is profoundly important and that for this purpose we must investigate very closely the details of the arithmetic. The present tendency of error analysis, in all branches of numerical analysis, largely refrains from following through the effects on the solution of each individual error, but accepts these errors and tries to determine what problem we have actually solved. Our methods are then evaluated according to our ability to perform the appropriate analysis and to the size of the upper bounds of the perturbations.

This is considered in Chapter 6. Chapters 3 and 4 study various direct processes for solving linear equations based on elimination and triangular decomposition, and the close relations which exist between the various methods. Many of these, together with the orthogonalisation methods of Chapter 5, might well be discarded for practical purposes, but they have some mathematical interest and considerable literature, and I thought it desirable to collect in one place a summary of the relevant facts. In Chapter 7 I consider very briefly the work and storage requirements for some of the methods, with particular reference to automatic digital computers. Chapter 8 gives an introduction to a class of iterative methods for solving linear equations, whose recent developments, particularly for the large sparse matrices relevant to elliptic differential equations, have been brilliantly expounded by R. S. Varga in his 'Matrix Iterative Analysis' (Prentice-Hall, 1962).

Chapters 9 and 10 discuss the determination of the latent roots and vectors of general matrices, both by iterative methods and by the search for similarity transformations of various kinds, associated with the names of Jacobi, Givens, Householder, Lanczos, Rutishauser,

Francis and others. These have been further developed by J. H. Wilkinson, together with a systematic error analysis whose general features are indicated in Chapter 11, and which are described in comprehensive detail in his forthcoming 'The algebraic eigenvalue problem' (Oxford, this series, in press).

There are, of course, some omissions which will displease many students and teachers. For example I have said very little about computing machines, and have not made detailed distinction between the error analysis of 'fixed-point' and 'floating-point' machine arithmetic. Coding and programming are mentioned only briefly in the introductory chapter, with no details of languages like FORTRAN or ALGOL. My personal opinion is that these things, while relatively easy to learn and master, take much space to describe and the mathematical undergraduate needs essentially the principles expressed in a book which is reasonably short and correspondingly inexpensive. Those who teach ALGOL, moreover can easily use as exercises the algorithms of this book, all of which, I hope, are expressed unambiguously in the language of English and of standard mathematical notation. In a few cases the algorithmic language would simplify the description, and in these cases it is interesting to note that hand computation is relatively tedious; the method of § 30 in Chapter 4 is one example of this. There is, of course, some advantage in using digital computers at the undergraduate stage, and I hope to introduce this at Oxford when we acquire facilities which are not completely saturated by the demands of research.

With regard to notation I have used the prime rather than the super-script T to denote matrix transposition, and usually capital letters denote matrices and lower-case letters denote vectors, in ordinary italic type. Exceptions are the row or column vectors of a matrix, usually denoted respectively by $R_s(A)$ and $C_s(A)$, and I fear that consistency lapses for the residual vector, sometimes called r and sometimes R, I suspect for personal historical reasons. All my matrices, incidentally, have distinct latent roots (which word I use consistently instead of eigenvalues) and consequently a full set of independent latent vectors, with obvious simplifications in the theory and no considerable restriction in practice.

Most of the material is already published in learned journals, and most books on numerical analysis have some account of parts of it. Few similar books, however, are available in English. Predecessors not mentioned in the text include P. S. Dwyer's 'Linear Computations'

(Wiley, 1951), written before the advent of the digital computer and the advances in error analysis, and E. Bodewig's 'Matrix Calculus' (North Holland Publishing Company, Amsterdam, 1959) which has more and deeper theoretical treatment but perhaps fewer practical details. More advanced books include those of Varga, the imminent treatise of Wilkinson, and the latter's just published 'Rounding errors in algebraic processes' (HMSO, 1963), and I hope that my readers will be able subsequently to benefit more easily from these learned works.

It is a pleasure to record my debt to Dr. E. T. Goodwin, who read the proofs and made several valuable suggestions; to Professor A. H. Taub, who invited me to Illinois for a sabbatical semester in which I found time to write several chapters; to the Clarendon Press, who made a special and successful effort to produce this book in time for the 1964 examinations; and above all to Dr. J. H. Wilkinson, who read all the first draft, made important criticisms and suggestions, and from whom I have learnt much.

L. Fox
Oxford, January 1964

Contents

NOTE. † indicates that there is a further mention of the section in *Additional notes and bibliography*, given at the end of each Chapter.

1

Introduction

Numerical analysis

1. THIS book is concerned with topics in the field of linear algebra, in particular with the solution of linear equations and the inversion of matrices, and the determination of the latent roots and vectors of matrices. Before embarking on our exposition it is desirable to make some introductory remarks on the nature and general aims of numerical analysis, and on the computing equipment which will enable us, without undue fatigue and in reasonable time, to obtain numerical answers to our problems.

The numerical answer is our aim. The roots of the quadratic equation $x^2 + 2bx + c = 0$ are

$$x_1, x_2 = -b \pm (b^2 - c)^{\frac{1}{2}}, \tag{1}$$

but we are concerned with the evaluation of x_1 and x_2 for given numerical values of b and c. We might, as here, have a 'closed expression' for the answer, in which we merely have to substitute the given numbers, the data of the problem. More commonly there is no simple formula, but there may be an algorithm, represented by an ordered sequence of numerical operations, additions, subtractions, multiplications and divisions, which is known to give the required result. The construction of such algorithms is one of the research activities of numerical analysis.

2. But we must be careful with the phrase 'required result'. An answer is rarely obtainable exactly as an integer or the ratio of two integers. Even for a simple problem like that represented by equation (1) we shall have to compute an irrational number, or non-terminating decimal, for most values of b and c. For example if $b = 1$ and $c = -1$ the required roots are $-1 \pm \sqrt{2}$, and if we want this as a single number we have to specify in advance the precision of our result, that is the number of figures which we should like to have correct. In the decimal scale the number $\sqrt{2}$ is $1 \cdot 41421356...$, and if we specify a precision of p decimals we have to *round* the number appropriately, and in such a way that the error committed is as small as possible.

To do this we *truncate* the number to the precision required, increasing by unity the last digit retained if the first neglected digit is

5, 6, 7, 8 or 9. We thereby ensure that the maximum error committed is not more than five units in the first neglected place, or half a unit in the last figure given, or 0.5×10^{-p}. To three decimals $\sqrt{2} = 1.414$, to seven decimals it is 1.4142136, and so on.

3. Even if the computation can in theory be performed with exact integers, moreover, we shall find that our computing machine cannot usually handle the large numbers involved in the arithmetic processes. For example, if we are solving simultaneous linear algebraic equations in n unknowns, in which the coefficients and right-hand sides are given as p-figure integers, an exact process could give the results as the ratios of two integers, both of which would contain np digits. In the more practicable methods which we discuss in this book the integers might contain $p \times 2^{n-1}$ digits. If n is 20, which is by no means large in practical problems, and p is say four, this number is of the order of 2×10^6, and no computing machine can store numbers of this size without complicating prohibitively the task of 'programming' and increasing prohibitively the time of operation.

If the coefficients are given as rational fractions, such as $\frac{1}{7}$, or as irrational numbers like e, π, $\sqrt{2}$ or sin 0.72 (radians), we shall have to round them to a given number of digits. The problem we are solving is then not quite the original problem, and one of our tasks will be to decide how many figures we need to keep in the original data, and also in the process of the computation, to obtain the required precision in the results.

4. Problems in which the data are known exactly, either as integers, rational or irrational numbers, I call *mathematical*. The author of such a problem has a perfect right to ask for any degree of precision which he needs for his purpose. On the other hand most problems with a scientific context will involve data obtained as a result of measurement, in some degree inaccurate, and our task now is to decide the *worthwhile* precision of the answers. Such problems are called *physical*, and it is self-deceptive to quote as answers more digits than those which remain unchanged however the data is varied within its limits of 'tolerance'. The 'required result' now becomes the 'meaningful result', and our methods should decide this for us.

As a trivial example, if we are asked to compute sin x, and a measurement of x gives the value $x = \frac{1}{4}\pi \pm 0.005$, we see that there is a range of values of the answer, from about 0.7036 to 0.7106, and a quoted result of 0.7071 has a possible error of ± 0.0035. It would clearly be stupid to quote more than three decimals in the result. We shall see later that the precision of the answer compared with that of the data varies considerably with the problem, and in complicated algorithms our work of determining this might be formidable and challenging.

5. We note also that we would often prefer to use an algorithm, rather than evaluate a closed solution, even when the latter exists. In the field of differential equations, for example, the solution of the first-order equation

$$\frac{dy}{dx} - \frac{2y}{1-x^4} = 0 \tag{2}$$

is

$$y = A\left(\frac{1+x}{1-x}\right)^{\frac{1}{2}} e^{\tan^{-1}x}, \tag{3}$$

where A is an arbitrary constant to be fixed by the specification of y for a particular value of x.

Now this is a useful formula for the computation of y for one or two particular values of x. But it is quite common to want a *graph*, or preferably a *table* of values of y for a set of (usually) equidistant values of x over a lengthy range. The calculation of the expression (3) is then not trivial, involving the evaluations of a square root, an inverse tangent, and an exponential function, in addition to one division and several multiplications. In the computation of these elementary functions, moreover, we shall either have to use some form of series or to interpolate in mathematical tables, and the whole operation is somewhat lengthy. We have numerical methods for solving such problems, though they belong to a field outside our present interest, which perform much less arithmetic and which produce successive values in the table without ever knowing the closed solution (3).

6. The closed solution, of course, is extremely valuable for many purposes, but unfortunately it can rarely be obtained in terms of the so-called 'elementary' functions. For example an apparently innocent change in (2), to the form

$$\frac{dy}{dx} - \frac{2y}{1-x^4} = x, \tag{4}$$

produces the more formidable-looking solution

$$y = \left(\frac{1+x}{1-x}\right)^{\frac{1}{2}} e^{\tan^{-1}x}\left(\int x\left(\frac{1-x}{1+x}\right)^{\frac{1}{2}} e^{\tan^{-1}x}\,dx + A\right). \tag{5}$$

This can hardly be called a solution at all, since we have no analytical methods for evaluating the indefinite integral in terms of elementary functions, and some *numerical* process has to be used for this purpose. We might just as well use our algorithmic numerical method for the equation (4) without recourse to (5), and in fact the extra numerical work in (4) compared with that of (2) is almost negligible.

7. Again, however, we should not ignore the possibility of obtaining a closed solution, and it is very important that we should understand the mathematics and mathematical methods for our problems, as well as the numerical analysis and possible algorithms. In particular we should try to decide in advance whether our given problem really has a solution, that is whether there is an *existence theorem* for it. With the development of automatic computing machines the mathematical analysis is increasingly important, and it should never be thought that the machine will do the mathematics for us.

Our algorithm may sometimes decide for us whether or not our problem has a solution, or at least a unique solution. For example it is usually the case that a set of simultaneous linear algebraic equations has a unique solution when the number of equations is equal to the number of unknowns. But it is clear that the equations

$$\left.\begin{aligned} x+y &= 3\\ 2x+2y &= 6 \end{aligned}\right\} \tag{6}$$

do not define a unique solution, the second equation being effectively a restatement of the first. If in the second of (6) the right-hand side were a number other than six it is clear, moreover, that the equations would have no solution at all. This is less obvious with the equations

$$\left.\begin{aligned} x+y+z &= \alpha\\ x-y-z &= \beta\\ 2x+4y+4z &= \gamma \end{aligned}\right\}, \tag{7}$$

which have no unique solution for any α, β and γ, and no solution at all unless $\gamma = 3\alpha - \beta$.

With many equations, and with more digits in the coefficients, we may have some trouble in this context, and the necessity for rounding may produce a solution from our computing machine when in fact no solution exists. We shall give examples of this in a later chapter and show how our algorithm can help to decide the questions. In other fields, notably in the solution of differential equations, our algorithm may be less valuable in the determination of existence, and mathematical analysis is essential.

8. Summarizing, we can say that numerical analysis is concerned with the production of numerical solutions to scientific and mathematical problems. Our aim is to find methods which are economic in time, which produce the results to the accuracy requested in mathematical problems, and which tell us how many figures are worth quoting in physical problems. To the numerical analysis we should add any mathematical knowledge we have or can find about the existence of solutions, and in some sense our methods, like those of mathematics itself, should be elegant!

As a rather trivial example of elegance we might consider the formula (1) for the solution of quadratic equations. If $b^2 - c$ is reasonably small, and we compute its square root to a given number of decimal places, the formula gives roughly the same number of correct digits in both roots. But if c is small, so that $(b^2 - c)^{\frac{1}{2}} = b + \epsilon$, where ϵ is small, then $x_2 = -2b - \epsilon$, $x_1 = \epsilon$, and x_2 is given accurately with many more digits than x_1. To avoid computing the square root to more figures we use our mathematics to note that $x_1 x_2 = c$, so that $x_1 = c/x_2$ and can be computed from this formula with a relative accuracy similar to that of x_2.

The loss of significant digits in subtracting large numbers is a common phenomenon, and we use all possible methods to avoid or mitigate the consequences thereof.

Computer arithmetic

9. There are two methods in common use for operating with numbers in a computing machine. In both cases the numbers are stored in registers of fixed length, so that we can retain only p digits say, in any given number, and a number containing more than p digits must be truncated or, with extra effort, stored in two or more such registers. In what follows we assume that we are working in the common decimal system.

With 'single-length' arithmetic, with p digits, we have either the *fixed-point* or the *floating-point* method of operation. In the fixed-point method it is customary to limit the size of numbers which may occur to the range -1 to $+1$, and any number outside this range must be scaled appropriately by dividing by a power of 10. The programmer must take definite steps to keep track of these scale factors so that the correct result can finally be obtained.

Since our machine can only store digits we must turn the positive and negative signs into quasi-digital form, and this we do with the

convention that all positive numbers have their first digit zero. The
decimal point will normally be thought to follow this digit, so that in
a four-digit register we can effectively store three figures. The number
0·924 will actually appear in that form, and the largest positive
number we can store is 0·999, the integer after the decimal point
being 10^p-1 in a $(p+1)$ register machine.

For a negative number x we store the complement $10^{p+1}-|x|$, so that
the first digit is always 9, and the number $-0·924$ appears as 9·076.
All negative numbers have nine as the first digit, and the largest
negative number we can store is 9·000, which is -1 in the 'signed'
convention, the 'fractional part' representing the integer 10^p.

It is easy to see that addition and subtraction, using the complements
of negative numbers, will always give the correct answers in the 'signed'
convention provided that the result is in the allowed range. In fact in a
sequence of such operations the intermediate results are allowed to exceed the range. For example $0·126-0·125 = 0·126+9·875 = 10·001$.
The first digit is 'lost' and we are left with 0·001, the true result. Again,
$0·125-0·126 = 0·125+9·874 = 9·999 = -0·001$, again correct. The
sum $0·986+0·125 = 1·111$ cannot be allowed, however, and we would
have to store this in the rounded form $0·111 \times 10^1$, remembering the
power of 10 involved. But

$$0·986+0·125-0·389 = 0·986+0·125+9·611 = 10·722 = 0·722,$$

and this is correct.

When we multiply together two permissible numbers the result is
certain to be within range. But the exact product of two numbers of p
digits has $2p$ digits, and we need two registers to store it exactly, a
so called 'double-length' accumulator. If we have to round it to single
length we commit an error of maximum amount $0·5 \times 10^{-p}$. The
division a/b is out of range if $a > b$, but otherwise we can perform the
calculation. In a 'single-length' register the stored result will have a
maximum error of $0·5 \times 10^{-p}$, unless the resulting decimal number
terminates in at most p digits.

10. In the floating-point system our numbers can be of almost any
size, and we store them in the form $10^a \times b$, making space in our
register for both a and b. This representation is not unique, but we
standardize by choosing b in the range $0·1 < |b| < 1$. For example the
number 1562 is stored as $0·1562 \times 10^4$, 0·001562 is given as $0·1562 \times
10^{-2}$. Both a and b can be negative, and are stored with the signed

convention, though a is always an integer and we can forget about the decimal point in its register.

Here the user is not worried by scaling problems and the machine automatically keeps track of the relevant powers of ten. 'Overflow' of the accumulator is now almost solely restricted to the case of division by zero, and otherwise the size of allowable numbers is governed by the size of the register we allow for the representation of the exponent a. We shall mention some other relevant facts about arithmetic in the appropriate contexts.

Simple error analysis

11. The fixed-point and floating-point representations introduce the ideas of *decimal places* and *significant figures*. Both the numbers 0·9246 and 0·0002 have four decimal places and would be stored in this form in the fixed-point method. The first number, however, has four significant figures whereas the second has only one significant figure. The point about the word 'significant' is that, if these numbers were obtained as a result of rounding with a possible maximum error of half a unit in the last place retained, each has a possible *absolute* error of $\pm 0\cdot00005$, but the former has a much smaller *relative* error. It is correct to approximately one part in 20,000, while the number 0·0002 is correct only to one part in 4.

In the floating-point representation these numbers are stored respectively as $0\cdot9246 \times 10^0$ and $0\cdot2000 \times 10^{-3}$. Here the number of non-zero digits in the fractional part represents the number of significant figures present, the three zeros in the second example being inserted to fill up the register. If we had more *significant* information about this value, for example that it was 0·0002329..., or 0·0002000 where the last three zeros are known to be correct, we could store it in a floating-point form like $0\cdot2329 \times 10^{-3}$ with a small relative error, whereas the rounded fixed-point number 0·0002 has a small absolute error but a large relative error. This, incidentally, does not imply that the floating-point representation is superior. There are many factors involved, some of which we shall mention later. We note immediately, however, that in an addition like $0\cdot9246 \times 10^0 + 0\cdot2329 \times 10^{-3}$ we have first to express the smaller number in the rounded form $0\cdot0002 \times 10^0$ in order to add it to the first, and we have had to discard its last three digits.

12. We shall need rules for assessing both types of error in simple operations, so that we can extend them to complicated situations.

Consider first the case of absolute error. If x is the true value, and $x \pm \delta x$ an approximation, the absolute error of $x \pm \delta x$ is just δx. In general, for instance after rounding, the error will normally have equal possibility of being positive or negative.

We can then assert quite obviously that the maximum absolute error in a sequence of additions or subtractions

$$x = \pm a \pm b \pm c \pm d \ldots \tag{8}$$

is just the *arithmetic* sum of the individual absolute errors, given by

$$|\delta x| = |\delta a| + |\delta b| + \ldots . \tag{9}$$

For a product ab we actually form $(a \pm \delta a)(b \pm \delta b)$, and the absolute error is

$$|b\delta a| + |a\delta b| + |\delta a \delta b|, \tag{10}$$

the last term usually being negligible, a quantity of 'second order', in relation to the others.

13. In fact we can use the differential calculus and say that, if

$$y = f(x_1, x_2, \ldots), \tag{11}$$

and $x_1, x_2, \ldots$ have absolute errors $|\delta x_1|, |\delta x_2|, \ldots$, then that of y is

$$|\delta y| = \left| \delta x_1 \frac{\partial f}{\partial x_1} \right| + \left| \delta x_2 \frac{\partial f}{\partial x_2} \right| + \ldots, \tag{12}$$

provided that the individual absolute errors are sufficiently small. For example, if $y = \sin x$, then $\delta y = \cos x \, \delta x$, and the absolute error in y is not greater than that in x.

Again, if $y = x^p$, we have

$$|\delta y| = px^{p-1} |\delta x|, \tag{13}$$

and the ratio $|\delta y/\delta x|$ will depend both on x and on p. If p and x both exceed unity the error in y is greater than that in x, but if $x > 1$ and $p < 1$, so that we are taking a fractional power, then $|\delta y| < |\delta x|$. The statement in many books that we cannot get a result with more correct figures than are contained in the data is clearly false. For example if

$$y = 2^{0 \cdot 01}, \tag{14}$$

and all we know about the '2' is that it is correctly rounded, we can certainly quote $y = 1 \cdot 007$ with a maximum absolute error of $0 \cdot 003$, or maximum relative error of 1 in 300.

14. We shall in fact more often be concerned with relative error, the dimensionless quantity $|\delta x/x|$. The relative error of a sum or difference has no simple expression, but corresponding to (8) and (9) we have the rule that if

$$x = a^{\pm 1}b^{\pm 1}c^{\pm 1}..., \tag{15}$$

then

$$\left|\frac{\delta x}{x}\right| = \left|\frac{\delta a}{a}\right| + \left|\frac{\delta b}{b}\right| + ..., \tag{16}$$

that is the maximum relative error of the result is the sum of the individual relative errors. This is proved immediately by taking the logarithmic derivative of (15).

This result will give us valuable information about the number of meaningful figures in the number x derived from an operation like (15). For example, if

$$x = \frac{0 \cdot 833 \times 22 \cdot 5}{0 \cdot 225}, \tag{17}$$

and all we know about the factors is that they are correctly rounded, to how many digits can we reasonably quote the result? From (16) we have

$$\left|\frac{\delta x}{x}\right| = \frac{0 \cdot 0005}{0 \cdot 833} + \frac{0 \cdot 05}{22 \cdot 5} + \frac{0 \cdot 0005}{0 \cdot 225} = 0 \cdot 0050 \tag{18}$$

to sufficient accuracy. The error in x is therefore one part in two hundred. From (17) our estimate of x is 83·3, and this therefore has a possible error of $\pm 0 \cdot 4$, and only two significant figures of x are worthwhile.

Computing machines, programming and coding

†**15.** There are about five steps in any computational task, though some of them may not always be needed. They are as follows:

(i) Expressing the scientific problem as a mathematical problem.
(ii) Finding the 'best' numerical method for solving the mathematical problem.
(iii) Expressing this method in algorithmic form, that is as a sequence of numerical operations, recordings, and so on.
(iv) Turning this sequence into the language of the machine.
(v) Performing the computation.

These items are in some sense independent of the nature of our computing equipment, but the latter will influence our choice in (ii), and to some extent the work of (iv).

Before about 1950 most computation was carried out on a desk machine which can perform arithmetic and store a few numbers. For example, if we want to compute ab, we can put a in the *setting* register, tap out the number b on the *multiplication* register, and obtain the result in the *product* register. The numbers a, b and ab are all visible—they are *stored* in the machine. In a long computation, however, in spite of various tricks that we can play, we shall have to record with pen and ink the results of many such intermediate calculations. We use an *auxiliary* storage medium, in this case the 'registers' on a sheet of paper.

The 'best' numerical method is then to some extent conditioned by our desire to avoid overmuch recording, which is tedious and error-provoking. On the other hand our auxiliary store is unlimited, and the point of this remark will become apparent later.

In item (iii) for our desk-machine work we write down, in considerable detail, the precise nature and order of the operations we wish to perform, and possibly present it to an assistant who will then perform item (v). In other words we give him a *programme of instructions*. The language we use in (iv) is the national tongue, with words and mathematical symbols. Our helper then operates the machine as in (v), records the intermediate results and produces the final answer, recorded on his sheet of paper.

16. The modern high-speed electronic digital computer (the machine) differs in several important respects, but our use of it has analogies with that of the desk machine, and we can use the same type of vocabulary. First, the machine has large storage capacity, its arithmetic speeds are very great, and intermediate calculations can be transferred to the registers of the machine rapidly and accurately. The registers in the machine are numbered, are given *addresses*, and we can ask the machine to put a number in a particular register, or to fetch it from that register, just as we used to ask our assistant to copy a number into a particular location on the computing sheet, or to take such a particular number and perform some numerical operation with it.

Second, the machine has an *arithmetic unit*, the operative part of which is the *accumulator*, corresponding directly to the product register of our desk machine. There is, however, one significant difference. In the desk machine our registers have a fixed length, that is we can store numbers to a certain precision, but it is perfectly possible, and indeed easier, to perform our arithmetic with fewer digits. In the electronic machine our registers can store a fixed number of digits (the *word length* of the machine) and there is no economy of effort in trying to use a smaller number of digits. We can, as in the desk machine, work with more digits at a cost in time and extra storage, but again the number of digits is an integral multiple of the word length, and we speak of work in *single*-length, *double*-length, and in general in *multiple*-length arithmetic.

17. Third, we can write our programme of instructions in a coded language which the machine can interpret, and we can store the programme inside the machine, effectively in registers reserved for this purpose, and a *control unit* in the machine will see to it that the instructions are performed, one after the other, in the desired sequence.

We can interrupt this sequence, and go along different paths in the programme, by asking the machine to perform a test, for example on the sign of a number, and to take its next instruction from one part or other, as appropriate, of the programme store. For example many computational tasks are repetitive. A common problem is the evaluation of a function $f(x)$ for a range of equidistant values of x. The number of instructions is kept to a minimum, and our store is not overloaded, if in such a case we have a single set of instructions, a *sub-routine*, for evaluating $f(x)$, and each time we enter this subroutine we change the value of x appropriately. But we want to do this only a finite number of times, so that we have to test each value of x to see whether we should stop this operation and proceed to other work. We might of course ask alternatively that the computation of $f(x)$ should cease whenever $f(x)$ reaches or exceeds a certain value, not knowing in advance the corresponding value of x. The test is similar, and of course this is exactly what our human operator does subconsciously when deciding to stop his paper and ink manipulations.

18. It has become customary to represent the sequence of operations in the form of a flow diagram, largely independent of any particular machine, which can subsequently be 'coded' for a particular machine. For example if we want to compute and print the value of $f(x)$ for $x = 0(1)50$ (which means start with $x = 0$, then proceed at intervals of unity until $x = 50$), our flow diagram would have the appearance of Fig. 1.

19. At the present time the coding depends on the particular machine, but the topic is developing fast and it is not impossible that by the time this is printed there will be general acceptance of a universal language. Here we trace very briefly the development of coding systems.

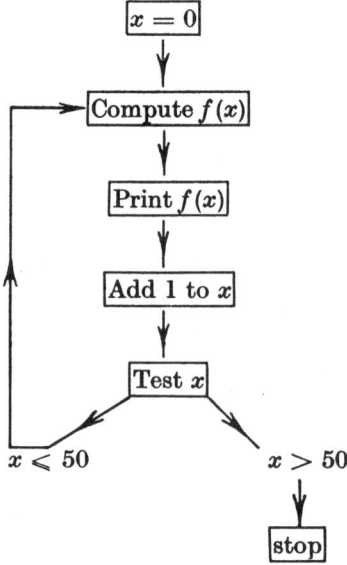

FIG. 1. This is an example with a single *loop* of instructions, and we will often have double or multiple loops of this kind. For example the flow diagram for computing $f(x, y)$ for $x = 0(1)50$, $y = 1(2)21$ would look like Fig. 2.

In the first place our machines prefer to work with the digits 0 and 1, in virtue of the two-state properties of electronic equipment. This means that the binary scale has replaced the decimal scale for computing purposes. All our numbers are represented as a sequence of zeros and ones, so that the binary integer 101 is $1 \times 2^2 + 0 \times 2^1 + 1 \times 2^0 = 5$ in the decimal scale, and the binary fraction 0·101 is

$$1 \times 2^{-1} + 0 \times 2^{-2} + 1 \times 2^{-3} = 0{\cdot}625$$

in the decimal scale. A given number needs more binary digits than decimal digits, by the factor $\log_2 10$, for representation to the same precision, so that our word lengths may be around 32 or 48 binary digits, roughly 10 or 15 decimal digits respectively.

Since the machine recognises only the digits 0 and 1 all our instructions are also represented by a string of binary digits, and the context in which such a 'word' appears, and the operation of the control unit, enables the machine to distinguish between a number and an instruction. Common basic instructions might be

(i) transfer to the accumulator the number stored in a specified address;

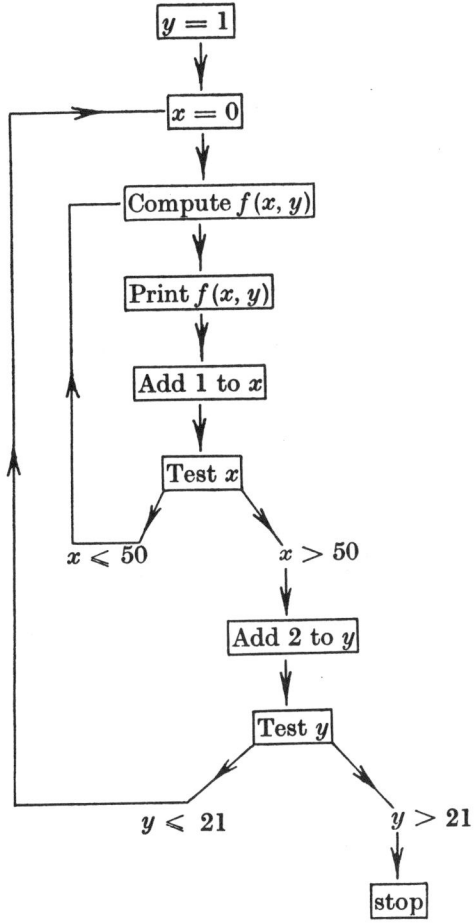

FIG. 2. In each case, of course, there may be a considerable number of instructions involved in the various boxes, particularly in the computations of $f(x)$, and somebody will have to give them in detail either just before or during the coding of the flow diagram into machine language.

    (ii) add into the accumulator (that is to what is already there) the number stored in a specified address;

    (iii) transfer the content of the accumulator to a specified address.

In this case the instruction word will include several blocks of digits, one giving the binary code number for the operation, one giving the binary number of the address, and possibly others for special purposes of a control nature.

Or our basic instruction may be more complicated, as for example 'add together the numbers from registers x and y and store the result in register z', for which four blocks of digits would be needed.

20. There are many variations and it is not our purpose to explore them in any detail. It is clear, however, that this type of coding is not easy to master and that there are many sources of blunder. The next step, roughly speaking, was to permit the use of decimal digits, or even letters and other symbols, and to associate this with a routine, living permanently in the store of the machine, which decoded the information it received into its own binary language. For example the three previous instructions might read

CA 149 ('clear accumulator', and put there the content of register 149);

Add 150 (add to the content of the accumulator the number in register 150);

Store 151 (transfer the result to register 151).

21. A further basic step is the development of Autocodes and similar languages. Important features of such languages are the ability to combine many arithmetic instructions into a single instruction, to use symbols like x, or even x_i, to refer to the contents of the various registers, and to use English statements like print, jump (to a particular numbered instruction) etc., when we want to see a result, to alter the sequence of operations, and so on. The three previous instructions would then be replaced by the single command $x = y + z$, where x, y and z refer to the contents of registers 151, 149 and 150 respectively. We can also write something like $x = \sin x$, and the content of register x will be replaced by the sine of itself.

The advantages of autocode language are that we need fewer instructions, our statements are mathematical and relatively easy to check, and the user of another machine can more easily understand our programme. Indeed universal languages are proposed and might well come into use, in which we might not only publish algorithms but also write our instructions. Each different machine will have its own translation routine, and though this is involved and the writing thereof is an expert process the user of the machine need not be concerned with it.

Finally, we note that we do not wish to submit our data to the machine, or to get it from the machine, in the unfamiliar binary form, and our translation routine includes instructions which convert our decimal numbers to binary on input, and the machine's binary numbers from binary to decimal on output. Moreover our translator will include many subroutines which we shall need to use very frequently for things like finding square roots, evaluating elementary functions and so on.

22. One further comment about the storage capacity is necessary in relation to the remark in § 15 about the unlimited capacity of the paper store of the desk-machine operator. Most modern machines have a 'high-speed' store which contains the instructions and data of the whole problem if these are small enough. For large problems more data and instructions may be contained in a 'backing store', and transferred at the appropriate time to the high-speed store. Such transfers are generally slow in relation to the other machine operations, and this may cause a significant time-difference in the solution of similar problems of slightly different size. For example the solution of n algebraic equations may be appreciably slower than that of $n-1$ such equations, and we often use ingenuity of various kinds to avoid too frequent transfers between the fast store and the backing store.

23. To complete this sketchy picture, we may remark that the input medium is commonly punched card or punched paper tape, the position of the holes in a row of card or tape representing a single digit or symbol in the appropriate code, and the output medium is either a punch, which produces card or tape which can be fed to a printer for the final answers, or fed back to the machine if we need a further, much slower, backing store. There may also be a fast printer which produces typed results directly from the machine.

The backing store might consist of tracks on magnetic tape or a magnetically-coated drum, and its capacity is usually much greater than that of the fast store. In particular the backing store might even contain instructions for the solution of complete problems, such as linear equations, and many such 'library routines' might be stored, at various sections of say a magnetic tape, and fed to the main store when required.

Checking
24. The necessity for checking our work cannot be overemphasized. Even if our method is good and will produce the required result there are many other sources of error. First, the data of the problem may be copied wrongly from some basic original source. Second, if the data is correct it may still be wrongly punched on the input cards or tape.

Third, our programme of instructions may be wrongly punched, either by omitting or changing numerical factors, or changing a plus sign into a minus sign, or in many other ways which will provide the machine with a problem which, though quite meaningful, is not the one we want to solve. Fourth, our machine may make blunders which we fail to detect, either in the arithmetic or in input or output. Finally, the results of the output may be transcribed wrongly for submission to the author of the problem.

We can do little about the first, though in some problems we can apply a type of 'sum check', making our machine add together all the data it has received and comparing this with the correct sum obtained from the earliest version of the data. This also checks our second source of error and the input part of the fourth. With regard to the third possibility it is often practicable to solve a particular case, with particularly simple data, for which the answer is known in advance. We must be careful here, however, that our choice of data checks all possibilities of error; the use of numbers like 0 and 1 is particularly dangerous. In other cases, generally in routines of a repetitive nature, we might use a desk machine to check the first step.

It is fortunate that, in the fourth class of error, our machines are less prone than the human desk-machine operator to make undetected errors. If the machine is making blunders it is usually obvious. The peripheral equipment (input and output) is however less safe and our final transcription might likewise be faulty. It pays to check our printed results, those that are transmitted to the client, by whatever means we have available, and these will vary with the problem.

It is a bore to talk about checking, and most workers find it such a bore to listen, and to act on the advice, that checking is unfortunately rarely practised. As a result I estimate that some 50 per cent of all computed results have bigger errors than the author of the problem will know about or admit. In the desk-machine era of Comrie and Peters, the great compilers of mathematical tables, the possibility of errors and the perpetual need for guarding against them were fully realised. They would be appalled by the self-confidence of their successors!

Additional notes

§ 15. The words 'programming' and 'programmer' are used almost indiscriminately and they mean different things to different people. They have developed from the less ambiguous word 'programme' which, at least originally, meant the set of coded instructions which was submitted to the machine with the data. I shall generally use 'programme' in this sense and avoid the other two altogether. Item (i) is the duty of the mathematician, often but not always an applied mathematician. Item (ii) is performed by a 'numerical analyst', often a mathematician with interests and experience in the field of 'numerical analysis'. Item (iii) might also be performed by a numerical analyst, though here the skills are less mathematical than logical, and item (iv) needs an expert on the particular machine available, with mathematical skills which are still less searching. The performance of the computation, by putting into the machine the appropriate tapes or cards, pressing the appropriate switches and carrying out any other necessarily manual acts, is the work of the 'machine operator', for whom intelligence rather than knowledge is the prime virtue.

Matrix Algebra

Introduction

1. THE solution of simultaneous linear algebraic equations and numerical operations with matrices, in particular the evaluation of their determinants, inverses, and latent roots and vectors, are included in the general field of 'Linear Algebra'. It is unnecessary, in a book on numerical methods, to stress the part which these problems play in practical computation, but experience in various computing services indicates that linear algebra is involved, wholly or in part, in about 75 per cent of all scientific problems.

Apart from problems involving a finite number of degrees of freedom, the continuous cases represented by ordinary or partial differential equations are commonly transformed, through the medium of the calculus of finite differences, into algebraic problems. Integral equations are often treated in a similar fashion. Approximations to non-linear problems are frequently corrected by a 'linearizing' process, thereby also entering the domain of linear algebra. 'Linear programming', a relatively new field concerned with the minimization of the cost and effort of relatively complicated operations, involves the solution of sets of linear equations subject to linear inequalities.

2. Linear algebra is therefore a wide field, and in this book we shall be concerned mainly with the problem of solving general sets of linear algebraic equations, inverting matrices, evaluating determinants and finding latent roots and vectors. The equations and matrices are general, in the sense that little attention is paid to the special sets arising, for example, from differential equations. Such problems are treated elsewhere in this series.

Linear equations. General considerations

3. A general set of m linear equations in n unknowns can be written in the form

$$\left. \begin{array}{l} a_{11}x_1 + a_{12}x_2 + \ldots + a_{1n}x_n = b_1 \\ a_{21}x_1 + a_{22}x_2 + \ldots + a_{2n}x_n = b_2 \\ \cdots\cdots\cdots\cdots\cdots\cdots\cdots\cdots\cdots \\ a_{m1}x_1 + a_{m2}x_2 + \ldots + a_{mn}x_n = b_m \end{array} \right\}. \tag{1}$$

The *unknowns* we seek to determine are the quantities x_1, x_2,..., x_n, the numbers a_{rs} are the *coefficients* in the equations, and b_1, b_2,..., b_m are the *right-hand sides* or *constant* terms.

In general there will be no unique solution x_1,..., x_n unless the number of equations is equal to that of the unknowns. We may have too few equations, as for example in the attempt to determine two unknowns from the single equation

$$2x_1 - 3x_2 = 1. \tag{2}$$

In this case there is an infinity of solutions, since either x_1 or x_2 may be chosen arbitrarily and the other obtained by direct substitution in (2). Or we may have too many equations, as for example in the calculation of x_1 and x_2 from the three equations

$$\left. \begin{aligned} 2x_1 - 3x_2 &= 1 \\ x_1 + x_2 &= 0 \\ 5x_1 - x_2 &= 3 \end{aligned} \right\}. \tag{3}$$

In this case there is in general no solution at all. For example the results $x_1 = \frac{1}{5}$, $x_2 = -\frac{1}{5}$ are the only possibilities for the satisfaction of the first two equations, and the third is then not satisfied. The equations

$$\left. \begin{aligned} 2x_1 - 3x_2 &= 1 \\ x_1 + x_2 &= 0 \\ 3x_1 - 2x_2 &= 1 \end{aligned} \right\} \tag{4}$$

are in fact all satisfied, uniquely, by $x_1 = \frac{1}{5}$, $x_2 = -\frac{1}{5}$, but this result can be obtained from any two equations of the set (4), the remaining equation being redundant, providing no additional information. It is, however, *consistent* with the other two (and this information might be 'useful') whereas in (3) the third equation gives *conflicting* information.

†4. It is apparent that the case 'more unknowns than equations' will generally give solutions but not unique solutions, whereas the case 'more equations than unknowns' gives solutions, which are then generally unique, only when the 'extra' equations are restatements of the previous equations. In (4), for example, the third equation is just the sum of the first two.

We are left with the case in which the number of unknowns is equal to the number of equations, and simple examples suggest that there is generally one and only one set of values of the unknowns which will satisfy all the equations. The single equation

$$a_{11}x_1 = b_1, \tag{5}$$

for example, has generally the unique solution $x_1 = b_1/a_{11}$. The exceptional cases are when $a_{11} = 0$, $b_1 \neq 0$, for which the solution is 'infinite' (does not 'exist'), and when $a_{11} = b_1 = 0$, so that x_1 is indeterminate.

Similarly the two equations

$$\left.\begin{aligned} a_{11}x_1 + a_{12}x_2 &= b_1 \\ a_{21}x_1 + a_{22}x_2 &= b_2 \end{aligned}\right\}, \tag{6}$$

in the two unknowns x_1 and x_2, can be reduced to a single equation in x_1 by taking a_{22} times the first equation and adding it to $-a_{12}$ times the second, giving

$$(a_{11}a_{22} - a_{21}a_{12})x_1 = (b_1a_{22} - b_2a_{12}), \tag{7}$$

and a similar process yields for x_2 the single equation

$$(a_{11}a_{22} - a_{21}a_{12})x_2 = (b_2a_{11} - b_1a_{21}). \tag{8}$$

The solution is generally unique, but does not exist if

$$a_{11}a_{22} - a_{21}a_{12} = 0,$$

unless $b_1a_{22} - b_2a_{12}$ (and therefore $b_2a_{11} - b_1a_{21}$) is also zero, in which case the solution is indeterminate.

The case of indeterminacy is equivalent to 'too little information', since if the three quantities mentioned vanish it is clear that the second equation is a restatement of the first. Similarly, if $a_{11}a_{22} - a_{21}a_{12}$ alone vanishes, the equations can be written as

$$\left.\begin{aligned} x_1 + kx_2 &= c_1 \\ x_1 + kx_2 &= c_2 \end{aligned}\right\}, \tag{9}$$

where $k = a_{12}/a_{11} = a_{22}/a_{21}$, $c_1 = b_1/a_{11}$, $c_2 = b_2/a_{21}$, and with $c_1 \neq c_2$ the equations give conflicting statements.

Homogeneous equations

†5. When the right-hand sides are zero the equations are said to be *homogeneous*. No unique solution is possible, since each unknown can be multiplied by the same constant without disturbing the satisfaction

of the equations. We might, however, determine uniquely the ratios of the unknowns. For example the equations

$$\left.\begin{array}{l} x_1+2x_2-x_3 = 0 \\ 2x_1-8x_2+x_3 = 0 \\ 4x_1-4x_2-x_3 = 0 \end{array}\right\} \qquad (10)$$

have the solution $x_1 = 2k$, $x_2 = k$, $x_3 = 4k$ for any value of k, so that the ratios $x_1/x_3 = \frac{1}{2}$, $x_2/x_3 = \frac{1}{4}$ are unique. These ratios, however, can be found from any two of the three equations (10), say from the first pair written in the form

$$\left.\begin{array}{l} (x_1/x_3)+2(x_2/x_3) = 1 \\ 2(x_1/x_3)-8(x_2/x_3) = -1 \end{array}\right\}. \qquad (11)$$

The third equation is therefore redundant, and for a possible solution it should not contradict the first two. In fact it must be possible to construct it from a linear combination of the first two. Here the third equation of (10) is formed by adding twice the first to the second.

The three equations

$$\left.\begin{array}{l} x_1+2x_2-x_3 = 0 \\ 2x_1-8x_2+x_3 = 0 \\ 4x_1-4x_2+x_3 = 0 \end{array}\right\} \qquad (12)$$

have no solution at all of this kind, since the only possibility for the first two is given by $x_1/x_3 = \frac{1}{2}$, $x_2/x_3 = \frac{1}{4}$, and this does not satisfy the third of (12). There is always, however, the so-called 'trivial' solution $x_1 = x_2 = x_3 = 0$.

Homogeneous equations are particularly important in the theory, and in practice it is often desired to calculate one or more values of a parameter, occurring in the coefficients, for which non-trivial solutions exist.

Linear equations and matrices

6. We find it convenient at this stage to introduce the elements of matrix algebra, partly to avoid cumbersome 'long-hand' expressions and partly to simplify the proofs of certain elementary results.

Referring to the linear algebraic equations (1) we say that the coefficients a_{rs} form a *matrix*, denoted by the letter† A, and written as

$$A = \begin{bmatrix} a_{11} & a_{12} & \cdots & a_{1n} \\ a_{21} & a_{22} & \cdots & a_{2n} \\ \cdots\cdots\cdots\cdots\cdots\cdots \\ a_{m1} & a_{m2} & \cdots & a_{mn} \end{bmatrix}. \tag{13}$$

We may call the quantities a_{rs} the *elements* of the matrix A. The elements $a_{r1}, a_{r2}, \ldots, a_{rn}$ form the rth *row* of A, and $a_{1s}, a_{2s}, \ldots, a_{ms}$ form the sth *column* of A. The matrix in (13) has m rows and n columns: it is an $(m \times n)$ matrix.

The unknowns x_1 to x_n form a *column matrix* or *vector* denoted by x, and expressed as

$$x = \begin{bmatrix} x_1 \\ x_2 \\ \cdot \\ \cdot \\ \cdot \\ x_n \end{bmatrix}. \tag{14}$$

The quantities x_1 to x_n are the *components* of the vector x. The constants b_1 to b_m similarly form a *vector* b of order m, and we can then write (1) in the form

$$\begin{bmatrix} a_{11} & a_{12} & \cdots & a_{1n} \\ a_{21} & a_{22} & \cdots & a_{2n} \\ \cdots\cdots\cdots\cdots\cdots\cdots \\ a_{m1} & a_{m2} & \cdots & a_{mn} \end{bmatrix} \begin{bmatrix} x_1 \\ x_2 \\ \cdot \\ \cdot \\ \cdot \\ x_n \end{bmatrix} = \begin{bmatrix} b_1 \\ b_2 \\ \cdot \\ \cdot \\ \cdot \\ b_m \end{bmatrix}, \tag{15}$$

or, more concisely, as

$$Ax = b. \tag{16}$$

Matrix addition and multiplication

7. In many respects it is convenient to regard a matrix as an operator, so that (16) might be interpreted as 'the operation on the vector x by the matrix A produces the vector b'. We shall, however, refer to Ax as a matrix-by-vector 'multiplication', the result of which is another vector b. Similarly matrices may be multiplied together, or

---

† We shall almost always use capital letters to denote matrices, whose elements are represented by the same lower-case letter with double suffixes. Vectors are almost always denoted by lower-case letters and their elements have a single suffix.

added or subtracted, to produce another matrix, whenever such operations are permitted by the rules of matrix algebra.

Since equations (15) and (16) are both equivalent to the set of equations (1), it follows that they represent a set of linear algebraic equations, of which the rth is

$$a_{r1}x_1 + a_{r2}x_2 + \ldots + a_{rn}x_n = b_r, \tag{17}$$

and is obtained by summing products of corresponding terms in the rth row of A and the column x, and equating the result to the rth element of b. We call the left-hand side a *scalar* product, and may write (17) in the form

$$R_r(A)C(x) = b_r, \tag{18}$$

that is 'rth row of matrix A multiplied by column x = rth element of column b'.

We note that matrix-by-vector multiplication conflicts with the 'rules' unless the number of elements in the vector x is equal to the number of columns in the matrix A, for otherwise no unambiguous meaning can be given to the scalar product. The operation is possible only if the matrix and vector are *conformable* in this sense.

8. We turn now to consider other possible elementary matrix operations. If each element of b in equations (1) is multiplied by a constant k, we write the resulting vector as kb. The equations have the same solution, that is the unknowns x_r are unchanged, only if every element a_{rs} is multiplied by k. This gives the rule for multiplication of a matrix, or a vector, by a constant: *every element of the matrix or vector is multiplied by that constant.*

Suppose next that in addition to (16) we have a second set of equations represented by the matrix equation

$$Bx = c, \tag{19}$$

and we add together corresponding equations of the sets (16) and (19). The result can clearly be expressed in the form

$$Px = q, \tag{20}$$

where $P = A + B$, $q = b + c$, and the *rule for addition* is that

$$p_{rs} = a_{rs} + b_{rs}, \qquad q_r = b_r + c_r, \tag{21}$$

that is all corresponding elements are added to form the resulting matrix and vector. The addition is possible without ambiguity only if the matrices A and B have the same 'shape', that is are both $(m \times n)$ matrices, and the vectors b and c have the same number m of elements.

Subtraction of matrices and vectors is performed in a similar way. We note in particular that the operation $A - A$ gives a matrix, of order $(m \times n)$, all of whose elements are zero. This is the *null matrix* (denoted by O). Similarly $b - b$ gives a *null vector* (denoted by o) of order m. As a corollary the equation $A = B$ implies that $a_{rs} = b_{rs}$ for all r, s, that is $A - B$ is the null matrix.

9. Multiplication of two matrices is a rather more complicated operation. It would arise, for example, if we made in (16) the transformation $x = By$, so that we would like to write

$$Ax = ABy = b, \tag{22}$$

and give meaning to the 'product' AB.

From our previous work it is clear that the operation $x = By$ is possible only if B has the same number of rows as the number of elements in x. The number of elements in y is then equal to the number of columns in B. Mathematically, we have

$$\left. \begin{aligned} A = (m \times n), \quad x = (n \times 1), \quad B = (n \times p) \\ y = (p \times 1), \quad b = (m \times 1) \end{aligned} \right\}, \tag{23}$$

where we use for a vector of order n the alternative notion of *column-matrix*, that is a matrix of n rows and 1 column.

The product AB is clearly a matrix, since we can carry out the algebraic long-hand operations involved and produce a set of linear equations for the components of y. We also deduce from the equation $ABy = b$ that AB has m rows, the same number as b, and p columns, the same as the number of rows of y, and the result can be exhibited pictorially as

$$\tag{24}$$

and

$$\tag{25}$$

The elements of C clearly involve those of A and B, and simple experiments show that the rule is given by the equation

$$c_{rs} = R_r(A)C_s(B), \tag{26}$$

that is the (r, s) element of the matrix represented by the product AB is equal to the scalar product of the rth row of A and the sth column of B. The products exist only if the number of columns of A is the same as the number of rows of B, so that (23) is verified. *Conformable* matrices A $(m \times n)$ and B $(n \times p)$ give rise to the result $AB = C$, where C is $(m \times p)$.

The result (26) is easily verified by considering the equations

$$\left. \begin{aligned} a_{11}x_1 + a_{12}x_2 &= b_1 \\ a_{21}x_1 + a_{22}x_2 &= b_2 \end{aligned} \right\}, \tag{27}$$

and the transformation

$$\left. \begin{aligned} x_1 &= b_{11}y_1 + b_{12}y_2 + b_{13}y_3 \\ x_2 &= b_{21}y_1 + b_{22}y_2 + b_{23}y_3 \end{aligned} \right\}. \tag{28}$$

Direct substitution yields the equations

$$\left. \begin{aligned} (a_{11}b_{11} + a_{12}b_{21})y_1 + (a_{11}b_{12} + a_{12}b_{22})y_2 + (a_{11}b_{13} + a_{12}b_{23})y_3 &= b_1 \\ (a_{21}b_{11} + a_{22}b_{21})y_1 + (a_{21}b_{12} + a_{22}b_{22})y_2 + (a_{21}b_{13} + a_{22}b_{23})y_3 &= b_2 \end{aligned} \right\}, \tag{29}$$

so that $Cy = b$, with C obtained from A and B by rule (26).

10. We recall that matrices are best regarded as operators, and the expression ABx means 'operate on vector x with matrix B, and then on the resulting vector with matrix A'. The *order* of the operations may be important, a fact of frequent occurrence in other branches of mathematics. For example in the differential calculus, if $D_1 = x\, d/dx$ and $D_2 = x^2\, d/dx$, then $D_1 D_2 y = 2x^2\, dy/dx + x^3\, d^2y/dx^2$, while $D_2 D_1 y = x^2\, dy/dx + x^3\, d^2y/dx^2$.

Similarly, in general the operations ABx and BAx will not produce the same result. Indeed AB might have meaning, when for example A is $(m \times n)$, B is $(n \times p)$, whereas BA might not, since here we lack conformability except in the special case where $p = m$. In this case the product AB gives a matrix of order $(m \times m)$, that is a *square* matrix of order m, while BA gives a square matrix of order n. These cannot possibly be the same unless $m = n$, that is unless both A and B are square matrices of the same order.

Even in this case, however, it is generally true that $AB \neq BA$. For example

$$\begin{bmatrix} 2 & 1 \\ 1 & 2 \end{bmatrix}\begin{bmatrix} 1 & 3 \\ 3 & 2 \end{bmatrix} = \begin{bmatrix} 5 & 8 \\ 7 & 7 \end{bmatrix}, \tag{30}$$

while

$$\begin{bmatrix} 1 & 3 \\ 3 & 2 \end{bmatrix}\begin{bmatrix} 2 & 1 \\ 1 & 2 \end{bmatrix} = \begin{bmatrix} 5 & 7 \\ 8 & 7 \end{bmatrix}, \tag{31}$$

and in general, for the respective products AB and BA of square matrices of order n, we have

$$\begin{bmatrix} R_1(A)C_1(B) & R_1(A)C_2(B) & \dots & R_1(A)C_n(B) \\ R_2(A)C_1(B) & R_2(A)C_2(B) & \dots & R_2(A)C_n(B) \\ \hdotsfor{4} \\ R_n(A)C_1(B) & R_n(A)C_2(B) & \dots & R_n(A)C_n(B) \end{bmatrix} \neq$$

$$\neq \begin{bmatrix} R_1(B)C_1(A) & R_1(B)C_2(A) & \dots & R_1(B)C_n(A) \\ R_2(B)C_1(A) & R_2(B)C_2(A) & \dots & R_2(B)C_n(A) \\ \hdotsfor{4} \\ R_n(B)C_1(A) & R_n(B)C_2(A) & \dots & R_n(B)C_n(A) \end{bmatrix}. \tag{32}$$

We refer to AB as B *premultiplied* or multiplied *on the left* by A, or as A *postmultiplied* or multiplied *on the right* by B.

The matrix product is a natural extension of the matrix-by-vector product discussed in § 7. Indeed we can regard B as being composed of a succession of column vectors, on which operation with A produces another set of column vectors comprising the matrix AB. Thus sets of equations $Ax = b$, $Ay = c$, $Az = d$,..., can be represented by the single matrix equation $AX = B$, where X has columns x, y, z,..., and B has columns b, c, d,... .

Inversion and solution. The unit matrix

11. If A and B are matrices the expression A/B literally has no meaning, just as the differential operation D_1/D_2, where D_1 and D_2 were defined in § 10, is strictly meaningless. We may, however, write the solution of (16), a concise statement of equations (1), in the form

$$x = A^{-1}b, \tag{33}$$

and give meaning to the expression A^{-1}.

We have already seen that unique and definite solutions are likely only if the matrix A is square, so that the operation of inversion, the formation and use of A^{-1}, will relate here solely to square matrices. The concept of inversion is analogous to that of the inverse of a differential operator, for example in the solution $y = D^{-1}x$ of the differential equation $Dy = x$.

The first noteworthy point is that A^{-1} is a matrix of order n, the same as that of A, b and x. Second, the elements of A^{-1} depend only on those of A. These remarks become intuitively obvious by considering simple cases. Equations (7) and (8), for example, express the solution of equations (6) in the form

$$\left.\begin{aligned} x_1 &= \alpha_{11}b_1 + \alpha_{12}b_2 \\ x_2 &= \alpha_{21}b_1 + \alpha_{22}b_2 \end{aligned}\right\},\tag{34}$$

where

$$\left.\begin{aligned} \alpha_{11} &= a_{22}/D, \quad \alpha_{12} = -a_{12}/D, \quad \alpha_{21} = -a_{21}/D, \quad \alpha_{22} = a_{11}/D \\ D &= a_{11}a_{22} - a_{21}a_{12} \end{aligned}\right\},\tag{35}$$

so that (34) can be written in matrix form (33), and we have

$$A^{-1} = (a_{11}a_{22} - a_{21}a_{12})^{-1}\begin{bmatrix} a_{22} & -a_{12} \\ -a_{21} & a_{11} \end{bmatrix},\tag{36}$$

remembering the rule for multiplication of a matrix by a constant. A^{-1} exists and is unique unless $a_{11}a_{22} - a_{21}a_{12}$ vanishes, the special case quoted in § 4 for lack of solution of the two equations. If A^{-1} exists, A is *non-singular*. A *singular* matrix has no inverse.

12. We note that the elements of A^{-1} may be calculated by solving successively the equation $Ax = b$ for different special values of the elements of the vector b, that is for different right-hand sides. If we take $b = \begin{bmatrix} 1 \\ 0 \end{bmatrix}$, for example, we see from (34) that the resulting x_1 and x_2 are the elements of the first column of A^{-1}, and the right-hand side $b = \begin{bmatrix} 0 \\ 1 \end{bmatrix}$ similarly produces the elements of the second column of A^{-1}. In general the rth column of A^{-1} is composed of the elements of the solution of $Ax = b$, where b has unity for its rth element and zero elsewhere.

Recalling the remark at the end of § 10, it follows that the matrix A^{-1} is related to A through the equation

$$AA^{-1} = I,\tag{37}$$

where I is a square matrix of order n whose elements are all zero except in the *main diagonal*, the line sloping from top left to bottom right, in which the elements are all unity. The matrix I is called the *unit matrix*, and is given a suffix when the order is important and not obvious. For order three, for example, we have

$$I_3 = \begin{bmatrix} 1 & 0 & 0 \\ 0 & 1 & 0 \\ 0 & 0 & 1 \end{bmatrix}. \tag{38}$$

The unit matrix may also be called the *identity* matrix, since the operation Ix produces the vector x, so that Ix and x are different expressions for the same quantity.

The result (37) follows also from equation (33). Premultiplication of both sides of (33) with A gives

$$Ax = AA^{-1}b, \tag{39}$$

and since $Ax = b = Ib$ it follows that $AA^{-1} = I$.

13. The unit matrix has the special property, easily verified, that its inclusion in any matrix expression involving multiplication leaves that expression unchanged. In particular the order of multiplication with I is unimportant, so that

$$AI = IA = A. \tag{40}$$

From (40) and (37) we can prove another useful result. Premultiplication of (37) with A^{-1} gives $A^{-1}AA^{-1} = A^{-1}I = A^{-1}$, from which it follows that

$$A^{-1}A = I. \tag{41}$$

The order of multiplication of a matrix and its inverse is therefore also unimportant, both products giving the unit matrix.

14. If the inverse matrix is known or obtained the solution of sets of equations like (1), with the same A but different b, can obviously be effected by first computing A^{-1} and then performing the operations $A^{-1}b$ in the formal solutions (33). In practice, however, it is found that the determination of A^{-1} is a lengthy process, not necessary for the solution of $Ax = b$, and A^{-1} will not generally be calculated unless it is needed in contexts other than the solution of linear equations.

Transposition and symmetry. Inversion of products

15. Continuing our elementary discussion of matrix calculus, we introduce here the *transposed matrix*† of A, denoted by A', and related to A by the formula

$$R_s(A') = C_s(A). \tag{42}$$

An alternative definition is that the (r, s) element of A' is the (s, r) element of A.

A vector can be transposed similarly. The transpose of the vector or column-matrix x is the *row-matrix* x', whose elements, written in a row, are those of x written in a column. A *null row* is denoted by o'.

We must establish rules for transposing matrix expressions. For example the matrix product AB is a matrix C, and it is clearly desirable to know the corresponding relation between A', B' and C'. The relation follows simply from the rule (26) for the formation of c_{rs} in terms of a_{rs} and b_{rs}. We write (26) in the form

$$c_{sr} = R_s(A)C_r(B), \tag{43}$$

use (42) to replace this by

$$c_{sr} = C_s(A')R_r(B') = R_r(B')C_s(A'), \tag{44}$$

and deduce from (44) the required result

$$C' = B'A', \tag{45}$$

so that the transpose of a product AB is $B'A'$, each matrix being transposed and the order of multiplication reversed. More generally, we find that

$$(ABC \dots X)' = X' \dots C'B'A'. \tag{46}$$

16. Matrix-vector operations may be transposed in a similar way. For example transposition of the equation $Ax = b$ gives $x'A' = b'$, and represents the same set of algebraic equations. We note in passing that $A'x'$ is meaningless, lacking the necessary conformability for multiplication.

Transposition of vectors makes possible their multiplication. If x has m components, and y has n components, the expression xy is meaningless unless $m = n = 1$, but $x'y$ gives a single number, a *scalar*, for any value of $m = n$. Moreover xy' gives a matrix of shape

† The notation A^T (T for 'transpose') is also used, particularly in America.

$(m \times n)$. For example if $x' = [1, 2, 3]$, $y' = [1, -1, 1]$ then $x'y = 2$, and if $x' = [1, 2, 3]$, $y' = [1, -1]$ we find

$$xy' = \begin{bmatrix} 1 \\ 2 \\ 3 \end{bmatrix} \begin{bmatrix} 1 & -1 \end{bmatrix} = \begin{bmatrix} 1 & -1 \\ 2 & -2 \\ 3 & -3 \end{bmatrix}. \tag{47}$$

17. The matrix A is called *symmetric* when it is identical with its transpose A'. In this case $a_{rs} = a_{sr}$ so that elements symmetrically disposed (mirror images) about the principal diagonal are equal. Both matrices on the left of (30) are symmetric, and this equation suffices to point the warning that the product of symmetric matrices gives a matrix which is not in general symmetric. That is, if $A = A'$, $B = B'$, *it does not follow* that $(AB)' = AB$. In fact $(AB)' = B'A' = BA$.

We may, however, note that when A is symmetric the matrix AA is also symmetric. The matrix AA is denoted by A^2, with similar notation for higher *powers* of the matrix A, and it is easy to see that when $A = A'$ it follows that $(A^n) = (A^n)'$ for any positive power of A.

The inverse of a symmetric matrix is also symmetric. For if $A = A'$ we have $I = AA^{-1} = A'A^{-1}$, and taking the transpose we find $(A^{-1})'A = I$, so that $(A^{-1})' = A^{-1}$.

18. Two other properties of matrices are important with respect to the order of multiplication. First, we have seen the special cases $AI = IA$, $A^{-1}A = AA^{-1}$, in which this order is irrelevant. Another case involves the powers of a matrix, and we can write

$$A^m A^n = A^n A^m = A^{n+m}, \tag{48}$$

where m and n are any integers, positive or negative. The proof is obvious with the remark that A^{-m} is defined as $(A^{-1})^m$.

19. Second, we derive a result for the inversion of a product, analogous to that of (46) for the transposition of a product. Writing

$$AB \ldots Z = C \tag{49}$$

we find, by premultiplication of (49) with C^{-1}, followed by successive postmultiplications by $Z^{-1}, \ldots, B^{-1}, A^{-1}$, the result

$$C^{-1} = Z^{-1} \ldots B^{-1}A^{-1}. \tag{50}$$

No meaning, of course, can be given to the expression $(Ax)^{-1}$, since only square matrices have inverses in the sense defined.

Some special matrices

20. We have mentioned already two special cases, that of a symmetric matrix and that of a unit matrix, and two others will be needed.

First there is the *diagonal matrix*, often denoted by D, which has zeros everywhere except in the principal diagonal. It is therefore similar in form to the unit matrix. Noteworthy operations with D include pre- and post-multiplication with a general A. The rules for multiplication indicate that the operation DA produces a B in which

$$R_s(B) = d_s R_s(A), \tag{51}$$

where d_s is the diagonal element of D in row and column s. Similarly the operation AD produces a matrix C in which

$$C_s(C) = d_s C_s(A). \tag{52}$$

The diagonal matrix D may therefore be used conveniently in an extension of the result already noted at the end of § 10, valuable in many aspects of linear algebra. The result is that the set of equations

$$Ax = k_1 b, \qquad Ay = k_2 c, \qquad Az = k_3 d,\ldots, \tag{53}$$

where the k_r are constants, can be represented by the single matrix equation

$$AX = BD, \tag{54}$$

where X has columns x, y, z,..., B has columns b, c, d,..., and D is a diagonal matrix with elements k_1, k_2, k_3,... .

Important and obvious results involving D are that the transpose D' is identical with D, the inverse D^{-1} is a diagonal matrix whose elements are the reciprocals of those of D, and the product of diagonal matrices is a diagonal matrix.

21. The next important special matrix is the *triangular matrix*. We denote by L a *lower triangular matrix* in which all elements *above* the principal diagonal are zero. Similarly the *upper triangular matrix* U has all its elements zero *below* the principal diagonal. The shape $\diagdown$ will denote an L, and $\diagup$ will denote a U. As examples we might have

$$L = \begin{bmatrix} 1 & 0 & 0 \\ -1 & 3 & 0 \\ 2 & 1 & 4 \end{bmatrix}, \qquad U = \begin{bmatrix} 2 & 3 & 4 \\ 0 & -1 & 0 \\ 0 & 0 & 3 \end{bmatrix}, \tag{55}$$

for triangular matrices of order three. The elements which are always zero are often omitted in the recording of such matrices.

Some triangular matrices may have all their diagonal terms unity and these are called *unit triangular matrices*.

The following properties of triangular matrices follow easily from the rules so far obtained:

(i) The transpose of an L is a U, and vice versa.

(ii) The product $L_1L_2L_3 \ldots$ is an L, the product $U_1U_2U_3 \ldots$ is a U.

(iii) The product LU or UL has no special significance: it is a general A.

Triangular matrices. The decomposition theorem

22. The importance of triangular matrices springs from the fact that linear equations are particularly easy to solve when the associated matrix is triangular. Consider, for example, the equation $Lx = b$, when L has the form of (55). The linear equations are given by

$$\left.\begin{aligned}
x_1 \qquad\qquad &= b_1 \\
-x_1 + 3x_2 \qquad &= b_2 \\
2x_1 + x_2 + 4x_3 &= b_3
\end{aligned}\right\} , \qquad (56)$$

and for given right-hand sides the unknowns are calculable in turn from successive equations by a process of *forward substitution*. The equation $Ux = b$, where U is defined in (55), represents the equations

$$\left.\begin{aligned}
2x_1 + 3x_2 + 4x_3 &= b_1 \\
-x_2 + 0x_3 &= b_2 \\
3x_3 &= b_3
\end{aligned}\right\} , \qquad (57)$$

and the unknowns are now calculable, in reverse order, by *back substitution* in successive equations starting with the last.

23. Consideration of equations (56) and (57) will provide another important result relating to the inversion of triangular matrices. The solutions of (56) and (57) are respectively

$$\begin{bmatrix} x_1 \\ x_2 \\ x_3 \end{bmatrix} = \begin{bmatrix} l_{11} & l_{12} & l_{13} \\ l_{21} & l_{22} & l_{23} \\ l_{31} & l_{32} & l_{33} \end{bmatrix} \begin{bmatrix} b_1 \\ b_2 \\ b_3 \end{bmatrix} , \quad \begin{bmatrix} x_1 \\ x_2 \\ x_3 \end{bmatrix} = \begin{bmatrix} u_{11} & u_{12} & u_{13} \\ u_{21} & u_{22} & u_{23} \\ u_{31} & u_{32} & u_{33} \end{bmatrix} \begin{bmatrix} b_1 \\ b_2 \\ b_3 \end{bmatrix} , \quad (58)$$

where l_{rs} and u_{rs} are here respective general elements of the inverse matrices L^{-1} and U^{-1}. Now it is clear from (56) that the component x_1 of the result depends only on b_1, and would be the same number whatever values were associated with b_2 and b_3. It follows, from the matrix-vector multiplication rule applied to the determination of x_1, that the multipliers l_{12} and l_{13} of b_2 and b_3 must both vanish. Again, x_2 is clearly independent of b_3, so that l_{23} is zero. Similar consideration of (57) gives the rule for inversion, that *the inverse of a lower (upper) triangular matrix is a lower (upper) triangular matrix.*

†24. The fact that the product LU of triangular matrices gives a matrix of no special form suggests the converse question—can a general matrix A be written as the product LU of a lower and an upper triangle?

It can in fact be shown in general, and illustrated by simple examples, that a general square matrix A is expressible in the form

$$A = LDU, \tag{59}$$

in which L and U are unit lower and upper triangles, D a diagonal, and that the expression is unique. For a matrix of order three we seek to determine the elements l_{rs}, d_r and u_{rs} in the equation

$$
\begin{bmatrix} a_{11} & a_{12} & a_{13} \\ a_{21} & a_{22} & a_{23} \\ a_{31} & a_{32} & a_{33} \end{bmatrix} = \begin{bmatrix} 1 & & \\ l_{21} & 1 & \\ l_{31} & l_{32} & 1 \end{bmatrix} \begin{bmatrix} d_1 & & \\ & d_2 & \\ & & d_3 \end{bmatrix} \begin{bmatrix} 1 & u_{12} & u_{13} \\ & 1 & u_{23} \\ & & 1 \end{bmatrix}. \tag{60}
$$

The matrix multiplication law gives, in relation to the first row of A, the equations

$$a_{11} = d_1, \quad a_{12} = d_1 u_{12}, \quad a_{13} = d_1 u_{13}, \tag{61}$$

which determine d_1, u_{12} and u_{13} in terms of the elements of A. For the second row of A we find the equations

$$a_{21} = l_{21} d_1, \quad a_{22} = l_{21} d_1 u_{12} + d_2, \quad a_{23} = l_{21} d_1 u_{13} + d_2 u_{23}, \tag{62}$$

from which we determine successively the quantities l_{21}, d_2 and u_{23}. Finally we determine in succession the elements l_{31}, l_{32} and d_3 from the equations relevant to the last row of A, given by

$$a_{31} = l_{31} d_1, \quad a_{32} = l_{31} d_1 u_{12} + l_{32} d_2, \quad a_{33} = l_{31} d_1 u_{13} + l_{32} d_2 u_{23} + d_3. \tag{63}$$

The elements of L, D and U are clearly unique, difficulties arising only in the case in which any of the d_r are zero, since these terms are used as divisors in the determination of l_{rs} and u_{rs}. The significance of this result will become apparent later.

We call (59) the *decomposition theorem* for a matrix. In practice we often combine D with either L or U, so that $A = L_1 U$ or LU_1, in which L_1 and U_1 are triangles but not unit triangles. We might also write $D = D_1 D_2$, with special choice of D_1 and D_2, associate D_1 with L and D_2 with U, giving $A = L_2 U_2$ in which neither triangle is a unit.

The determinant

25. We found in § 4 that the solutions of the equations

$$\left.\begin{array}{l} a_{11}x_1 + a_{12}x_2 = b_1 \\ a_{21}x_1 + a_{22}x_2 = b_2 \end{array}\right\} \tag{64}$$

can be written as

$$\frac{x_1}{\Delta_1} = \frac{x_2}{\Delta_2} = \frac{1}{\Delta}, \tag{65}$$

where

$$\Delta_1 = a_{22}b_1 - a_{12}b_2, \quad \Delta_2 = a_{11}b_2 - a_{21}b_1, \quad \Delta = a_{11}a_{22} - a_{12}a_{21}. \tag{66}$$

This is a special case of a general theorem (Cramer's rule) that the solution of a general set of equations is given by

$$\frac{x_1}{\Delta_1} = \frac{x_2}{\Delta_2} = \dots = \frac{1}{\Delta}, \tag{67}$$

in which the quantity Δ depends only on the elements of A in the equation $Ax = b$. This quantity is called the *determinant* of the square matrix A and denoted by $|A|$ or det. A. We write

$$|A| = \begin{vmatrix} a_{11} & a_{12} & \dots & a_{1n} \\ a_{21} & a_{22} & \dots & a_{2n} \\ \multicolumn{4}{c}{\dotfill} \\ a_{n1} & a_{n2} & \dots & a_{nn} \end{vmatrix}. \tag{68}$$

We would expect, from considerations of symmetry, that the quantities Δ_r are determinants, and indeed Δ_r is the determinant of the matrix obtained from A by replacing its rth column by the column b of right-hand sides.

A sufficient rule for evaluating determinants is contained in the results

$$|a_{11}| = a_{11}, \quad \begin{vmatrix} a_{11} & a_{12} \\ a_{21} & a_{22} \end{vmatrix} = a_{11}|a_{22}| - a_{12}|a_{21}|, \tag{69}$$

$$\begin{vmatrix} a_{11} & a_{12} & a_{13} \\ a_{21} & a_{22} & a_{23} \\ a_{31} & a_{32} & a_{33} \end{vmatrix} = a_{11}\begin{vmatrix} a_{22} & a_{23} \\ a_{32} & a_{33} \end{vmatrix} - a_{12}\begin{vmatrix} a_{21} & a_{23} \\ a_{31} & a_{33} \end{vmatrix} + a_{13}\begin{vmatrix} a_{21} & a_{22} \\ a_{31} & a_{32} \end{vmatrix}, \tag{70}$$

the determinants on the right of (70) being obtained by omitting from the original determinant the row and column containing their multipliers. Further expansion gives

$$|A| = a_{11}(a_{22}a_{33} - a_{32}a_{23}) - a_{12}(a_{21}a_{33} - a_{31}a_{23}) +$$
$$+ a_{13}(a_{21}a_{32} - a_{31}a_{22}). \quad (71)$$

It is clear that each element of A appears in only one term of the expansion (71) for $|A|$, and that no elements are multiplied together which occur in the same row. We could therefore rewrite the result in other ways, by expansion 'along' different rows. For example

$$|A| = -a_{21}(a_{12}a_{33} - a_{13}a_{32}) + a_{22}(a_{11}a_{33} - a_{13}a_{31}) -$$
$$- a_{23}(a_{11}a_{32} - a_{12}a_{31}), \quad (72)$$

$$|A| = a_{31}(a_{12}a_{23} - a_{22}a_{13}) - a_{32}(a_{11}a_{23} - a_{21}a_{13}) +$$
$$+ a_{33}(a_{11}a_{22} - a_{21}a_{12}) \quad (73)$$

The quantities in brackets in (71), (72) and (73) are called *first minors*. The first minor of a_{rs} is denoted by M_{rs}, and is a determinant of order one less than that of A. We can then write

$$|A| = \quad a_{11}M_{11} - a_{12}M_{12} + a_{13}M_{13}$$
$$= -a_{21}M_{21} + a_{22}M_{22} - a_{23}M_{23} \Big\}, \quad (74)$$
$$= \quad a_{31}M_{31} - a_{32}M_{32} + a_{33}M_{33}$$

and similarly for a general matrix of order n.

It is also easy to see that no quantities in the same column ever appear in a product, so that we can 'expand by columns', obtaining the results

$$|A| = \quad a_{11}M_{11} - a_{21}M_{21} + a_{31}M_{31}$$
$$= -a_{12}M_{12} + a_{22}M_{22} - a_{32}M_{32} \Big\}. \quad (75)$$
$$= \quad a_{13}M_{13} - a_{23}M_{23} + a_{33}M_{33}$$

26. From these expressions for $|A|$ we can verify, for order three, some important properties concerning determinants of general order n.

(i) $|A'| = |A|$. This follows from any two corresponding expressions in (74) and (75).

(ii) If any two rows or columns of A are identical, then $|A| = 0$, since the first minors of the other row or column are all zero.

(iii) If any row or column of A is null, then $|A| = 0$, which follows from expansion by that particular row or column.

(iv) If any two rows or columns are interchanged, the determinant changes its sign.

27. Other properties come from consideration of equation (67) for the general solution of the linear equations $Ax = b$. For example if all the elements of b are zero, then in general all the elements of x are zero. The vector b appears as a column in each of the determinants $\Delta_1, \Delta_2,...,$ of (67), so that we have here a special case of property (iii). The exceptional case, for which a non-trivial solution may exist, arises when $|A|$ is also zero, giving quantitative significance to the discussion on homogeneous equations in § 5. That discussion suggests also that $|A|$ is zero if any row is obtainable by a linear combination of any other rows, and this fact is a special case of the following property:

(v) The determinant is unchanged if to any row is added any multiples of any other rows.

Again, from the linear equations, we see that if the right-hand sides are multiplied by a constant k the solutions are also multiplied by this factor, so that

(vi) If a column is multiplied by a constant k, the determinant is multiplied by that constant.

28. Property (i) implies that 'row' and 'column' are interchangeable terms in most statements concerning determinants, including (v) and (vi). Concerning linear equations, however, we should note that multiplication by k of a row of A and the corresponding element of b has no effect on the solution, whereas multiplication of the column b changes the solution by that factor, and if a column of A is multiplied by k the corresponding unknown is affected by a factor k^{-1}.

29. Finally we have an important result relating to the product of matrices which we infer from (67) and the equations $Ax = b$, $x = By$, $Cy = b$:

(vii) If $C = AB$, then $|C| = |A|\,|B|$.

This result, which can be verified in simple cases by direct evaluation of $|A|$, $|B|$ and $|C|$, will have many important applications in our numerical work.

Cofactors and the inverse matrix
30. We are now able to find a formal expression for the inverse of A in terms of certain determinants relevant to A. For this purpose we return to (67) for the solution of $Ax = b$, recalling that the element x_r is obtained as Δ_r/Δ, where Δ is $|A|$, and Δ_r is the

determinant of the matrix obtained from A by replacing the rth column of A by the column b. We note also that if we replace b in turn by the columns of the unit matrix, denoting by $x^{(1)}$, $x^{(2)}$,..., the vector solutions of the corresponding equations, then that of $Ax = b$ is given by

$$x = b_1 x^{(1)} + b_2 x^{(2)} + ... + b_n x^{(n)}. \tag{76}$$

From expressions like (75) for expansion of a determinant by columns we easily see, for example, that

$$\left.\begin{array}{ll} x_1^{(1)} = A_{11}/|A|, & x_2^{(1)} = A_{12}/|A|,... \\ x_1^{(2)} = A_{21}/|A|, & x_2^{(2)} = A_{22}/|A|,... \end{array}\right\}, \tag{77}$$

in which $A_{rs} = (-1)^{r+s} M_{rs}$ is called the *cofactor* of a_{rs}. We then have, for $Ax = b$, the solution

$$\left.\begin{array}{l} x_1 = |A|^{-1}(A_{11}b_1 + A_{21}b_2 + ...) \\ x_2 = |A|^{-1}(A_{12}b_1 + A_{22}b_2 + ...) \\ \cdots\cdots\cdots\cdots\cdots\cdots\cdots\cdots\cdots \end{array}\right\}. \tag{78}$$

Comparing this result with the formal solution $x = A^{-1}b$ we find an expression for the matrix A^{-1}, in terms of the determinant and cofactors of A, given by

$$|A|\,A^{-1} = \begin{bmatrix} A_{11} & A_{21} & ... & A_{n1} \\ A_{12} & A_{22} & ... & A_{n2} \\ \multicolumn{4}{c}{\cdots\cdots\cdots\cdots\cdots} \\ A_{1n} & A_{2n} & ... & A_{nn} \end{bmatrix}. \tag{79}$$

This matrix is called the *adjugate* matrix of A, and we can write

$$\mathrm{Adj}\,A = |A|\,A^{-1}. \tag{80}$$

We note that in the adjugate A_{rs} is in *column* r and *row* s. We note also that A is singular if and only if $|A| = 0$.

Determinants of special matrices

31. We can now record the determinants of special matrices. Using the expansion rules (74) and (75), for example, we find that

  (i) the determinant of the unit matrix is unity;
  (ii) the determinant of a diagonal matrix, or an upper or lower triangle, is the product of the diagonal terms.

From the equation $A^{-1}A = I$ we find that

(iii) the determinant of A^{-1} is $|A|^{-1}$.

From equation (79), finally, we observe that $|A^{-1}| = |A|^{-n} |\text{Adj } A|$, since a determinant, unlike a matrix, is multiplied by a constant whenever any row or column is multiplied by that constant. This result and property (iii) produces

(iv) the determinant of Adj A is $|A|^{n-1}$.

Partitioned matrices

32. We note now some results involving the partitioning of matrices and vectors into smaller units. Consider for example the set of equations

$$\left.\begin{aligned}
a_{11}x_1+a_{12}x_2+b_{11}y_1+b_{12}y_2 &= p_1 \\
a_{21}x_1+a_{22}x_2+b_{21}y_1+b_{22}y_2 &= p_2 \\
c_{11}x_1+c_{12}x_2+d_{11}y_1+d_{12}y_2 &= q_1 \\
c_{21}x_1+c_{22}x_2+d_{21}y_1+d_{22}y_2 &= q_2
\end{aligned}\right\}. \tag{81}$$

We can represent (81) in matrix language in the form

$$\left[\begin{array}{c|c} A & B \\ \hline C & D \end{array}\right]\left[\begin{array}{c} x \\ \hline y \end{array}\right] = \left[\begin{array}{c} p \\ \hline q \end{array}\right], \tag{82}$$

where A, B, C and D are square matrices of order two, x, y, p and q are vectors of the same order, and the apparently strange notation in (81) becomes comprehensible.

The internal lines indicate *partitioning* of the original matrix and vector into matrices and vectors of lower order. The expansion of (82) is performed as if the matrix and vector elements were numerical quantities, so that we replace (82) by the two equations

$$\left.\begin{aligned}
Ax+By &= p \\
Cx+Dy &= q
\end{aligned}\right\}. \tag{83}$$

The 'coefficients' of these equations are themselves matrices, and the 'unknowns' and 'constant' terms are themselves vectors, so that the order in which the multiplied terms in (83) are recorded is important.

The equations (83) can of course be expanded further, and each of them gives two equations of the set (81), verifying the consistency of the definition of expansion of (82).

33. We can partition in any way for which every multiplication involved is conformable. In particular, in (82) C may be a single row, B a column, and D a single element, in which case both y and q are single elements and the second of (83) is then a relation between *scalars*, pure numerical quantities.

For example the equations

$$\begin{aligned}
a_{11}x_1 + a_{12}x_2 + b_1 y &= p_1 \\
a_{21}x_1 + a_{22}x_2 + b_2 y &= p_2 \\
c_1 x_1 + c_2 x_2 + dy &= q
\end{aligned}\quad\right\}\tag{84}$$

can be written as

$$\left[\begin{array}{c|c} A & b \\ \hline c' & d \end{array}\right]\left[\begin{array}{c} x \\ \hline y \end{array}\right] = \left[\begin{array}{c} p \\ \hline q \end{array}\right],\tag{85}$$

where c' is the row vector with components (c_1, c_2), b, p and x are column vectors of order two, and the quantities d, y and q are scalars. Expansion of (85) gives

$$\begin{aligned}
Ax + by &= p \\
c'x + dy &= q
\end{aligned}\quad\right\},\tag{86}$$

the second of (86) representing a single equation, the third of (84).

†34. Finally, it is not essential that any of the matrices in (82) or (85) should be square. For example we could replace the equations

$$\begin{aligned}
a_{11}x_1 + a_{12}x_2 + b_{11}y_1 + b_{12}y_2 &= p_1 \\
a_{21}x_1 + a_{22}x_2 + b_{21}y_1 + b_{22}y_2 &= p_2 \\
a_{31}x_1 + a_{32}x_2 + b_{31}y_1 + b_{32}y_2 &= p_3 \\
c_1 x_1 + c_2 x_2 + d_1 y_1 + d_2 y_2 &= q
\end{aligned}\quad\right\}\tag{87}$$

by the partitioned matrix equation

$$\left[\begin{array}{c|c} A & B \\ \hline c' & d' \end{array}\right]\left[\begin{array}{c} x \\ \hline y \end{array}\right] = \left[\begin{array}{c} p \\ \hline q \end{array}\right].\tag{88}$$

We must, however, avoid the temptation of trying to invert any matrices, arising from partitioning, which are not square.

Latent roots and vectors

35. We remarked at the end of § 5 that a common problem in linear algebra is the determination of parameters such that a set of homogeneous linear equations should have non-trivial solutions. Problems of this kind arise frequently in the theory of vibrations, of 'small oscillations', and are commonly presented in the first place as one or more differential equations. For example we may have a pair of such equations in the form

$$\left.\begin{array}{l} \dfrac{d^2x}{dt^2}+ax+by = 0 \\[3mm] \dfrac{d^2y}{dt^2}+cx+dy = 0 \end{array}\right\}, \tag{89}$$

and the assumption that the vibrations of x and y have the same period and phase but different amplitudes, so that

$$x = \alpha \cos(pt+\epsilon), \qquad y = \beta \cos(pt+\epsilon), \tag{90}$$

leads to the linear homogeneous equations

$$\left.\begin{array}{l} (a-p^2)\alpha+b\beta = 0 \\[2mm] c\alpha+(d-p^2)\beta = 0 \end{array}\right\}, \tag{91}$$

which have non-trivial solutions for those values of p^2 for which the determinantal equation

$$\begin{vmatrix} a-p^2 & b \\ c & d-p^2 \end{vmatrix} = 0 \tag{92}$$

is satisfied. Here the parameter p^2 enters linearly, and we shall consider this case only.

36. We start with the general equations

$$\left.\begin{array}{l} (a_{11}-\lambda)x_1+a_{12}x_2+\ldots+a_{1n}x_n = 0 \\[2mm] a_{21}x_1+(a_{22}-\lambda)x_2+\ldots+a_{2n}x_n = 0 \\[1mm] \cdots\cdots\cdots\cdots\cdots\cdots\cdots\cdots\cdots\cdots\cdots \\[1mm] a_{n1}x_1+a_{n2}x_2+\ldots+(a_{nn}-\lambda)x_n = 0 \end{array}\right\}, \tag{93}$$

for which the corresponding form in matrix notation is given by

$$(A-I\lambda)x = o. \tag{94}$$

We have already seen that the values of λ for which non-trivial solutions exist are the roots of the determinantal equation

$$|A-I\lambda| = 0, \tag{95}$$

and since the determinant can obviously be expanded as a polynomial of degree n, whose coefficient of λ^n is $(-1)^n$ and whose constant term is $|A|$, the fundamental theorem of algebra says that there are n roots, in general either real or complex and not necessarily distinct.

These roots $\lambda_1, \lambda_2, ..., \lambda_n$ are called the *latent roots* (or *eigenvalues*) of the matrix A, and the vector $x^{(r)}$, obtained by solving (93) with $\lambda = \lambda_r$, is the *latent vector* (or *eigenvector*) corresponding to λ_r. The vector is of course not unique, since if $x^{(r)}$ is any solution then $kx^{(r)}$, where k is any constant, is also a solution. In practice we find it convenient to *normalize* $x^{(r)}$, making it unique, either by arranging that its largest element is unity, or that the sum of the squares of its elements is unity ($x'x = 1$), or in some other convenient way.

37. It is easy to see that if all the roots are distinct then so are all the vectors. Certainly the vector corresponding to λ_r cannot be identical with that corresponding to λ_s, for otherwise subtraction of corresponding equations of the set (93) would make every component zero, giving the trivial solution. Moreover we cannot have any relation of the form $\sum\limits_{r=1}^{n} \alpha_r x^{(r)} = o$ unless every $\alpha_r = 0$. For this would also imply $A^p \sum\limits_{r=1}^{n} \alpha_r x^{(r)} = o = \sum\limits_{r=1}^{n} \alpha_r \lambda_r^p x^{(r)}$. If we now take $p = 0, 1, 2, ..., n-1$, and consider the resulting set of homogeneous equations involving the first components $x_1^{(r)}$ of these vectors, we find that every $\alpha_r x_1^{(r)} = 0$ unless

$$\begin{vmatrix} 1 & 1 & \cdots & 1 \\ \lambda_1 & \lambda_2 & \cdots & \lambda_n \\ \lambda_1^2 & \lambda_2^2 & \cdots & \lambda_n^2 \\ \cdots\cdots\cdots\cdots\cdots\cdots \\ \lambda_1^{n-1} & \lambda_2^{n-1} & \cdots & \lambda_n^{n-1} \end{vmatrix} = 0. \tag{96}$$

This cannot be true since the determinant is proportional to $\prod\limits_{r \neq s} (\lambda_r - \lambda_s)$, and $\lambda_r \neq \lambda_s$. The same argument gives $\alpha_r x_k^{(r)} = 0$ for every value of k, and since no latent vector is null we must have every $\alpha_r = 0$.

If two or more roots are equal, however, we may or may not have independent vectors. For example the unit matrix has all roots equal to unity, and the corresponding latent vectors can be taken as the vectors $e^{(r)}$, $r = 1, 2, ..., n$, where $e^{(r)}$ is the rth column of the unit matrix. Any linear combination of the $e^{(r)}$, of course, is also a solution. But the matrix $\begin{bmatrix} 1 & 1 \\ 0 & 1 \end{bmatrix}$ has roots $\lambda_1 = 1$, $\lambda_2 = 1$, but only the single vector with components 1, 0.

In this book we shall assume throughout that the roots are distinct and that there are therefore n different latent vectors. By 'different' we mean that they are linearly independent in the sense quoted, that is there are no constants α_r, other than zero, such that

$$\sum_{r=1}^{n} \alpha_r x^{(r)} = o. \tag{97}$$

This restriction, while not serious in practice, simplifies our mathematics. We state here, without proof, that a *symmetric* matrix always has n distinct latent vectors even if some of the latent roots are equal.

Similarity transformations

38. The full set of equations (94), for all vectors $x^{(r)}$ and roots λ_r, can be written as the single matrix equation

$$AX = XD, \tag{98}$$

where the matrix X, the *modal matrix*, has the latent vectors as its columns, and D is a diagonal matrix with coefficients λ_r in the corresponding columns. Moreover if the vectors are linearly independent the matrix X is non-singular, so that we can multiply (98) on the right by X^{-1} and produce

$$A = XDX^{-1}. \tag{99}$$

The matrices A and D, with the same latent roots, are called *similar*, and if we write (99) in the form

$$D = X^{-1}AX \tag{100}$$

we have performed a *similarity* transformation of A, producing a matrix D whose latent roots are easy to find, being the diagonal elements of D.

We can also deduce from (99) that the latent roots of a power A^m of A are the same power of those of A. For

$$A^m = (XDX^{-1})(XDX^{-1}) \ldots (m \text{ factors}) = XD^mX^{-1}, \tag{101}$$

and if m is negative the same result holds. In particular the latent roots of A^{-1} are the reciprocals of those of A.

The transformation (100) is a special case of the more general similarity transformation given by

$$B = Y^{-1}AY, \tag{102}$$

where Y is any non-singular matrix. B and A have the same latent roots, for if we substitute in (98) for A obtained from (102) we find

$$BZ = ZD, \qquad Z = Y^{-1}X, \tag{103}$$

so that the vectors of B are related to those of A through the second of (103). Many of our methods of finding latent solutions seek a transformation (102) which, through a suitable choice of Y, produces a matrix B whose latent roots are relatively easy to find.

†39. As a first corollary we notice that AB and BA, though not in general the same, have the same latent roots. For if AB has modal matrix X and diagonal matrix D of roots, we can write

$$ABX = XD, \qquad BA(BX) = (BX)D, \tag{104}$$

and the vectors of BA form the matrix BX, with the corresponding elements of D as latent roots.

As a second corollary we can verify a condition on the matrices A and B which allows the two products AB and BA to give the same matrix. For if the roots and vectors of A form the respective matrices D_1 and X, and D_2 and Y have the same significance for the matrix B, then (99) and the assumption $AB = BA$ gives the equation

$$XD_1X^{-1}YD_2Y^{-1} = YD_2Y^{-1}XD_1X^{-1}. \tag{105}$$

Premultiplying by Y^{-1} and postmultiplying by X, and writing $Y^{-1}X = C$, we find

$$CD_1C^{-1}D_2C = D_2CD_1. \tag{106}$$

Diagonal matrices *commute*, that is $D_1D_2 = D_2D_1$, and it follows that one solution of (106) is $C = C^{-1} = I$, that is $X = Y$, and the matrices A and B have the same latent vectors.

Finally, we note that if the latent vectors are linearly independent, then any other arbitrary vector z can be expressed uniquely as a linear combination of them, in the form

$$z = \sum_{r=1}^{n} \alpha_r x^{(r)}. \tag{107}$$

For if we denote by α the vector whose components are $\alpha_1, \alpha_2, ..., \alpha_n$, equation (107) can be written in matrix notation in the form

$$X\alpha = z, \tag{108}$$

and since X is non-singular it follows that α exists and is unique. This result will be needed in many applications and proofs.

Orthogonality

40. We can now investigate the properties of the latent vectors of a general matrix A. If A is unsymmetric we have to consider simultaneously the transpose A'. Since $|A - I\lambda| = |A' - I\lambda|$ their roots are the same, but the modal matrices X and Y are in general different. We then have

$$\left.\begin{aligned} AX &= XD \\ A'Y &= YD \end{aligned}\right\}, \tag{109}$$

the columns of X being the *right* latent vectors of A, those of Y being the *left* latent vectors of A since the second of (109) is equivalent to $Y'A = DY'$. If we use the second of (109) in its transposed form, and multiply the two equations respectively by Y' on the left and X on the right, we find

$$Y'AX = (Y'X)D = D(Y'X). \tag{110}$$

The only possibilities for the matrix product to be independent of the order, with one of them a diagonal, is that the other is also a diagonal or null. But in the latter case, $Y'X$ null, we have $|Y'||X| = 0$, so that either Y or X is singular, contrary to our assumption.

Hence $Y'X$ is diagonal, and in fact we can conveniently normalize the individual vectors so that

$$Y'X = I. \tag{111}$$

This equation includes the *orthogonality* relations of the latent vectors, that is

$$y^{(s)'}x^{(r)} = 0, \qquad r \neq s, \tag{112}$$

and we have normalized here so that

$$y^{(r)'}x^{(r)} = 1. \tag{113}$$

The transformation (100) can then be written in the form

$$D = Y'AX. \tag{114}$$

These orthogonality relations facilitate the computation of the coefficients α_r in the expansion (107) for an arbitrary vector, if we know the latent vectors of both A and A'. For the solution of (108) is $\alpha = X^{-1}z$, and we easily see from (111) that $X^{-1} = Y'$ so that

$$\alpha = Y'z, \tag{115}$$

and the solution of linear equations is replaced by a matrix-by-vector multiplication.

Symmetry, Rayleigh's Principle. Hermitian matrices

41. If the matrix is symmetric, so that $A = A'$, the foregoing analysis is simplified considerably. For now there is only one modal matrix X whose columns, the latent vectors of A, are mutually *orthogonal*. We have

$$X^{-1} = X', \tag{116}$$

and X is called an *orthogonal* matrix. The transformation (114) becomes

$$D = X'AX, \tag{117}$$

and this is an *orthogonal similarity transformation*. The individual orthogonality and normalizing relations are included in the equations

$$x^{(r)'}x^{(s)} = 0, \qquad r \neq s; \qquad x^{(r)'}x^{(r)} = 1. \tag{118}$$

We note also that $XX' = I$, so that the *rows* of X are also mutually orthogonal.

42. Real symmetric matrices are very common in practice, but if A is complex it is more useful to consider with A its complex conjugate $\bar{A}$, and if $\bar{A} = A'$ the matrix is called *Hermitian*. We can deduce properties of real symmetric matrices from those of Hermitian type simply by dropping the 'bar'.

· We show first that the roots of a Hermitian matrix are real. For if λ is a root, and x the corresponding vector, we have $Ax = \lambda x$, and premultiplication with $\bar{x}'$ gives for λ the expression

$$\lambda = \frac{\bar{x}'Ax}{\bar{x}'x}. \tag{119}$$

Now the denominator is real, being the sum of squares of the moduli of the components of x. We show also that the numerator is real. For if we write $x = \alpha + i\beta$, $\bar{x} = \alpha - i\beta$, where α and β are real vectors, and $A = C + iD$, where C and D are real matrices with

$C' = C$, $D' = -D$ in the Hermitian case, we find

$$\bar{x}'Ax = (\alpha' - i\beta')(C + iD)(\alpha + i\beta)$$
$$= (\alpha'C\alpha + \beta'D\alpha - \alpha'D\beta + \beta'C\beta) +$$
$$+ i(\alpha'D\alpha - \beta'C\alpha + \alpha'C\beta + \beta'D\beta). \quad (120)$$

The coefficient of the imaginary part can be simplified with the observation that

$$\alpha'C\beta = \beta'C'\alpha = \beta'C\alpha, \quad (121)$$

by transposition of a scalar quantity and the Hermitian property. Also, by transposition,

$$\alpha'D\alpha = \alpha'D'\alpha = 0 = \beta'D\beta, \quad (122)$$

since $D = -D'$. The quantity $\bar{x}'Ax$ is therefore purely real, and hence λ is real. If A is real the corresponding vector is also obviously real, but in the complex Hermitian case the vectors are generally complex.

Second, we note that if A is Hermitian the vectors of A' are the complex conjugates of those of A. For if $AX = XD$, then $\bar{A}\bar{X} = \bar{X}\bar{D}$, and since $\bar{A} = A'$ and $\bar{D} = D$ we have the required result. The similarity transformation then becomes

$$\bar{X}'AX = D, \quad (123)$$

and the orthogonality and normalizing relations are given by

$$\bar{X}'X = I = X'\bar{X}. \quad (124)$$

The matrix X, orthogonal when A is real, is called a *unitary* matrix in the Hermitian case, and the similarity transformation is a *unitary transformation*.

†43. Returning to the real symmetric case, the formula for λ in (119) becomes

$$\lambda = \frac{x'Ax}{x'x}, \quad (125)$$

and we have seen that λ and x are real. The root is also positive if the matrix A is *positive definite*, that is

$$x'Ax > 0 \quad (126)$$

for any real vector x. These matrices arise often in practice, and a particularly important case is given by $A = B'B$. For then

$$x'Ax = (Bx)'(Bx),$$

and is the sum of the squares of the components of the vector Bx.

In this case the quantity

$$\lambda_R = \frac{y'Ay}{y'y},\tag{127}$$

where y is not a latent vector, is a number called the *Rayleigh quotient*, and has important properties embodied in *Rayleigh's Principle*.

With our basic assumption that the vectors are linearly independent we can express y in the form

$$y = \sum_{r=1}^{n} \alpha_r x^{(r)},\tag{128}$$

and if we substitute in (127) and use the orthogonality and normalizing properties we find

$$\lambda_R = \frac{y'Ay}{y'y} = \frac{\sum\limits_{r=1}^{n} \alpha_r^2 \lambda_r}{\sum\limits_{r=1}^{n} \alpha_r^2},\tag{129}$$

since $Ax^{(r)} = \lambda_r x^{(r)}$ and $x^{(s)'}Ax^{(r)} = \lambda_r x^{(s)'}x^{(r)} = 0$ for $r \neq s$; $= \lambda_r$ for $r = s$. Now suppose the roots are arranged in algebraic order $\lambda_1 > \lambda_2 > \dots > \lambda_n$. Then

$$\lambda_R - \lambda_n = \sum_{r=1}^{n-1} \alpha_r^2 (\lambda_r - \lambda_n) / \sum_{r=1}^{n} \alpha_r^2.\tag{130}$$

In the numerator every number is positive, and the denominator is clearly positive, so that $\lambda_R > \lambda_n$ and the *Rayleigh quotient, for an arbitrary vector, overestimates the smallest latent root*. A similar argument shows that it *underestimates the largest latent root* in algebraic terms.

Suppose further that y is not quite arbitrary, but is an approximation with 'first-order' error to a latent vector $x^{(k)}$, which means that in (128) all the coefficients α_r are of order ϵ, except α_k which is of order unity. Then

$$\lambda_R - \lambda_k = \sum_{r=1}^{n} \alpha_r^2 (\lambda_r - \lambda_k) / \sum_{r=1}^{n} \alpha_r^2,\tag{131}$$

and the right-hand side is of order ϵ^2 since the coefficient of α_k^2 is zero. We therefore have the stationary property of Rayleigh's principle, that the Rayleigh quotient, for a vector with *first-order approximation* to a latent vector, gives a *second-order approximation* to the corresponding latent root.

For an unsymmetric A there is an analogous stationary property, but we have here to consider two arbitrary vectors, x for A and y for

A'. In analogy with the Rayleigh quotient we consider the ratio $y'Ax/y'x$, and if

$$x = \sum_{r=1}^{n} \alpha_r x^{(r)}, \quad y = \sum_{r=1}^{n} \beta_r y^{(r)}, \tag{132}$$

we find corresponding to (129) the result

$$\frac{y'Ax}{y'x} = \sum_{r=1}^{n} \alpha_r \beta_r \lambda_r / \sum_{r=1}^{n} \alpha_r \beta_r. \tag{133}$$

If x and y are good approximations to $x^{(k)}$ and $y^{(k)}$ the quotient (133) will give a better approximation to λ_k, but we cannot say anything about underestimation or overestimation in the case of arbitrary vectors.

Limits, series and norms

†44. We shall need to consider infinite series of vectors and matrices, and to discuss the idea of limits of sequences and convergence of series. For this purpose we also need some method of measuring the magnitude of vectors and matrices.

A *sequence* of vectors $x^{(1)}, x^{(2)}, ..., x^{(k)}, ...$ is said to have a limit x if each sequence of corresponding components has a finite limit, and the components of x are the limiting values of the separate components of the vectors $x^{(k)}$. The *series* of vectors

$$x^{(1)} + x^{(2)} + ... + x^{(k)} + ...$$

converges to the vector x if the sequence

$$y^{(k)} = (x^{(1)} + x^{(2)} + ... + x^{(k)})$$

has a limit as $k \to \infty$. The corresponding result for matrices is obtained by changing x to A and examining every sequence of corresponding elements in the matrices.

We can measure the size of a vector in various ways, three of which are in common use. For 'size' we use the word *norm*, with notation $\|x\|$. The norm $\|x\|$ is a non-negative number with the properties

$$\left.\begin{array}{l} \|x\| \neq 0 \text{ unless } x = o, \text{ and } \|o\| = 0 \\[4pt] \|cx\| = |c|\,\|x\|, \text{ where } c \text{ is a constant} \\[4pt] \|x+y\| < \|x\| + \|y\| \end{array}\right\}. \tag{134}$$

From the third of (134) we can easily deduce, by writing

$$x = x - y + y,$$

that

$$\|x - y\| > \|x\| - \|y\|. \tag{135}$$

For matrices we shall use norms similar to those of (134), with A replacing x, together with a property involving multiplication given by

$$\|AB\| < \|A\| \, \|B\|. \tag{136}$$

The importance of the norm lies in the obvious fact that $x^{(k)} \to x$ if and only if $\|x^{(k)} - x\| \to 0$, with a similar property for matrices.

45. Three different vector norms are in common use. They are

$$\left. \begin{array}{l} \|x\|_\infty = \max_r |x_r| \\[2mm] \|x\|_1 = \sum_{r=1}^{n} |x_r| \\[2mm] \|x\|_2 = \left\{ \sum_{r=1}^{n} |x_r|^2 \right\}^{\frac{1}{2}} \end{array} \right\}, \tag{137}$$

the third norm having many of the properties of length in Euclidean space.

Since matrices and vectors occur together in our subject, we want to have matrix norms which are connected with vector norms. For this purpose we choose

$$\|A\| = \max_{\|x\|=1} \|Ax\|, \tag{138}$$

that is $\|A\|$ is the largest of the vector norms $\|Ax\|$ with respect to vectors x of unit norm. Such a norm satisfies the definitions of a matrix norm and the compatibility condition

$$\|Ax\| < \|A\| \, \|x\|, \tag{139}$$

which is (136) with B replaced by x.

We can then deduce the matrix norms corresponding (*subordinate*) to the vector norms of (137). For the first we have

$$\|A\|_\infty = \max_{\|x\|_\infty = 1} \|Ax\|_\infty = \max_i \left| \sum_{j=1}^{n} a_{ij} x_j \right| < \max_i \sum_{j=1}^{n} |a_{ij}|,$$

since $\max_r |x_r| = 1$. We can reach equality by choosing x suitably. For if $x_r = 1$ for $a_{pr} > 0$, -1 for $a_{pr} < 0$, where p is the value of i which gives the maximum of $\sum_{j=1}^{n} |a_{ij}|$, then $\|x\|_\infty = 1$ and $\left| \sum_{j=1}^{n} a_{pj} x_j \right| = \sum_{j=1}^{n} |a_{pj}|$.

Thus we have the first matrix norm

$$\|A\|_\infty = \max_i \sum_{j=1}^{n} |a_{ij}|, \tag{140}$$

the largest row sum of absolute values.

For the second norm we have

$$\|A\|_1 = \max_{\|x\|_1=1} \sum_{i=1}^{n} \left| \sum_{j=1}^{n} a_{ij}x_j \right| < \sum_{j=1}^{n} \sum_{i=1}^{n} |a_{ij}| \, |x_j| < \max_j \sum_{i=1}^{n} |a_{ij}|,$$

since $\sum |x_j| = 1$. Again we can reach equality by choosing x to be zero except in the element corresponding to the value of j for which $\sum |a_{ij}|$ is largest, and to have unity in this component. Then $\|x\|_1 = 1$ and

$$\|A\|_1 = \max_j \sum_{i=1}^{n} |a_{ij}|, \tag{141}$$

the largest column sum of absolute values.

Finally, for the third norm, we have $\|x\|_2 = (x'x)^{\frac{1}{2}}$, so that

$$\|A\|_2 = \max_{\|x\|_2=1} (x'A'Ax)^{\frac{1}{2}}.$$

Now $A'A$ is symmetric and positive definite, its latent roots μ_r are positive and it has n independent vectors $x^{(r)}$. Then for a vector x of unit norm but otherwise arbitrary we have

$$x = \sum \alpha_r x^{(r)}, \qquad x'A'Ax = \sum \alpha_r^2 \mu_r, \tag{142}$$

and $\sum \alpha_r^2 = 1$ since $x'x = 1$. Then

$$x'A'Ax < \mu_1 \sum \alpha_r^2 < \mu_1, \quad \text{where} \quad \mu_1 > \mu_2 > \ldots > \mu_n. \tag{143}$$

We can achieve equality by taking x to be the latent vector $x^{(1)}$, so that

$$\|A\|_2 = \mu_1^{\frac{1}{2}}. \tag{144}$$

46. We can use these norms to estimate the rate of convergence of sequences and series, particularly of matrices. First, however, we note from (136) that if A can be expressed in the form (99), with corollary (101), then $A^m \to O$ if and only if $D^m \to O$, that is if all the latent roots of A are less than unity in absolute value. The latter statement is in fact true, though not proved here, even if A is not expressible in the form (99).

Second, we see from (136) that

$$\|A^m\| < \|A^{m-1}\| \, \|A\| < \ldots < \|A\|^m, \tag{145}$$

so that $\|A^m\| \to 0$, and therefore $A^m \to O$, if $\|A\| < 1$.

Third, we find that, for all the norms considered,

$$\|A\| > |\lambda_r|, \tag{146}$$

where here λ_r is any latent root of A. For taking norms of both sides of the equation $\lambda x = Ax$ we find

$$|\lambda|\,\|x\| = \|Ax\| < \|A\|\,\|x\|, \tag{147}$$

and the result follows since $\|x\| \neq 0$.

47. Consider now the series $I+A+A^2+\ldots+A^m+\ldots$. For convergence it is clearly necessary that $A^m \to O$. This condition is also sufficient, for if $A^m \to O$ it follows that $|\lambda_r| < 1$, and therefore $|I-A|$ does not vanish and the inverse $(I-A)^{-1}$ exists. But

$$(I+A+A^2+\ldots+A^m)(I-A) = I-A^{m+1}, \tag{148}$$

and postmultiplication with $(I-A)^{-1}$ gives

$$I+A+A^2+\ldots+A^m = (I-A)^{-1}-A^{m+1}(I-A)^{-1} \to (I-A)^{-1} \tag{149}$$

as $m \to \infty$, so that

$$I+A+A^2+\ldots+A^m+\ldots = (I-A)^{-1}, \tag{150}$$

proving the sufficiency and establishing the sum of the series.

We also have

$$\|(I-A)^{-1}-(I+A+A^2+\ldots+A^m)\| = \|A^{m+1}+A^{m+2}+\ldots\|$$
$$< \|A\|^{m+1}+\|A\|^{m+2}+\ldots$$
$$= \frac{\|A\|^{m+1}}{1-\|A\|}, \tag{151}$$

giving an estimate of the rate of convergence of the series in (150).

As a by-product we have

$$\|(I\pm A)^{-1}\| = \|I\mp A+A^2\mp\ldots\|$$
$$< \|I\|+\|A\|+\|A\|^2+\ldots = (1-\|A\|)^{-1}, \tag{152}$$

since $\|I\| = 1$. This and similar results will be needed later in estimating the rate of convergence of iterative processes.

Numerical methods

48. We find it convenient to distinguish two classes of methods for numerical operations in linear algebra. In the so-called *direct* methods we normally perform a sequence of operations once only, and the results obtained are an approximation to the true results. If all arithmetic operations had been performed exactly the results would be exact, and they are approximate only because multiplications and divisions, in particular, are recorded and used subsequently with errors which depend on the computing equipment. An important problem in numerical analysis is the consideration of such rounding errors, to estimate their combined effect on the accuracy of the answers.

The indirect methods attempt solution by a process of successive approximation. The same sequence of operations, usually shorter than that of a direct method, is repeated several times, the results either increasing steadily in accuracy (converging) or showing no definite trend to constancy (diverging). Here the numerical analyst is interested in questions of convergence, and particularly in securing the most rapid rate of convergence.

Finally, we may combine the two types of method. For example a direct method which yields a few accurate figures might be included as part of an iterative scheme, which may yield better accuracy with a minimum of effort.

49. Before proceeding to discussion of these topics we note here some methods of checking which may be used with advantage, and which are all based on the *linear* nature of many of our operations. They can all be included under the general heading of *sum checks*.

When we operate on rows and columns and on complete matrices, square or rectangular, it is often convenient to calculate each *row sum*, that is the sum of the elements in a row, each *column sum*, and occasionally the *matrix sum*, the sum of all the matrix elements. For a matrix of shape $(m \times n)$ the row sums form a column vector denoted here by $s^{(1)}$, with components $s_1^{(1)}$, $s_2^{(1)}$, ..., $s_m^{(1)}$, the column sums form a row vector denoted here by $s^{(2)'}$, with components $s_1^{(2)}$, $s_2^{(2)}$, ..., $s_n^{(2)}$, and the matrix sum is a single element s. The use of these quantities for checking purposes is illustrated by the following examples. First we have

$$s = \sum_{r=1}^{m} s_r^{(1)} = \sum_{r=1}^{n} s_r^{(2)}. \tag{153}$$

Second, if $A = B+C$, then $a_{ij} = b_{ij}+c_{ij}$, and the calculation of each element can be checked by computing from the a_{ij} the sum terms, and using any of the formulae

$$\left.\begin{aligned}
s_r^{(1)}(A) &= s_r^{(1)}(B)+s_r^{(1)}(C) \\
s_r^{(2)}(A) &= s_r^{(2)}(B)+s_r^{(2)}(C) \\
s(A) &= s(B)+s(C)
\end{aligned}\right\}. \tag{154}$$

Only compensating errors can escape detection, and we may note that in the event of error the first or second of (154) gives more information than the third about the location of the offending element.

Third, if $A = BC$, then $a_{ij} = \sum b_{ik}c_{kj}$, and the calculation of all the elements in a particular row k is checked by the formula

$$s_k^{(1)}(A) = R_k(B)s^{(1)}(C), \tag{155}$$

and the complete calculation can be checked from

$$s(A) = s^{(2)'}(B)s^{(1)}(C), \tag{156}$$

the quantities on the right of (155) and (156) being row-by-column (scalar) products.

Equation (157) shows an example of matrix multiplication, $A = BC$.

	B			$s^{(1)}(B)$		C		$s^{(1)}(C)$
	3	1	2	6		-1	2	1
	-4	1	6	3		3	5	8
	0	2	-1	1		2	1	3
$s^{(2)'}(B)$	-1	4	7	$s = 10$	$s^{(2)'}(C)$	4	8	$s = 12$

	A		$s^{(1)}(A)$
	4	13	17
	19	3	22
	4	9	13
$s^{(2)'}(A)$	27	25	$s = 52$

(157)

For the second row of A equation (155) gives

$$R_2(B)s^{(1)}(C) = 22 = s_2^{(1)}(A),$$

which checks the elements in this row, and equation (156) gives $s(A) = s^{(2)'}(B)s^{(1)}(C) = 52$, which checks all the elements of A.

In the particular case in which C is alleged to be the inverse B^{-1} of B, and we wish to check that the product $BB^{-1} = I$, equation (155) should give unity for each scalar product, and equation (156) should give the number n, the order of B.

Finally, a common calculation is the formation of a matrix A by combining linearly certain rows of another matrix B. If the elements of the ith row of A satisfy the equation

$$R_i(A) = R_i(B) + kR_j(B), \tag{158}$$

then the single check

$$s_i^{(1)}(A) = s_i^{(1)}(B) + ks_j^{(1)}(B) \tag{159}$$

will suffice for all the elements of the row.

50. The nature of our computing equipment affects to some degree the details of the computing procedure and in particular our attitude to checking. With desk-machine operation human errors are not uncommon, and frequent checks of the kind illustrated are almost essential. With high-speed digital machines isolated undetected errors are increasingly rare, and the speed of computation is so great that sum checks can reasonably be omitted. We shall in fact often omit them in the illustrative examples.

Still important, however, are the 'terminal' checks. Our machine must receive into its store the true data of our problem, and even if the reading mechanism is perfect there is still the human fallibility in punching on card or tape the information required. The sum check can be used with advantage, comparing the machine-calculated sum of all the coefficients or those of each row or column with the true values computed, say by desk machine, from the initial data.

A check on the final results is also important, both if the 'writing' equipment is prone to error and particularly if the results are transcribed further.

51. Finally, the use of desk machines requires a different emphasis on the arrangement of the figures on the computing paper. For example row-by-column multiplication of matrices is not particularly convenient for matrices of even moderate order; the terms in each product are well separated and occur neither in the same row nor column. A more convenient arrangement is that of row-by-row or column-by-column multiplication, since partners in each product are then more easily identified. This can be achieved by transposing one of the matrices, for then

$$a_{ij} = R_i(B)C_j(C) = R_i(B)R_j(C') = C_i(B')C_j(C). \tag{160}$$

In the problem of equation (157), for example, we could write

$$
\begin{array}{cc}
B' & C
\end{array}
$$

$$
\begin{bmatrix} 3 & -4 & 0 \\ 1 & 1 & 2 \\ 2 & 6 & -1 \end{bmatrix}
\begin{bmatrix} -1 & 2 \\ 3 & 5 \\ 2 & 1 \end{bmatrix}
$$

and use column-by-column multiplication to produce A. The check equations of types (155) and (156) are also more convenient now for the arrangement of the numbers involved.

This arrangement we shall use from time to time in the examples, but only for convenience. The considerations for automatic-machine computation are of course of a quite different nature.

Additional notes and bibliography

§§ 4, 5. We do not wish to discuss at length all the factors which are involved in the question of existence of solutions of m linear equations in n unknowns, since in most practical problems the tentative ideas of the text will suffice. The ideas can, however, be put on a rigorous basis, and the theorem can be proved that the homogeneous system of linear equations $Ax = o$, where A is an $(m \times n)$ matrix, has a non-trivial solution, that is a solution for which at least one component of x is not zero, if $n > m$ (more unknowns than equations). Moreover if $n \leqslant m$ such a solution exists if every $(n \times n)$ determinant formed from n rows of A is zero. A proof of the theorem is given, for example, in

BELLMAN, R., 1960, *Introduction to matrix analysis*. McGraw-Hill, New York.

The theorem may with caution be extended to the inhomogeneous case $Ax = b$ by treating b as an extra column of A, and introducing a new unknown. Equations (4), for example, can then be written as

$$
\left.\begin{array}{l}
2x_1 - 3x_2 - x_3 = 0 \\
x_1 + x_2 = 0 \\
3x_1 - 2x_2 - x_3 = 0
\end{array}\right\},
$$

and a non-trivial solution of these last three equations exists because the single determinant of order 3 obtained from the (3×3) matrix does in fact vanish. There is a corresponding solution of equations (4) if x_3 is not zero. But there is no solution if x_3 is zero. For example the equations $2x_1 + 3x_2 = 5$, $4x_1 + 6x_2 = 9$ have no solution, even though the homogeneous equations have a non-trivial solution. The latter is $k(-3, 2, 0)$, where k is any non-zero constant, and x_3 is zero.

For the linear equations (4) we can give the value unity to the component x_3 of the corresponding homogeneous equations, and the solution for x_1 and x_2 is then unique. This cannot, however, be guaranteed from the vanishing of the determinant. Indeed the equations

$$
\left.\begin{array}{l}
x_1 - x_2 = 1 \\
2x_1 - 2x_2 = 2 \\
3x_1 - 3x_2 = 3
\end{array}\right\}
$$

also have a solution, since

$$\begin{vmatrix} 1 & -1 & -1 \\ 2 & -2 & -2 \\ 3 & -3 & -3 \end{vmatrix} = 0,$$

and in the corresponding homogeneous equations x_3 can be given a non-zero value. But either x_1 or x_2 can then be given any value, and the other is obtained from any one of the equations. In fact the rows of the matrix have more than one degree of dependence, since not only does the second row equal twice the first, but the third is the sum of the first two. All first minors of the matrix also vanish, and only one of the equations gives 'independent' information about x_1 and x_2, that is there is only one 'independent' row.

If a square matrix of order n has no row dependencies it is said to have *rank n*. If there are k row dependencies it has rank $n - k$.

A simple square matrix with only one independent row, that is a matrix of rank unity, is given by the expression xy', where x and y are vectors. For the previous example we have

$$\begin{bmatrix} 1 \\ 2 \\ 3 \end{bmatrix} \begin{bmatrix} 1 & -1 & -1 \end{bmatrix} = \begin{bmatrix} 1 & -1 & -1 \\ 2 & -2 & -2 \\ 3 & -3 & -3 \end{bmatrix},$$

and the second and third rows are just multiples of the first. Similarly the product $AB = C$, where A has n rows and k independent columns, and B has k independent rows and n columns, with $n > k$, is a $(n \times n)$ square matrix of rank k, with $n - k$ dependencies. For example in the product

$$C = \overset{A}{\begin{bmatrix} a_1 & a_2 \\ b_1 & b_2 \\ c_1 & c_2 \\ d_1 & d_2 \end{bmatrix}} \overset{B}{\begin{bmatrix} \alpha_1 & \beta_1 & \gamma_1 & \delta_1 \\ \alpha_2 & \beta_2 & \gamma_2 & \delta_2 \end{bmatrix}},$$

we have

$$\left. \begin{array}{l} R_1(C) = a_1 R_1(B) + a_2 R_2(B) \\ R_2(C) = b_1 R_1(B) + b_2 R_2(B) \end{array} \right\},$$

and these are independent. But

$$\left. \begin{array}{l} R_3(C) = c_1 R_1(B) + c_2 R_2(B) \\ R_4(C) = d_1 R_1(B) + d_2 R_2(B) \end{array} \right\},$$

and we find that

$$R_3(C) = \frac{c_1 b_2 - c_2 b_1}{a_1 b_2 - a_2 b_1} R_1(C) + \frac{a_1 c_2 - a_2 c_1}{a_1 b_2 - a_2 b_1} R_2(C),$$

a linear combination of R_1 and R_2, and for $R_4(C)$ we can find a similar result.

Conversely any square matrix of order n and rank k can be expressed as the product of two rectangular matrices of respective shapes (n, k) and (k, n). We shall see in §§ 33–39 of Chapter 3 how to determine the rank of a square matrix and the corresponding factors.

Returning to the case of more equations than unknowns, we find that the equations

$$\left.\begin{aligned} 2x_1 - 3x_2 &= 1 \\ x_1 + x_2 &= 0 \\ 3x_1 - 2x_2 &= 1 \\ 5x_1 - x_2 &= 3 \end{aligned}\right\}$$

have no solution, since although the determinant vanishes which comes from the first three equations, no such determinant vanishes which involves the coefficients of the fourth equation. Similarly the homogeneous equations (12) have no non-trivial solution since $|A| \neq 0$.

In the general case, either with $m > n$ or $m < n$, we can sometimes decide fairly easily whether a system of m linear equations in n unknowns has a solution which may or may not be unique. If the equations are written in matrix form $Ax = b$, and we consider also the system $A'y = c$, where A' is the transpose of A, we have

$$Ax = b, \qquad x'A' = b', \qquad x'A'y = b'y$$

from the first equation, and

$$x'A'y = x'c$$

from the second. Then $x'c = b'y$, and if we take c to be the null vector we have $b'y = 0$. There is therefore no solution of $Ax = b$ unless the equations $A'y = o$ have only a trivial solution or independent non-trivial solutions which are orthogonal to the vector b. It can also be shown that a solution exists if these conditions are satisfied, and this we state without proof.

For example the equations

$$\left.\begin{aligned} 2x_1 - 3x_2 &= 1 \\ x_1 + x_2 &= 0 \\ 3x_1 - 2x_2 &= 1 \end{aligned}\right\}$$

have a solution because the equations

$$\left.\begin{aligned} 2y_1 + y_2 + 3y_3 &= 0 \\ -3y_1 + y_2 - 2y_3 &= 0 \end{aligned}\right\}$$

have only one non-trivial solution, $y' = k(1, 1, -1)$ for any k, and this vector is orthogonal to $(1, 0, 1)$. On the other hand, for the equations

$$\left.\begin{aligned} 2x_1 - 3x_2 &= 1 \\ x_1 + x_2 &= 0 \\ 3x_1 - 2x_2 &= 1 \\ 5x_1 - x_2 &= 3 \end{aligned}\right\}$$

we must examine the homogeneous system

$$\left.\begin{aligned} 2y_1 + y_2 + 3y_3 + 5y_4 &= 0 \\ -3y_1 + y_2 - 2y_3 - y_4 &= 0 \end{aligned}\right\},$$

and this has two non-trivial solutions

$$y^{(1)\prime} = k(1, 1, -1, 0), \qquad y^{(2)\prime} = K(1 \cdot 2, \ 2 \cdot 6, \ 0, \ -1).$$

Although $y^{(1)}$ is orthogonal to $(1, 0, 1, 3)$ the second vector $y^{(2)}$ does not have this property, so that the equations have no solution.

We might ask similarly whether the equations

$$\left.\begin{aligned} 2x_1 + x_2 + 3x_3 + 5x_4 &= 8 \\ -3x_1 + x_2 - 2x_3 - x_4 &= -3 \end{aligned}\right\},$$

with more unknowns than equations, have any solution, and for this purpose we examine the 'transposed' homogeneous set

$$\left.\begin{aligned} 2y_1 - 3y_2 &= 0 \\ y_1 + y_2 &= 0 \\ 3y_1 - 2y_2 &= 0 \\ 5y_1 - y_2 &= 0 \end{aligned}\right\}.$$

These equations have only the trivial solution, and hence there is some solution of the original set whatever the right-hand sides may be.

But for the system

$$\left.\begin{aligned} x_1 + x_2 + x_3 &= 1 \\ 2x_1 + 2x_2 + 2x_3 &= 3 \end{aligned}\right\}$$

we find that the corresponding homogeneous equations

$$y_1 + 2y_2 = 0, \qquad y_1 + 2y_2 = 0, \qquad y_1 + 2y_2 = 0$$

have a non-trivial solution $y' = k(-2, 1)$, and this is not orthogonal to the vector $(1, 3)$ so that the system has no solution.

We give further discussion of these points in § 40 of Chapter 3, and they are treated also, with examples, in

LANCZOS, C., 1957, *Applied analysis*. Prentice-Hall, New York; Pitman, London.

§§ 24, 34. The use of partitioned matrices often simplifies the proof of various matrix properties. For example we can easily show by induction that we can find the unique triangles in the decomposition $A = LU$, with L a unit lower triangle, provided that all the leading submatrices of A have non-zero determinant.

For suppose we have succeeded in this decomposition for the sub-matrix formed from the first p rows and columns of A, and we try to extend the property to the sub-matrix 'bordered' by the next row and column. If this can be done we can write

$$\left[\begin{array}{c|c} A_p & C_{p+1} \\ \hline R'_{p+1} & a_{p+1,p+1} \end{array}\right] = \left[\begin{array}{c|c} L_p & o \\ \hline l'_{p+1} & 1 \end{array}\right]\left[\begin{array}{c|c} U_p & u_{p+1} \\ \hline o' & u_{p+1,p+1} \end{array}\right],$$

where R'_{p+1}, l'_{p+1} are rows, C_{p+1}, u_{p+1} are columns, $a_{p+1,p+1}$, $u_{p+1,p+1}$ are single elements, and the arrangement on the right preserves the triangularity of L and U and the unity of L. Then

$$\left.\begin{aligned} A_p &= L_p U_p \\ C_{p+1} &= L_p u_{p+1} \\ R'_{p+1} &= l'_{p+1} U_p \\ a_{p+1,p+1} &= l'_{p+1} u_{p+1} + u_{p+1,p+1} \end{aligned}\right\}.$$

The first of these equations represents the triangulation of A_p, which by our hypothesis is possible and is such that $|A_p| \neq 0$. Then A_p and therefore U_p is non-singular, so that L_p and U_p have inverses. We can therefore compute u_{p+1} and l'_{p+1} from the second and third equations, and $u_{p+1,p+1}$ follows from the last equation.

We have therefore succeeded in decomposing A_{p+1}, and U_{p+1} is non-singular provided that $u_{p+1,p+1}$ is not zero. In this case we can proceed to the next stage. The proof also shows that $|A_p| = |U_p|$, so that the product of the first p diagonal elements of the final U is equal to the determinant of the sub-matrix A_p. These elements of U are of course the diagonal elements of D in the expression $A = LDU$, with both L and U unit triangles, discussed in § 24, and we see the significance of the 'difficulties' mentioned at the end of that section.

We note also that the theorem still holds if the last term $u_{n,n}$ turns out to be zero, since then A is singular but the inverses exist of its leading sub-matrices.

§ 39. The deduction from (106) that $C = I$ is perhaps a little unconvincing. The following is an alternative proof that if $AB = BA$ the separate matrices have common latent vectors, at least if A and B are non-singular.

If $AB = BA$, then $B = A^{-1}BA = ABA^{-1}$, so that if B has a modal matrix Z it has also the modal matrices $A^{-1}Z$ and AZ. Therefore, since we may have a different normalizing factor in each case, we can write $A^{-1}ZD = AZ$, so that $A^2Z = ZD$. This implies that the vectors of A^2, which are the same as those of A, form the matrix Z, and we have the required result.

§ 43. In practice we may have to solve the more general equation

$$(A - B\lambda)x = o,$$

where neither A nor B is a diagonal. If B is not singular we can reduce this to the simpler form $(C - I\lambda)x = o$ by premultiplication with B^{-1}, so that $C = B^{-1}A$.

Commonly, however, both A and B are symmetric. There is some virtue in preserving symmetry in the reduction to the simpler form, and $B^{-1}A$ is not necessarily symmetric. We shall see in Chapter 4 that a symmetric B can be decomposed into the product $B = LL'$, where L is a lower triangle, and that L is not singular if B is positive definite.

We can show, by arguments similar to those of § 43, that if A and B are symmetric, and B is positive definite, then the latent roots λ are real, and if A is also positive definite the roots are positive. In either case we can write

$$(A - B\lambda)x = (A - LL'\lambda)x = \{A(L')^{-1} - L\lambda\}(L'x),$$

and premultiplication with L^{-1} produces the simple form

$$(C - I\lambda)y = o,$$

where $C = L^{-1}A(L')^{-1}$, which is symmetric, and $y = L'x$.

§ 44. Most of this discussion on norms is taken from

FADDEEVA, V. N., 1959, *Computational methods of linear algebra.* Dover Publications, New York.

3

Elimination Methods of Gauss, Jordan and Aitken

Introduction

1. THE simplest of all methods for solving linear equations is the elimination method attributed to Gauss, which is effectively a disciplined form of the school-book method. Consider, for example, the equations

$$\left.\begin{aligned} 4x_1 - 9x_2 + 2x_3 &= 5\\ 2x_1 - 4x_2 + 6x_3 &= 3\\ x_1 - x_2 + 3x_3 &= 4 \end{aligned}\right\} . \tag{1}$$

By forming suitable combinations of pairs of these equations we can produce two independent equations which do not contain x_1, reducing the problem to the solution of two equations in the two unknowns x_2 and x_3. A further elimination produces one equation containing x_3 only, and direct solution is then possible.

For example if we add -0.5 times the first equation to the second, and -0.25 times the first equation to the third, we find that (1) can be replaced by the new set

$$\left.\begin{aligned} 4x_1 - 9x_2 + 2x_3 &= 5\\ 0.5x_2 + 5x_3 &= 0.5\\ 1.25x_2 + 2.5x_3 &= 2.75 \end{aligned}\right\} . \tag{2}$$

Ignoring the first equation of (2) we can now eliminate x_2 from the third by adding -2.5 times the second to the third, so that (1) has finally been replaced by the *triangular* set

$$\left.\begin{aligned} 4x_1 - 9x_2 + 2x_3 &= 5\\ 0.5x_2 + 5x_3 &= 0.5\\ -10x_3 &= 1.5 \end{aligned}\right\} . \tag{3}$$

A process of *back-substitution* now produces the required result. The third equation of (3) gives directly the value $x_3 = -0.15$, from the second of (3) we then have $0.5x_2 = 0.5 - 5x_3 = 1.25$, so that $x_2 = 2.5$, and final substitution in the first of (3) produces the result

$$4x_1 = 5 + 9x_2 - 2x_3 = 27.8,$$

so that $x_1 = 6.95$. Substitution in the second and third of the original set (1) verifies the accuracy of our results.

2. The solution is unique, but there are obviously other methods of elimination which produce different 'derived' sets of equations, and in the artificial school examples the coefficients were often arranged so that some ingenuity could be used to lighten the arithmetic. For example we might notice that a combination of -0.5 times the second added to the third of (1) produces an equation from which both x_1 and x_3 are eliminated simultaneously, so that x_2 is obtained without further ado. Instead of the set (2), for example, we could produce the equivalent set

$$\left. \begin{aligned} 4x_1 - 9x_2 + 2x_3 &= 5 \\ 0.5x_2 + 5x_3 &= 0.5 \\ x_2 &= 2.5 \end{aligned} \right\}, \qquad (4)$$

and the process of back-substitution can be performed without further elimination, the unknowns now being calculated in the order x_2, x_3, x_1.

In practical problems the search for such simplifications is virtually useless, since such possibilities are rare in equations in up to say 100 unknowns, with coefficients which are not simple integers but perhaps ten-decimal numbers. Moreover the use of high-speed electronic computing machines is becoming increasingly common and economic in problems in linear algebra, and for these machines we prefer standard programmes which are straightforward and independent of accidental and random relationships between the coefficients.

We therefore carry out the elimination in a sequence of operations which is relatively independent of the coefficients of the given equations, the qualification 'relatively' becoming apparent in the sequel.

The Gauss methods

3. In the elimination which produced equations (2) the first of the original set (1) was involved in the production of all of (2), and in these circumstances we call this the *pivotal* equation, though it is more common to stick to matrix ideas and to refer to the *pivotal row*. The coefficient in the pivotal row of the unknown which is currently to be eliminated is called the *pivot*. Once the pivot has been selected the sequence of operations in this stage of the elimination is well-defined and unique.

In the elimination resulting in equations (4), however, there was no pivot in this sense, since the derived equations were obtained respectively from the first, first and second, and second and third of the original set (1), and no member of (1) is involved directly in every member of (4). In a large set of equations there are obviously many elimination possibilities of this kind, but in general no particular selection will have significant advantage and the method is therefore less systematic.

4. In the example of (1) the pivots were taken in order down the principal diagonal of the matrix of coefficients, and in each case multiples of the pivotal row were added to all other rows from which the particular column elements were to be eliminated, that is to all rows *below* the pivotal row. A less rigid choice of pivotal rows is possible, and the phrase 'all rows below the pivotal row' is then replaced by 'all rows which have not previously been pivotal'.

†5. In calculation by desk machine the work and arrangement of the computing sheet in the first method would have the appearance of Table 1, with all unnecessary symbols suppressed.

TABLE 1

m	A			b	$s^{(1)}$
	4	−9	2	5	2
−0·5	$\overline{2}$	−4	6	3	7
−0·25	1	−1	3	4	7
	$\underline{0·5}$	5		0·5	6
−2·5		1·25	2·5	2·75	6·5
			−10	1·5	−8·5
x'	6·95	2·5	−0·15		
$s^{(2)\prime}$	7	−14	11	12	

The factors m, multiplying the pivotal rows, are recorded on the same lines as the rows containing the respective terms to be eliminated, and the sum column $s^{(1)}$ is used to check the elimination. The pivotal rows are recorded once only, and the pivots are underlined to indicate the equations which are used in the back substitution. The results are given in the row labelled x', the prime denoting 'row-vector' in normal matrix notation. The row $s^{(2)\prime}$ is formed from the column sums of A and b, and the scalar product of the elements of x with the corresponding elements of $s^{(2)}$ should be equal to the last element of $s^{(2)}$, the sum of the elements of b. This checks the back-substitution and is equivalent to substituting the alleged solution into the equation formed from the sum of the original equations. A final check of this kind is essential, and must use *all* the original equations: substitution into $(n-1)$ of the original equations does not suffice, since it is possible for these equations to be satisfied by a result containing an error made at some stage, while the remaining equation is not satisfied.

6. As a by-product of this work we can easily calculate the determinant of A. The assembly of pivotal rows forms the matrix

$$\begin{bmatrix} 4 & -9 & 2 \\ 0 & 0\cdot5 & 5 \\ 0 & 0 & -10 \end{bmatrix}, \tag{5}$$

and this has been obtained from A by operations which, according to determinantal rules, do not alter the value of $|A|$. The determinant of (5) is just the product of its diagonal terms, that is of the pivotal elements of the elimination, so that we have immediately the result $|A| = 4(0\cdot5)(-10) = -20$. Other determinants derived from A are also calculable from the pivots. It is clear, for example, that the last row and column of A affect the results only in the last column of the final derived form of type (5), so that the product of the first $(n-1)$ pivots is equal to the determinant derived from A by omitting its last row and column. In general the product of the first $(n-r)$ pivots is equal to the determinant of the matrix obtained from A by omitting the last r rows and columns of A, that is the leading sub-matrix of order $(n-r)$.

7. This choice of pivots is obviously very convenient both for desk-machine work and for programming in high-speed computation, but unfortunately it is not always possible. It breaks down, for example, whenever any pivot (except the very last), including perhaps the first term of the original equations, is zero, since the multipliers are then infinite or indeterminate. Again we wish to have a standard unique process which avoids this possibility, and a convenient method is to choose as pivot the coefficient of largest absolute value in the relevant column of the reduced matrix, taking the columns successively so that the unknowns are eliminated in natural order. If all the elements in a column become zero our method fails, but this is not surprising since in that case the determinant vanishes, the matrix is singular, and the equations have no unique solution.

With this method, called 'partial pivoting', the computing sheet has the appearance of Table 2.

TABLE 2

m	A			b	$s^{(1)}$
	$\underline{4}$	−9	2	5	2
−0·5	2	−4	6	3	7
−0·25	1	−1	3	4	7
−0·4		0·5	5	0·5	6
		$\underline{1·25}$	2·5	2·75	6·5
			$\underline{4}$	−0·6	3·4
x'	6·95	2·5	−0·15		
$s^{(2)\prime}$	7	−14	11	12	

We note several differences between the two methods. First, the search for the largest coefficient, to be used as pivot, needs extra programming for the automatic machine, and takes a time which compared with the search by eye of a desk-machine operator may not be negligible. Second, in this choice all the multipliers are less than unity in absolute value, a circumstance which is useful in keeping numbers within range, particularly for a fixed-point machine. Third, this method has definite advantages in accuracy, as we shall see later, when the results of any calculations are rounded, which happens in almost all practical problems.

8. Finally, we need extra programming or other effort in order to calculate $|A|$. In this example the order of pivotal rows, compared with that of the rows of A, is 1, 3 and 2, and we have reduced A, by operations which do not alter $|A|$, to the form

$$\begin{bmatrix} 4 & -9 & 2 \\ 0 & 0 & 4 \\ 0 & 1·25 & 2·5 \end{bmatrix}. \tag{6}$$

Unlike the result (5) this matrix is not triangular, and the determinant is not just the product of the pivots. We can transform (6) into an upper triangle by interchanging certain rows, but each interchange alters the sign of the determinant. We deduce that

$|A| = (-1)^p$ times the product of the pivots, where p is the number of such interchanges. Here we need only interchange rows 2 and 3, so that $|A| = (-1)(4)(4)(1 \cdot 25) = -20$ as before.

The extra programming is required to keep track of the selected order of pivotal rows, and to count the number of interchanges needed to transform the final assembly of these rows into an upper triangle. With automatic computing a convenient method is to interchange in the store the pivotal row, where necessary, with the top row of the current matrix, adding one to a count, initially zero, at every such operation. If the final count is odd the sign of the product of the pivots must be changed to give $|A|$.

In spite of the extra programming and the time used with the second method its advantages are so great in practice that the first method is used only in special circumstances, when for example it is known in advance that no 'diagonal pivot' can be small.

Jordan elimination

9. It is possible to extend the elimination processes of the Gauss method so that the equations are reduced to a form in which the matrix is diagonal, so that no back-substitution is required. The pivots are chosen as in the Gauss method, but elements are also eliminated from the previous pivotal rows, the actual pivots being unaffected since the other elements in previous pivotal columns are zero. Tables 3 and 4 show the calculations for each method corresponding respectively to those of Tables 1 and 2. In each case the final check could be completed as before.

TABLE 3

m	A			b	$s^{(1)}$
	4	−9	2	5	2
−0·5	2	−4	6	3	7
−0·25	1	−1	3	4	7
18·0	4	−9	2	5	2
	0	0·5	5	0·5	6
−2·5	0	1·25	2·5	2·75	6·5
9·2	4	0	92	14	110
0·5	0	0·5	5	0·5	6
	0	0	−10	1·5	−8·5
	4	0	0	27·8	31·8
	0	0·5	0	1·25	1·75
	0	0	−10	1·5	−8·5
x'	6·95	2·5	−0·15		

In the first method the operations on the matrix A have produced in the end a diagonal matrix. In the second method the final matrix is not immediately diagonal, but can be made so by suitable inter-

TABLE 4

m		A		b	$s^{(1)}$
	$\underline{4}$	-9	2	5	2
$-0{\cdot}5$	2	-4	6	3	7
$-0{\cdot}25$	1	-1	3	4	7
$7{\cdot}2$	4	-9	2	5	2
$-0{\cdot}4$	0	$0{\cdot}5$	5	$0{\cdot}5$	6
	0	$\underline{1{\cdot}25}$	$2{\cdot}5$	$2{\cdot}75$	$6{\cdot}5$
-5	4	0	20	$24{\cdot}8$	$48{\cdot}8$
	0	0	$\underline{4}$	$-0{\cdot}6$	$3{\cdot}4$
$-0{\cdot}625$	0	$1{\cdot}25$	$2{\cdot}5$	$2{\cdot}75$	$6{\cdot}5$
	4	0	0	$27{\cdot}8$	$31{\cdot}8$
	0	0	4	$-0{\cdot}6$	$3{\cdot}4$
	0	$1{\cdot}25$	0	$3{\cdot}125$	$4{\cdot}375$
x'	$6{\cdot}95$	$2{\cdot}5$	$-0{\cdot}15$		

changes of rows, each interchange being carried out in automatic computation as soon as a pivotal row has been selected. If $|A|$ only is required but not the solution x the extra work in Jordan is unnecessary, since the Gauss upper triangle provides all relevant information. Other relative advantages of Gauss and Jordan elimination, including the amounts of arithmetic involved, are mentioned in Chapter 7.

Calculation of the inverse

10. If we have more than one right-hand side with the same matrix A we can of course perform the elimination as before, extra columns on the right being treated together, and a separate back-substitution must then be carried out for each column. In particular we can find A^{-1} by replacing b by the columns of the unit matrix. The n solutions so obtained then form the n columns of A^{-1}. If the solutions are written in rows as in Tables 1–4 the aggregate of these rows forms the transposed inverse $(A^{-1})'$. The computation for the Gauss method, corresponding to that of Table 2, is shown in Table 5. A single sum column $s^{(1)}$ checks all the elimination. Each row of $(A^{-1})'$ could be checked as in Table 2, or alternatively a complete check is provided by the scalar product of $s^{(2)'}(A)$ and $s^{(2)}(A^{-1})'$, the column sums of the respective matrices, which should be equal to the order n of A. The latter check, of course, throws less light on the location of any error than does the former.

TABLE 5

m		A			I		$s^{(1)}$
	$\underline{4}$	-9	2	1	0	0	-2
-0.5	$\underline{2}$	-4	6	0	1	0	5
-0.25	1	-1	3	0	0	1	4
-0.4		0.5	5	-0.5	1	0	6
		$\underline{1.25}$	2.5	-0.25	0	1	4.5
			$\underline{4}$	-0.4	1	-0.4	4.2

	$(A^{-1})'$		
	0.3	0	-0.1
	-1.25	-0.5	0.25
	2.3	1.0	-0.1

$s^{(2)'}A$	7	-14	11	(Product $=$ 3)
$s^{(2)'}(A^{-1})'$	1.35	0.5	0.05	

11. The corresponding Jordan method is shown in Table 6, with the final check omitted.

TABLE 6

m		A			I		$s^{(1)}$
	$\underline{4}$	-9	2	1	0	0	-2
-0.5	$\underline{2}$	-4	6	0	1	0	5
-0.25	1	-1	3	0	0	1	4
7.2	4	-9	2	1	0	0	-2
-0.4	0	0.5	5	-0.5	1	0	6
	0	$\underline{1.25}$	2.5	-0.25	0	1	4.5
-5	4	0	20	-0.8	0	7.2	30.4
	0	0	$\underline{4}$	-0.4	1	-0.4	4.2
-0.625	0	1.25	2.5	-0.25	0	1	4.5
	4	0	0	1.2	-5	9.2	9.4
	0	0	4	-0.4	1	-0.4	4.2
	0	1.25	0	0.0	-0.625	1.25	1.875

	$(A^{-1})'$		
	0.3	0	-0.1
	-1.25	-0.5	0.25
	2.3	1.0	-0.1

It is also true that if the first quasi-diagonal form of A on the left of Table 6 is arranged by interchanges into a true diagonal, and thence into a unit matrix by dividing the rows by their diagonal terms, the corresponding operations on the final matrix on the right of Table 6 will in fact replace that matrix by the required A^{-1}, without transposition.

Matrix equivalent of elimination

12. This last fact becomes apparent when we consider some matrix equivalents of the various elimination methods described for a general set of equations given by

$$
\begin{array}{ccc}
A & x & b \\
\begin{bmatrix}
a_{11} & a_{12} & \cdots & a_{1n} \\
a_{21} & a_{22} & \cdots & a_{2n} \\
\cdot & & & \\
\cdot & & & \\
\cdot & & & \\
a_{n1} & a_{n2} & \cdots & a_{nn}
\end{bmatrix}
&
\begin{bmatrix}
x_1 \\
x_2 \\
\cdot \\
\cdot \\
\cdot \\
x_n
\end{bmatrix}
&=&
\begin{bmatrix}
b_1 \\
b_2 \\
\cdot \\
\cdot \\
\cdot \\
b_n
\end{bmatrix}.
\end{array}
\tag{7}
$$

In the first Gauss process, with pivots taken in order down the diagonal, it is easily verified that the first reduced set of equations has the matrix representation

$$
J_1 A x = J_1 b,
\tag{8}
$$

where J_1 has the form

$$
J_1 =
\begin{bmatrix}
1 & & & & & \\
m_{21} & 1 & & & & \\
m_{31} & 0 & 1 & & & \\
\cdot & & & & & \\
\cdot & & & & & \\
\cdot & & & & & \\
m_{n1} & 0 & 0 & \cdots & 0 & 1
\end{bmatrix},
\tag{9}
$$

and $m_{r1} = -a_{r1}/a_{11}$. The second reduction is equivalent to the matrix operation

$$
J_2 J_1 A x = J_2 J_1 b,
$$

where

$$
J_2 = \begin{bmatrix}
1 & & & & & \\
0 & 1 & & & & \\
0 & m_{32} & 1 & & & \\
\cdot & & & & & \\
\cdot & & & & & \\
\cdot & & & & & \\
0 & m_{n2} & 0 & \cdots & 0 & 1
\end{bmatrix}, \tag{10}
$$

the terms m_{r2} being the second set of multipliers, their number decreasing by unity in each successive operation.

13. The final set of equations is given by

$$
J_{n-1} \ldots J_2 J_1 A x = J_{n-1} \ldots J_2 J_1 b, \tag{11}
$$

and all the J_r are lower triangular matrices. The product J of the lower triangles is itself a lower triangle, so that our final equations are equivalent to the matrix equation

$$
JAx = Jb. \tag{12}
$$

We have seen that the final matrix operating on x is in fact an upper triangle U, so that on the left-hand side we have carried out a process equivalent to the operation

$$
JA = U, \tag{13}
$$

and have determined U and Jb by a definite sequence of computations, recording at each stage the relevant terms in $J_1 A$, $J_1 b$, $J_2 J_1 A$, $J_2 J_1 b$, etc.

14. Equation (13) implies that

$$
A = J^{-1} U, \tag{14}
$$

and since the inverse of a lower triangle is itself a lower triangle we have performed a process equivalent to the expression of A in terms of the product of a lower and upper triangle. The elements of J^{-1}, in terms of the multipliers m_{ij}, are easily obtained. For

$$
J^{-1} = J_1^{-1} J_2^{-1} \ldots J_{n-1}^{-1}, \tag{15}
$$

and we easily find

$$
J_1^{-1} =
\begin{bmatrix}
1 & & & & & & \\
-m_{21} & 1 & & & & & \\
-m_{31} & 0 & 1 & & & & \\
\cdot & & & & & & \\
\cdot & & & & & & \\
\cdot & & & & & & \\
-m_{n1} & 0 & 0 & \cdots & 0 & 1
\end{bmatrix},
$$

(16)

$$
J_2^{-1} =
\begin{bmatrix}
1 & & & & & \\
0 & 1 & & & & \\
0 & -m_{32} & 1 & & & \\
\cdot & & & & & \\
\cdot & & & & & \\
\cdot & & & & & \\
0 & -m_{n2} & 0 & \cdots & 0 & 1
\end{bmatrix},
$$

each J_r^{-1} being identical with J_r except that the sign of each m is changed. Then

$$
J_1^{-1}J_2^{-1} =
\begin{bmatrix}
1 & & & & & \\
-m_{21} & 1 & & & & \\
-m_{31} & -m_{32} & 1 & & & \\
-m_{41} & -m_{42} & 0 & 1 & & \\
\cdot & & & & & \\
\cdot & & & & & \\
\cdot & & & & & \\
-m_{n1} & -m_{n2} & 0 & 0 & \cdots & 0 & 1
\end{bmatrix},
$$

(17)

successive post-multiplications with $J_3^{-1}, \ldots, J_{n-1}^{-1}$ effectively replace the zero elements by the new multipliers, and we have finally

$$
J^{-1} = \begin{bmatrix}
1 & & & & & & \\
-m_{21} & 1 & & & & & \\
-m_{31} & -m_{32} & 1 & & & & \\
-m_{41} & -m_{42} & -m_{43} & 1 & & & \\
\cdot & & & & & & \\
\cdot & & & & & & \\
\cdot & & & & & & \\
-m_{n1} & -m_{n2} & -m_{n3} & \cdots & -m_{n,n-1} & 1
\end{bmatrix}.
\quad (18)
$$

In the problem of Table 1, for example, it is easily verified that

$$
\overset{A}{\begin{bmatrix} 4 & -9 & 2 \\ 2 & -4 & 6 \\ 1 & -1 & 3 \end{bmatrix}}
=
\overset{J^{-1}}{\begin{bmatrix} 1 & & \\ 0{\cdot}5 & 1 & \\ 0{\cdot}25 & 2{\cdot}5 & 1 \end{bmatrix}}
\overset{U}{\begin{bmatrix} 4 & -9 & 2 \\ & 0{\cdot}5 & 5 \\ & & -10 \end{bmatrix}}.
\quad (19)
$$

The possibility of expressing a matrix in this form is an important result, which will be used in Chapter 4 as the basis of 'compact' elimination methods.

†15. The analysis corresponding to the choice of largest elements in successive columns as pivots is more complicated. For the first operation, corresponding to the choice of the element a_{i1} as pivot, the corresponding J_1 has the form

$$
J_1 = \begin{bmatrix}
1 & 0 & 0 & \cdots & m_{1i} & 0 & \cdots & 0 \\
0 & 1 & 0 & \cdots & m_{2i} & 0 & \cdots & 0 \\
\cdot & & & & & & & \\
\cdot & & & & & & & \\
\cdot & & & & & & & \\
0 & 0 & 0 & \cdots & m_{i-1,i} & 0 & \cdots & 0 \\
0 & 0 & 0 & \cdots & 1 & 0 & \cdots & 0 \\
0 & 0 & 0 & \cdots & m_{i+1,i} & 1 & \cdots & 0 \\
\cdot & & & & & & & \\
\cdot & & & & & & & \\
\cdot & & & & & & & \\
0 & 0 & 0 & \cdots & m_{ni} & 0 & \cdots & 1
\end{bmatrix},
\quad (20)
$$

and again it is easily verified that J_1^{-1} has the form of J_1 with the signs changed of the coefficients m.

Detailed analysis in a particular case will illustrate the general arrangement. Suppose that for a matrix A of order four the pivotal rows are taken in order 3, 1, 4, 2. The matrices J_1, J_2, J_3 then have the respective forms

$$
\overset{J_1}{\begin{bmatrix} 1 & 0 & m_{13} & 0 \\ 0 & 1 & m_{23} & 0 \\ 0 & 0 & 1 & 0 \\ 0 & 0 & m_{43} & 1 \end{bmatrix}},
\overset{J_2}{\begin{bmatrix} 1 & 0 & 0 & 0 \\ m_{21} & 1 & 0 & 0 \\ 0 & 0 & 1 & 0 \\ m_{41} & 0 & 0 & 1 \end{bmatrix}},
\overset{J_3}{\begin{bmatrix} 1 & 0 & 0 & 0 \\ 0 & 1 & 0 & m_{24} \\ 0 & 0 & 1 & 0 \\ 0 & 0 & 0 & 1 \end{bmatrix}}, \quad (21)
$$

and inversion merely changes the signs of the off-diagonal terms. We find

$$
J_1^{-1}J_2^{-1}J_3^{-1} = \begin{bmatrix} 1 & 0 & -m_{13} & 0 \\ -m_{21} & 1 & -m_{23} & -m_{24} \\ 0 & 0 & 1 & 0 \\ -m_{41} & 0 & -m_{43} & 1 \end{bmatrix}, \quad (22)
$$

and the matrix $J_3J_2J_1A$, with pivots underlined, looks like

$$
J_3J_2J_1A = \begin{bmatrix} 0 & \underline{\times} & \times & \times \\ 0 & 0 & 0 & \underline{\times} \\ \underline{\times} & \times & \times & \times \\ 0 & 0 & \underline{\times} & \times \end{bmatrix}. \quad (23)
$$

At this stage we have carried out the operation $JA = C$, where $J = J_3J_2J_1$, and C is the matrix in (23). We note that C is a row-permutation of an upper triangle U, so that $I_rC = U$, where here

$$
I_r = \begin{bmatrix} 0 & 0 & 1 & 0 \\ 1 & 0 & 0 & 0 \\ 0 & 0 & 0 & 1 \\ 0 & 1 & 0 & 0 \end{bmatrix}, \quad (24)
$$

the position of the 1 in successive rows of I_r having the numbers

3, 1, 4, 2, the order of the successive pivotal rows. We then have

$$A = J^{-1}I_r^{-1}U, \tag{25}$$

where

$$I_r^{-1} = \begin{bmatrix} 0 & 1 & 0 & 0 \\ 0 & 0 & 0 & 1 \\ 1 & 0 & 0 & 0 \\ 0 & 0 & 1 & 0 \end{bmatrix} \quad \text{and} \quad J^{-1}I_r^{-1} = \begin{bmatrix} \times & 1 & 0 & 0 \\ \times & \times & \times & 1 \\ 1 & 0 & 0 & 0 \\ \times & \times & 1 & 0 \end{bmatrix}. \tag{26}$$

The latter is not a lower triangle, but is a row-permutation of a lower triangle, and we find that

$$I_r J^{-1} I_r^{-1} = \begin{bmatrix} 1 & 0 & 0 & 0 \\ -m_{13} & 1 & 0 & 0 \\ -m_{43} & -m_{41} & 1 & 0 \\ -m_{23} & -m_{21} & -m_{24} & 1 \end{bmatrix}, \tag{27}$$

a unit lower triangle. We have therefore effectively performed the operation of expressing $I_r A$, a row-permutation of A, as a product of lower and upper triangles. The row-permutation is of course equivalent to an interchange of the rows of A so that successive pivots lie on the main diagonal. Comparison of (27) with the separate J_r matrices in (21) shows the relation of the lower triangle to its original components.

16. In the Jordan elimination the equivalent effect is again successive premultiplication by matrices of type J_r. If pivots are taken down the diagonal the first multiplying matrix, J_1, is exactly as before, and given in (9), but in successive J_r there are terms m_{rs} in all positions in the pivotal column, including the elements of previous pivotal rows. For example

$$J_2 = \begin{bmatrix} 1 & m_{12} & 0 & \dots & 0 \\ 0 & 1 & 0 & \dots & 0 \\ 0 & m_{32} & 1 & \dots & 0 \\ \cdot & & & & \\ \cdot & & & & \\ \cdot & & & & \\ 0 & m_{n2} & 0 & \dots & 1 \end{bmatrix}, \tag{28}$$

and the product J of the J_r is such that JA is a diagonal matrix. The inverse of each J_r is like J_r with the signs changed of the multipliers, but there is no simple form for J^{-1} in terms of the m_{rs}.

If the order of pivotal rows is varied, it is also clear that the operation JA produces a row permutation of a diagonal matrix and, as in the Gauss process, final multiplication by I_r, a matrix of the form of (24), will produce a true diagonal D. We therefore find

$$D^{-1}I_r JA = I, \tag{29}$$

so that $D^{-1}I_r J = A^{-1}$, illustrating a remark at the end of § 11, that if operations on A produce I then the same operations on I produce A^{-1}.

17. As numerical illustration, for the example of Table 6 we have

$$
\overset{J_3}{\begin{bmatrix} 1 & -5 & 0 \\ 0 & 1 & 0 \\ 0 & -0\cdot625 & 1 \end{bmatrix}}
\overset{J_2}{\begin{bmatrix} 1 & 0 & 7\cdot2 \\ 0 & 1 & -0\cdot4 \\ 0 & 0 & 1 \end{bmatrix}}
\overset{J_1}{\begin{bmatrix} 1 & 0 & 0 \\ -0\cdot5 & 1 & 0 \\ -0\cdot25 & 0 & 1 \end{bmatrix}}
\overset{A}{\begin{bmatrix} 4 & -9 & 2 \\ 2 & -4 & 6 \\ 1 & -1 & 3 \end{bmatrix}}
= \overset{JA}{\begin{bmatrix} 4 & 0 & 0 \\ 0 & 0 & 4 \\ 0 & 1\cdot25 & 0 \end{bmatrix}}. \tag{30}
$$

Then

$$
I_r JA =
\overset{I_r}{\begin{bmatrix} 1 & 0 & 0 \\ 0 & 0 & 1 \\ 0 & 1 & 0 \end{bmatrix}}
\overset{JA}{\begin{bmatrix} 4 & 0 & 0 \\ 0 & 0 & 4 \\ 0 & 1\cdot25 & 0 \end{bmatrix}}
= \overset{I_r JA}{\begin{bmatrix} 4 & 0 & 0 \\ 0 & 1\cdot25 & 0 \\ 0 & 0 & 4 \end{bmatrix}}, \tag{31}
$$

and $D^{-1}I_r JA = $
$$
\overset{D^{-1}}{\begin{bmatrix} 0\cdot25 & 0 & 0 \\ 0 & 0\cdot8 & 0 \\ 0 & 0 & 0\cdot25 \end{bmatrix}}
\overset{I_r JA}{\begin{bmatrix} 4 & 0 & 0 \\ 0 & 1\cdot25 & 0 \\ 0 & 0 & 4 \end{bmatrix}}
= \overset{I}{\begin{bmatrix} 1 & 0 & 0 \\ 0 & 1 & 0 \\ 0 & 0 & 1 \end{bmatrix}}. \tag{32}
$$

Further calculation gives

$$
J = J_3 J_2 J_1 = \begin{bmatrix} 1\cdot2 & -5 & 9\cdot2 \\ -0\cdot4 & 1 & -0\cdot4 \\ 0 & -0\cdot625 & 1\cdot25 \end{bmatrix},
$$

$$
D^{-1}I_r J = \begin{bmatrix} 0\cdot3 & -1\cdot25 & 2\cdot3 \\ 0 & -0\cdot5 & 1\cdot0 \\ -0\cdot1 & 0\cdot25 & -0\cdot1 \end{bmatrix},
\tag{33}
$$

the first of (33) being the final matrix on the right of Table 6 and the second of (33) giving the inverse A^{-1}.

The method of Aitken

18. The methods of Gauss and Jordan can produce solutions X of the matrix equation

$$AX = B, \qquad (34)$$

where the matrices are conformable for multiplication. If B is a single column b then X is a column x, and we have solved one set of linear equations. If B is a matrix with several columns $b^{(r)}$ then X has the same number of columns $x^{(r)}$, and each $x^{(r)}$ is obtained from the solution of $Ax^{(r)} = b^{(r)}$; the total work involves one elimination process, applied simultaneously to A and B, and several back-substitutions, one for each column $x^{(r)}$. In particular if B is the unit I, then $X = A^{-1}$, the inverse of A. The solution of (34) is

$$X = A^{-1}B, \qquad (35)$$

and this general result includes the particular cases mentioned.

The method of Aitken allows the calculation of a more elaborate expression

$$Y = CA^{-1}B, \qquad (36)$$

where the matrices C, A and B are conformable for multiplication. The result (36) might follow from the elimination of X from the pair of matrix equations

$$\left. \begin{array}{l} AX = B \\ CX = Y \end{array} \right\}, \qquad (37)$$

in which, for example, A is $(m \times m)$, B is $(m \times n)$, C is $(p \times m)$, and Y is to be determined. Consider first the application of the Gauss elimination process to replace the first of (37) by the equation

$$UX = F, \qquad (38)$$

obtained by premultiplication with a suitable lower triangular matrix. Next we take the array

$$\left[\begin{array}{c|c} U & F \\ \hline C & O \end{array} \right], \qquad (39)$$

where O is a null matrix of order $(p \times n)$, and add multiples of the first row of U to successive rows of C so that the elements in the first column of C are replaced by zeros. Further addition of multiples of the second row of U to those of C can produce zeros in the second column of C, without affecting its first column since U is upper triangular. Ultimately C will be replaced by a null matrix, and similar operations with F, added to the rows of O in (39), will replace O by a matrix of the same shape. This matrix is in fact $-CU^{-1}F$.

To see this we observe that we have performed the matrix operation

$$\left[\begin{array}{c|c} I & O \\ \hline K & I \end{array}\right]\left[\begin{array}{c|c} U & F \\ \hline C & O \end{array}\right] = \left[\begin{array}{c|c} U & F \\ \hline P & Q \end{array}\right], \tag{40}$$

where

$$P = O = KU + C, \quad Q = KF. \tag{41}$$

The first of (41) gives $K = -CU^{-1}$, so that $Q = -CU^{-1}F$. Further, we have $X = U^{-1}F$ from (38), and this is the same as $A^{-1}B$ from (37), so that, finally, $Q = -CA^{-1}B$.

19. If we start with the matrix operation

$$\left[\begin{array}{c|c} I & O \\ \hline K & I \end{array}\right]\left[\begin{array}{c|c} A & B \\ \hline C & O \end{array}\right] = \left[\begin{array}{c|c} A & B \\ \hline KA+C & KB \end{array}\right], \tag{42}$$

and choose K so that $KA + C = O$, then $KB = -CA^{-1}B$ as before. The reason for first reducing A to an upper triangle U is to facilitate the operation of reducing $KA + C$ to O, a process, as we saw, very easy if A is a U. The reduction of A to U, followed by that of C to O can of course be performed simultaneously, treating the rows of C for the purposes of elimination as if they belonged to A, pivots being chosen only from A. The relevant matrix operation is then

$$\left[\begin{array}{c|c} J & O \\ \hline K & I \end{array}\right]\left[\begin{array}{c|c} A & B \\ \hline C & O \end{array}\right] = \left[\begin{array}{c|c} JA & JB \\ \hline KA+C & KB \end{array}\right], \tag{43}$$

where J is a unit lower triangle and $KA + C = O$.

20. As numerical illustration we take the example of Table 7, in which pivotal rows are discarded as soon as they have been used.

TABLE 7

m	A			B	
	4	−9	2	3	1
−0·5	2	−4	6	−1	4
−0·25	1	−1	3	2	1
	C			O	
−0·25	1	2	3	0	0
−1	4	−5	6	0	0
	4	−9	2	3	1
	0	0·5	5	−2·5	3·5
−2·5	0	1·25	2·5	1·25	0·75
−8·5	0	4·25	2·5	−0·75	−0·25
−8·0	0	4·0	4·0	−3·0	−1·0
	4	−9	2	3	1
	0	0·5	5	−2·5	3·5
	0	0	−10	7·5	−8·0
−4·0	0	0	−40	20·5	−30·0
−3·6	0	0	−36	17·0	−29·0
	4	−9	2	3	1
	0	0·5	5	−2·5	3·5
	0	0	−10	7·5	−8·0
	0	0	0	−9·5	2·0
	0	0	0	−10·0	−0·2

The (2×2) matrix in the bottom right corner is $-CA^{-1}B$, and it can be verified that the final array comes from the matrix multiplications

$$
\left[\begin{array}{ccc|cc} 1 & 0 & 0 & 0 & 0 \\ 0 & 1 & 0 & 0 & 0 \\ 0 & 0 & 1 & 0 & 0 \\ \hline 0 & 0 & -4\cdot0 & 1 & 0 \\ 0 & 0 & -3\cdot6 & 0 & 1 \end{array}\right]
\left[\begin{array}{ccc|cc} 1 & 0 & 0 & 0 & 0 \\ 0 & 1 & 0 & 0 & 0 \\ 0 & -2\cdot5 & 1 & 0 & 0 \\ \hline 0 & -8\cdot5 & 0 & 1 & 0 \\ 0 & -8\cdot0 & 0 & 0 & 1 \end{array}\right]
\left[\begin{array}{ccc|cc} 1 & 0 & 0 & 0 & 0 \\ -0\cdot5 & 1 & 0 & 0 & 0 \\ -0\cdot25 & 0 & 1 & 0 & 0 \\ \hline -0\cdot25 & 0 & 0 & 1 & 0 \\ -1\cdot0 & 0 & 0 & 0 & 1 \end{array}\right]
\left[\begin{array}{ccc|cc} 4 & -9 & 2 & 3 & 1 \\ 2 & -4 & 6 & -1 & 4 \\ 1 & -1 & 3 & 2 & 1 \\ \hline 1 & 2 & 3 & 0 & 0 \\ 4 & -5 & 6 & 0 & 0 \end{array}\right],
$$

$$(44)$$

where in each case the top left-hand matrix in the multipliers is the J_r of the simple Gauss process.

21. It is again desirable to choose as pivot the largest element in successive columns of A and its derived form, and the corresponding calculations are shown in Table 8, where, as in desk computing, only essential quantities are recorded and sum checks are included. We discuss this point further in § 31 of Chapter 4.

TABLE 8

m	A			B		$s^{(1)}$
	4	−9	2	3	1	1
−0·5	2	−4	6	−1	4	7
−0·25	1	−1	3	2	1	6
	C			O		
−0·25	1	2	3	0	0	6
−1·0	4	−5	6	0	0	5
−0·4		0·5	5·0	−2·5	3·5	6·5
		1·25	2·5	1·25	0·75	5·75
−3·4		4·25	2·5	−0·75	−0·25	5·75
−3·2		4·0	4·0	−3·0	−1·0	4·0
			4·0	−3·0	3·2	4·2
1·5			−6·0	−5·0	−2·8	−13·8
1·0			−4·0	−7·0	−3·4	−14·4
				−9·5	2·0	−7·5
				−10·0	−0·2	−10·2

Again the matrix $-CA^{-1}B$ finally replaces the original null matrix, and since the order neither of the rows of C nor the columns of B is altered the final result is not affected by the new choice of pivots from A.

22. Again there are useful special cases. For example if B is a column b and C a single row y', we produce $y'A^{-1}b = y'x$, that is a linear combination of the elements x of the solution of $Ax = b$, without actually evaluating the solution.

If B is I and C is I we replace the null matrix by $-A^{-1}$, and in no case need we perform any back-substitution, elimination with suitable choice of pivots being the only operation for which programming is required.

23. There may be advantages in applying a different version of the Aitken process. The equations

$$\left. \begin{matrix} A'X = C'' \\ B'X = Y' \end{matrix} \right\}$$ (45)

produce, on elimination of X, the result $Y' = B'(A')^{-1}C'$, the transpose of the result corresponding to (37). Instead of the former array we could therefore use

$$\left[\begin{array}{c|c} A' & C' \\ \hline B' & O \end{array}\right], \tag{46}$$

and eliminate as before to produce the required result in place of O. In particular if B' has fewer rows than C the second method involves the calculation of fewer multipliers than the first. We consider in Chapter 7 the respective amounts of arithmetic involved.

The symmetric case

24. In many practical cases the matrix A is symmetric, and if pivots can be taken down the diagonal each reduced matrix is symmetric, so that the labour of calculating the upper triangular transformation of A is almost halved. This result is easily proved by considering the set of equations (7) and evaluating the coefficients of the first reduction following the choice of a_{11} as pivot. We find

$$(a_{rs})_2 = (a_{rs})_1 + m_{r1}(a_{1s})_1, \tag{47}$$

for the new elements in rows and columns other than the first. Moreover

$$m_{r1} = -(a_{r1}/a_{11})_1 = -(a_{1r}/a_{11})_1, \tag{48}$$

so that

$$(a_{rs})_2 = (a_{rs})_1 - (a_{r1}a_{1s}/a_{11})_1, \tag{49}$$

and the symmetry of A, giving $(a_{pq})_1 = (a_{qp})_1$, shows that

$$(a_{rs})_2 = (a_{sr})_2,$$

so that the reduced matrix is symmetric in the rows and columns other than the first. Only the upper triangular part need be recorded, and the pivots are obtained from the second form of (48). Table 9 illustrates the computation.

TABLE 9

m		A		b	$s^{(1)}$
	4	5	2	1	12
$-1\cdot25$		6	-1	2	12
$-0\cdot50$			2	0	3
		$-0\cdot25$	$-3\cdot5$	$0\cdot75$	-3
$-14\cdot0$			$1\cdot0$	$-0\cdot50$	-3
			$50\cdot0$	$-11\cdot0$	39
x'		$0\cdot26$	$0\cdot08$	$-0\cdot22$	
$s^{(2)'}$		11	10	3	3

A point of the computation is the use of symmetry in the calculation of the sum checks. For any particular row we add all elements recorded both in the row and also in the corresponding column.

25. We saw in § 14 that the final upper triangle U is related to A through an equation of the form $A = LU$, where L is a lower triangle simply derived from the multipliers in the elimination process. In the present example we have, from equation (18),

$$L = \begin{bmatrix} 1 & & \\ 1{\cdot}25 & 1 & \\ 0{\cdot}50 & 14{\cdot}0 & 1 \end{bmatrix}, \qquad U = \begin{bmatrix} 4 & 5 & 2 \\ & -0{\cdot}25 & -3{\cdot}5 \\ & & 50{\cdot}0 \end{bmatrix}. \qquad (50)$$

Now it is clear from the symmetry of A and its reduced forms that the multipliers contained in L are calculable directly from the rows of U, and indeed the matrix L' is identical with U if the rows of the latter are divided by their diagonal elements. We then find the equation

$$A = LDL'. \qquad (51)$$

In the present example we have

$$\begin{array}{cccc} A & L & D & L' \end{array}$$
$$\begin{bmatrix} 4 & 5 & 2 \\ 5 & 6 & -1 \\ 2 & -1 & 2 \end{bmatrix} = \begin{bmatrix} 1 & & \\ 1{\cdot}25 & 1 & \\ 0{\cdot}50 & 14{\cdot}0 & 1 \end{bmatrix}\begin{bmatrix} 4 & & \\ & -0{\cdot}25 & \\ & & 50 \end{bmatrix}\begin{bmatrix} 1 & 1{\cdot}25 & 0{\cdot}50 \\ & 1 & 14{\cdot}0 \\ & & 1 \end{bmatrix},$$
$$(52)$$

the matrices L and L' being *unit* triangles, and the elements of D are the diagonal terms of U, the pivots in the elimination process.

The symmetric, positive-definite case

26. This method will fail if any of the pivots is zero and may be inaccurate if any pivot is small. For this reason it is quite common, in general cases, to ignore symmetry and use largest coefficients as pivots despite the extra time and labour involved.

In one particular case, however, this method is valuable, when it is known in advance that no diagonal element can vanish. In general this is not known, but in a certain class of problem, governed by the conditions that an essentially positive quadratic form shall have its minimum value, the diagonal terms will certainly be positive.

†27. This will arise, for example, with 'normal equations' derived from the method of 'least squares', in which, for a problem with more equations than unknowns, we seek that solution for which the sum of the squares of the *residuals*, the elements of the vector $Ax - b$, is a minimum.

Consider the equations

$$\left.\begin{aligned} a_{11}x_1 + a_{12}x_2 + a_{13}x_3 &= b_1 \\ a_{21}x_1 + a_{22}x_2 + a_{23}x_3 &= b_2 \\ a_{31}x_1 + a_{32}x_2 + a_{33}x_3 &= b_3 \\ a_{41}x_1 + a_{42}x_2 + a_{43}x_3 &= b_4 \end{aligned}\right\}. \tag{53}$$

The sum of the squares of the residuals is

$$S = (a_{11}x_1 + a_{12}x_2 + a_{13}x_3 - b_1)^2 + \dots + (a_{41}x_1 + a_{42}x_2 + a_{43}x_3 - b_4)^2, \tag{54}$$

and a necessary condition that S should be a minimum with respect to variations in the quantities x_1, x_2, x_3 is that

$$\frac{\partial S}{\partial x_1} = \frac{\partial S}{\partial x_2} = \frac{\partial S}{\partial x_3} = 0. \tag{55}$$

The first of (55) gives

$$a_{11}(a_{11}x_1 + a_{12}x_2 + a_{13}x_3 - b_1) + \dots + a_{41}(a_{41}x_1 + a_{42}x_2 + a_{43}x_3 - b_4) = 0, \tag{56}$$

and when the terms are combined we have a linear equation of the form

$$c_{11}x_1 + c_{12}x_2 + c_{13}x_3 = d_1, \tag{57}$$

in which

$$c_{11} = C_1(A)C_1(A),\ c_{12} = C_1(A)C_2(A),\ c_{13} = C_1(A)C_3(A),$$
$$d_1 = C_1(A)C(b). \tag{58}$$

Other similar equations are obtained from the last two of (55), and we derive finally a set of three equations, called the *normal equations*, which we can write as

$$Cx = d, \tag{59}$$

in which $C = A'A$, $d = A'b$. The matrix C is clearly symmetric, and it is also called *positive definite*, in virtue of the fact that the quadratic form $x'Cx$ is positive for any real vector x. It has the property that all the pivots in the elimination method, when chosen down the diagonal, are non-zero and positive.

28. This property can be deduced from the results of § 25. For any vector x the quadratic form

$$x'Cx = x'A'Ax = (a_{11}x_1+a_{12}x_2+a_{13}x_3)^2+\ldots+(a_{41}x_1+a_{42}x_2+a_{43}x_3)^2$$
$$(60)$$

is essentially positive. We can also write $C = LDL'$, from (51), and the transformation $y = L'x$ gives

$$x'Cx = x'LDL'x = y'Dy = d_1y_1^2+d_2y_2^2+\ldots+d_ny_n^2, \qquad (61)$$

where the d_r are the diagonal terms of D. Since (61) is essentially positive all the diagonal elements of D are positive and hence also those of U in the elimination process.

Exact and approximate solutions. Integer coefficients

29. In all the examples of this chapter the original coefficients in the matrices and in the right-hand sides of linear equations were taken as integers, and they were chosen so that all subsequent operations, including divisions, could be performed exactly with no rounding errors. In all cases the final results were exact and were capable of exact representation with only a few figures.

Practical examples of this kind are rare, and we shall deal in later chapters with cases in which rounded arithmetical operations are the rule. In some problems, however, in which the original coefficients are simple integers, it may be desired to obtain exact solutions, and the methods so far described are likely to lack this facility.

30. Consider, for example, the equations

$$\left.\begin{aligned}
7x_1+9x_2- x_3+2x_4 &= 1\\
4x_1-5x_2+2x_3-7x_4 &= 2\\
x_1+6x_2-3x_3-4x_4 &= 3\\
3x_1-2x_2- x_3-5x_4 &= 4
\end{aligned}\right\}. \qquad (62)$$

In the processes described we should choose the leading element as pivot, and the first multipliers are $-4/7$, $-1/7$ and $-3/7$. None of these quantities is expressible exactly as a terminating decimal fraction, and it is of course not convenient to work with proper fractions with any computing machine. Rounded multipliers would produce errors in a machine of fixed 'word-length', and the final results would not be exact.

In these circumstances we can perform the elimination in another way, retaining exact integers throughout, provided only that the number of digits in our largest number does not exceed the word-length of the machine.

We can eliminate x_1 from the second equation by adding -4 times the first to 7 times the second, the result being an equation all of whose coefficients are 7 times those of the standard Gauss process and which are exact integers. Similar procedures will eliminate x_1 from the other equations. This is equivalent to a process of cross-multiplication; for example the coefficient of x_2 in the new second equation is equal to the determinant $\begin{vmatrix} 7 & 9 \\ 4 & -5 \end{vmatrix}$, and has the value

$$(7)(-5)-(9)(4) = -71.$$

The first reduced set, with the first equation omitted, is then

$$\left.\begin{aligned} -71x_2+18x_3-57x_4 &= 10 \\ 33x_2-20x_3-30x_4 &= 20 \\ -41x_2- 4x_3-41x_4 &= 25 \end{aligned}\right\}. \tag{63}$$

Further 'cross-elimination' gives

$$\left.\begin{aligned} \underline{826}x_3+4011x_4 &= -1750 \\ 1022x_3+ 574x_4 &= -1365 \end{aligned}\right\}, \tag{64}$$

and the final equation is

$$-3625118x_4 = 661010. \tag{65}$$

The numbers can be reduced somewhat by the observation that all the coefficients in (64) are divisible exactly by the first pivot, which was 7, so that instead of (64) we could write

$$\left.\begin{aligned} \underline{118}x_3+573x_4 &= -250 \\ 146x_3+ 82x_4 &= -195 \end{aligned}\right\}, \tag{66}$$

and eliminate x_3 from (66) to produce

$$-73982x_4 = 13490. \tag{67}$$

This equation, similarly, is exactly divisible by -71, the second pivot, so that (67) can be replaced by

$$1042x_4 = -190. \tag{68}$$

The assembly of pivotal equations

$$\left.\begin{array}{rrrrr} 7x_1 + 9x_2 - & x_3 + & 2x_4 = & 1 \\ -71x_2 + 18x_3 - & 57x_4 = & 10 \\ 118x_3 + & 573x_4 = & -250 \\ 1042x_4 = & -190 \end{array}\right\} \tag{69}$$

has coefficients which are integers and for which no further reduction in size is generally possible.

31. Before completing the solution we shall examine equations (69), particularly the diagonal coefficients, and justify the remarks concerning the exact divisibility of equations like (64) by previous pivots. For this purpose we take the more general equations

$$\left.\begin{array}{l} a_{11}x_1 + a_{12}x_2 + \ldots + a_{1n}x_n = b_1 \\ a_{21}x_1 + a_{22}x_2 + \ldots + a_{2n}x_n = b_2 \\ \quad \cdot \quad \cdot \quad \cdot \quad \cdot \quad \cdot \quad \cdot \quad \cdot \quad \cdot \quad \cdot \quad \cdot \quad \cdot \\ a_{n1}x_1 + a_{n2}x_2 + \ldots + a_{nn}x_n = b_n \end{array}\right\}, \tag{70}$$

and choose pivots down the diagonal. The first reduced set, corresponding to (63), is

$$\left.\begin{array}{l} c_{22}x_2 + \ldots + c_{2n}x_n = d_2 \\ \quad \cdot \quad \cdot \quad \cdot \quad \cdot \quad \cdot \quad \cdot \quad \cdot \quad \cdot \\ c_{n2}x_2 + \ldots + c_{nn}x_n = d_n \end{array}\right\}, \tag{71}$$

where $\qquad c_{rs} = a_{11}a_{rs} - a_{r1}a_{1s}, \qquad d_r = a_{11}b_r - a_{r1}b_1.$ $\qquad$ (72)

A further reduction gives the equations

$$\left.\begin{array}{l} e_{33}x_3 + \ldots + e_{3n}x_n = f_3 \\ \quad \cdot \quad \cdot \quad \cdot \quad \cdot \quad \cdot \quad \cdot \quad \cdot \\ e_{n3}x_3 + \ldots + e_{nn}x_n = f_n \end{array}\right\}, \tag{73}$$

where $\qquad e_{rs} = c_{22}c_{rs} - c_{r2}c_{2s}, \qquad f_r = c_{22}d_r - c_{r2}d_2.$ $\qquad$ (74)

From (72) and (74) we find

$$\left.\begin{array}{l} e_{rs} = (a_{11}a_{22} - a_{21}a_{12})(a_{11}a_{rs} - a_{r1}a_{1s}) - (a_{11}a_{r2} - a_{r1}a_{12})(a_{11}a_{2s} - a_{21}a_{1s}) \\ f_r = (a_{11}a_{22} - a_{21}a_{12})(a_{11}b_r - a_{r1}b_1) - (a_{11}a_{r2} - a_{r1}a_{12})(a_{11}b_2 - a_{21}b_1) \end{array}\right\}. \tag{75}$$

In each case the two terms not containing a_{11} cancel, so that a_{rs} and f_r are divisible by a_{11}. Later results of this kind follow from similar arguments.

32. Consider now the relation between the original matrix A and the final upper triangular U. As in other elimination methods we have performed a sequence of matrix operations which has the representation $JA = U$ and, for a matrix of order four, JA is a product of the form

$$
\overset{J_5}{\begin{bmatrix} 1 & 0 & 0 & 0 \\ 0 & 1 & 0 & 0 \\ 0 & 0 & 1 & 0 \\ 0 & 0 & 0 & u_{22}^{-1} \end{bmatrix}}
\overset{J_4}{\begin{bmatrix} 1 & 0 & 0 & 0 \\ 0 & 1 & 0 & 0 \\ 0 & 0 & 1 & 0 \\ 0 & 0 & \times & u_{33} \end{bmatrix}}
\overset{J_3}{\begin{bmatrix} 1 & 0 & 0 & 0 \\ 0 & 1 & 0 & 0 \\ 0 & 0 & u_{11}^{-1} & 0 \\ 0 & 0 & 0 & u_{11}^{-1} \end{bmatrix}}
\overset{J_2}{\begin{bmatrix} 1 & 0 & 0 & 0 \\ 0 & 1 & 0 & 0 \\ 0 & \times & u_{22} & 0 \\ 0 & \times & 0 & u_{22} \end{bmatrix}} \times
$$

$$
\times \overset{J_1}{\begin{bmatrix} 1 & 0 & 0 & 0 \\ -a_{21} & a_{11} & 0 & 0 \\ -a_{31} & 0 & a_{11} & 0 \\ -a_{41} & 0 & 0 & a_{11} \end{bmatrix}}
\overset{A}{\begin{bmatrix} a_{11} & a_{12} & a_{13} & a_{14} \\ a_{21} & a_{22} & a_{23} & a_{24} \\ a_{31} & a_{32} & a_{33} & a_{34} \\ a_{41} & a_{42} & a_{43} & a_{44} \end{bmatrix}} =
$$

$$
= \overset{U}{\begin{bmatrix} u_{11} & u_{12} & u_{13} & u_{14} \\ & u_{22} & u_{23} & u_{24} \\ & & u_{33} & u_{34} \\ & & & u_{44} \end{bmatrix}}, \tag{76}
$$

in which the u_{rr} are the diagonal elements of the final U and the elements marked with a cross depend on elements of the reduced matrices in non-pivotal rows: their values are not required in the argument.

It follows that the product of the determinants of the J_r and A is equal to that of U. Then

$$
u_{22}^{-1} u_{33} u_{11}^{-2} u_{22}^2 a_{11}^3 \, |A| = u_{11} u_{22} u_{33} u_{44}, \tag{77}
$$

and since $u_{11} = a_{11}$ we find $|A| = u_{44}$, so that the last element in U is equal to the determinant of A. By an obvious analogy u_{33} is the determinant of the matrix obtained from A by omitting its last row and column, and so on.

33. We can now perform the back-substitution process in (69) and obtain the solution to our example. Recalling a result of Chapter 2, giving the components of x as the ratios of determinants, of which the denominator is $|A|$, we note that in our case, of integer coefficients, the numerators are also exact integers. We can therefore introduce the auxiliary vector $y = |A|x$, and the elements of y are integers.

In particular, $y_4 = -190$ from (69), and the other elements of y are obtained by back-substitution in the first three of (69), in which the term on the right is multiplied by $|A| = 1042$, and the final divisions give exact integers. We find

$$\left.\begin{aligned}
y_4 &= -190 \\
118y_3 &= 1042(-250)-573(-190), \text{ or } y_3 = -1285 \\
-71y_2 &= 1042(10)-18(-1285)+57(-190), \text{ or } y_2 = -320 \\
7y_1 &= 1042(1)-9(-320)+1(-1285)-2(-190), \text{ or } y_1 = 431
\end{aligned}\right\} ,$$

(78)

so that the required components of x are obtained as the ratios of integers, with $|A|$ in the denominator.

34. Various features of other methods can be applied here also. For example sum checks can be incorporated at various stages. Again, it is not essential to take pivots in order down the diagonal, and indeed the method fails, as usual, if any pivot is zero. For the symmetric case symmetry is maintained if pivots are taken on the diagonal, and in the positive-definite symmetric case no such pivot can vanish.

To avoid disaster in the general case we must select as pivot a non-zero coefficient in the relevant column, and for use with automatic computing equipment it is probably best, as usual, to exchange the current first row with the chosen pivotal row, recording the number of such changes if we are interested in $|A|$. It may also be preferable to choose the *smallest* non-zero element as pivot, since then the various integers remain as small as possible, and there is less danger of the numbers growing out of range before division by the relevant pivot.

Consider, for example, the computation of Table 10, in which on the left we evaluate $|A|$ by taking pivots down the diagonal and on the right by choosing successive smallest coefficients: in the latter case the results are given corresponding to the appropriate interchange of rows. The numbers in brackets are the values computed prior to the relevant divisions.

TABLE 10

9	1	5	6		9	1	5	6	
$\overline{1}$	4	1	3		$\frac{1}{8}$	4	1	3	
−8	7	9	2		−8	7	9	2	
2	6	7	4		2	6	7	4	
	$\frac{35}{71}$	4	21			−35	−4	−21
	71	121	66			39	17	26
	52	53	24			$\underline{-2}$	5	−2
		$\frac{439}{183}$	91	(3951 819)			−229	26
		183	−28	(1647 −252)			183	−28
			−827	(−28945)				827	(−1654)

In the second computation there are three row interchanges, so that the sign of $|A|$ is opposite to that of the final element.

Determination of rank

35. We saw in the notes on Chapter 2 that a square matrix of order n, with $n-m$ row dependencies, that is a matrix of rank m, can be expressed as the product of two rectangular matrices of respective shapes $(n \times m)$ and $(m \times n)$. For example a matrix C of order 5 with rank 3 is expressible in the form

$$C = \begin{bmatrix} a_{11} & a_{12} & a_{13} \\ a_{21} & a_{22} & a_{23} \\ a_{31} & a_{32} & a_{33} \\ a_{41} & a_{42} & a_{43} \\ a_{51} & a_{52} & a_{53} \end{bmatrix} \begin{bmatrix} b_{11} & b_{12} & b_{13} & b_{14} & b_{15} \\ b_{21} & b_{22} & b_{23} & b_{24} & b_{25} \\ b_{31} & b_{32} & b_{33} & b_{34} & b_{35} \end{bmatrix}. \tag{79}$$

where the left matrix is A and the right matrix is B.

If the matrix formed by the first three rows of A is non-singular, (79) effectively states that the first three rows of C are independent, whereas the last two are merely linear combinations of the first three. The matrix B, moreover, can be regarded as being formed from independent linear combinations of the first three rows of C.

In fact if we perform the Gauss elimination process with interchanges on a matrix with these properties we might find, after the third stage of the elimination, that the corresponding row-permuted C has been transformed to the shape

$$P = \begin{bmatrix} \times & \times & \times & \vline & \times & \times \\ - & \times & \times & \vline & \times & \times \\ - & - & \times & \vline & \times & \times \\ \hline - & - & - & \vline & 0 & 0 \\ - & - & - & \vline & 0 & 0 \end{bmatrix}, \tag{80}$$

where the dashes indicate zeros which have been produced deliberately by the elimination process, and the zeros given explicitly in (80) indicate the row dependence.

In matrix notation this implies that we have produced P by taking a row-permutation $I_r C$ of C and performing the operations

$$J_3 J_2 J_1 (I_r C),$$

where

$$J_1 = \begin{bmatrix} 1 & & & & \\ m_{21} & 1 & & & \\ m_{31} & & 1 & & \\ m_{41} & & & 1 & \\ m_{51} & & & & 1 \end{bmatrix}, \quad J_2 = \begin{bmatrix} 1 & & & & \\ & 1 & & & \\ & m_{32} & 1 & & \\ & m_{42} & & 1 & \\ & m_{52} & & & 1 \end{bmatrix},$$

$$J_3 = \begin{bmatrix} 1 & & & & \\ & 1 & & & \\ & & 1 & & \\ & & m_{43} & 1 & \\ & & m_{53} & & 1 \end{bmatrix}, \tag{81}$$

the m_{rs} being the multipliers in the elimination processes. It follows that

$$I_r C = LP, \quad L = \begin{bmatrix} 1 & & & \vline & & \\ -m_{21} & 1 & & \vline & & \\ -m_{31} & -m_{32} & 1 & \vline & & \\ \hline -m_{41} & -m_{42} & -m_{43} & \vline & 1 & \\ -m_{51} & -m_{52} & -m_{53} & \vline & 0 & 1 \end{bmatrix}, \tag{82}$$

and if we partition L and P in the ways shown in (80) and (82) we can write

$$I_r C = \left[\begin{array}{c|c} L_1 & O \\ \hline Q_1 & I \end{array}\right]\left[\begin{array}{c|c} U_1 & Q_2 \\ \hline O & O \end{array}\right] = \left[\begin{array}{c} L_1 \\ \hline Q_1 \end{array}\right][U_1 \mid Q_2], \qquad (83)$$

which gives the decomposition of $I_r C$ into rectangular matrices. The latter are special in the sense that L_1 and U_1 are triangular, but this decomposition is not unique since we can replace an equation of type (79) by

$$C = AXX^{-1}B, \qquad (84)$$

where X is any non-singular matrix of order m.

36. It follows that in this case we can determine the rank of a matrix by performing the elimination and stopping when the next reduced matrix is identically null, the rank being equal to the number of eliminations performed. The relevant rectangular matrices are formed from the multipliers and the pivotal rows which are not null. An example is given in Table 11, which shows the elimination and the rectangular matrices corresponding to the row-permuted square matrix. We work with rational fractions to ensure exact arithmetic.

TABLE 11

m			C			
$\frac{1}{2}$	1	1	1	1	1	(1)
	$\underline{-2}$	-1	0	1	3	(2)
$-\frac{1}{2}$	-1	0	1	2	4	(3)
0	0	1	2	3	5	(4)
$\frac{1}{2}$	1	2	3	4	6	(5)
$-\frac{1}{3}$		0·5	1	1·5	2·5	(1)
$-\frac{1}{3}$		0·5	1	1·5	2·5	(3)
$-\frac{2}{3}$		1	2	3	5	(4)
		$\underline{1\cdot5}$	3	4·5	7·5	(5)
			0	0	0	(1)
			0	0	0	(3)
			0	0	0	(4)

$$\underset{I_r C}{\begin{bmatrix} -2 & -1 & 0 & 1 & 3 \\ 1 & 2 & 3 & 4 & 6 \\ -1 & 0 & 1 & 2 & 4 \\ 0 & 1 & 2 & 3 & 5 \\ 1 & 1 & 1 & 1 & 1 \end{bmatrix}} = \underset{A}{\begin{bmatrix} 1 & 0 \\ -\frac{1}{2} & 1 \\ \frac{1}{2} & \frac{1}{3} \\ 0 & \frac{2}{3} \\ -\frac{1}{2} & \frac{1}{3} \end{bmatrix}} \underset{B}{\begin{bmatrix} -2 & -1 & 0 & 1 & 3 \\ 0 & 1\cdot5 & 3 & 4\cdot5 & 7\cdot5 \end{bmatrix}}$$

The rank is 2, and to find the decomposition of the original matrix we note that the row-permuting matrix is here

$$I_r = \begin{bmatrix} 0 & 1 & 0 & 0 & 0 \\ 0 & 0 & 0 & 0 & 1 \\ 0 & 0 & 1 & 0 & 0 \\ 0 & 0 & 0 & 1 & 0 \\ 1 & 0 & 0 & 0 & 0 \end{bmatrix}, \tag{85}$$

and we have decomposed $I_r C$ into the product AB shown in Table 11. The decomposition of C is then $I_r^{-1}AB = I_r'AB$, and $I_r'A$ is merely a row permutation of A, given here by

$$I_r'A = \begin{bmatrix} -\tfrac{1}{2} & \tfrac{1}{3} \\ 1 & 0 \\ \tfrac{1}{2} & \tfrac{1}{3} \\ 0 & \tfrac{2}{3} \\ -\tfrac{1}{2} & 1 \end{bmatrix}. \tag{86}$$

37. We might also ask what, in this case, are the combinations of the first two rows of $I_r C$ which produce the last three rows.

From the computation we find

$$\left. \begin{aligned} R_1(I_r C) &= R_1(B) \\ R_2(I_r C) &= -\tfrac{1}{2}R_1(B) + R_2(B), \qquad R_2(B) = \tfrac{1}{2}R_1(I_r C) + R_2(I_r C) \end{aligned} \right\} \tag{87}$$

Then

$$R_3(I_r C) = \tfrac{1}{2}R_1(B) + \tfrac{1}{3}R_2(B) = \tfrac{2}{3}R_1(I_r C) + \tfrac{1}{3}R_2(I_r C), \tag{88}$$

etc.

In general, with reference to equation (83), we have the m independent rows of $I_r C$ given by the equation

$$[R_1(I_r C), \ldots, R_m(I_r C)] = L_1[R_1(B), \ldots, R_m(B)], \tag{89}$$

and the remaining rows come from

$$[R_{m+1}(I_r C), \ldots, R_n(I_r C)] = Q_1[R_1(B), \ldots, R_m(B)]. \tag{90}$$

A combination of (89) and (90) gives the relevant linear combinations of the independent rows in the form

$$[R_{m+1}(I_r C), \ldots, R_n(I_r C)] = Q_1 L_1^{-1}[R_1(I_r C), \ldots, R_m(I_r C)], \tag{91}$$

and the matrix $S = Q_1 L_1^{-1}$ can be obtained by a simple process of back-substitution in the equations

$$L_1' S_1' = Q_1'. \tag{92}$$

Complete pivoting

38. The process of elimination in which we take pivots from *successive* columns is often called *partial pivoting*, and in our description of this process for solving linear equations we implied that it would terminate if every element of the current pivotal column were zero, since the matrix is then singular and this information is often sufficient. It is by no means certain, however, that the other columns of the reduced matrix are null, and we cannot yet determine the rank. Table 12, for example, shows partial pivoting for a case of this kind.

TABLE 12

0·5	1	1	1	0·4	1	(1)
	$\underline{-2}$	−1	0	1	3	(2)
−0·5	−1	0	1	1·7	4	(3)
0·5	1	1·4	1·8	1	3	(4)
0	0	1	2	3	5	(5)
−0·5		0·5	1	0·9	2·5	(1)
−0·5		0·5	1	1·2	2·5	(3)
−0·9		0·9	1·8	1·5	4·5	(4)
		$\underline{1}$	2	3	5	(5)
		0	−0·6	0		(1)
		0	−0·3	0		(3)
		0	$\underline{-1·2}$	0		(4)

At this stage we can ignore the first null column and proceed to eliminate if necessary from the next, choosing the largest element as pivot, and here the next reduced matrix is null so that the rank of the given matrix is three.

An alternative method is to use *complete pivoting*, in which at each stage we use as pivot the largest element in the whole of the relevant matrix. The process automatically terminates with exact arithmetic when the largest element is zero, and for the previous example the computation is shown in Table 13. In practice we shall usually commit some rounding errors, and we discuss this question in general in Chapter 6.

Table 13

−0·2	1	1	1	0·4	1	(1)
−0·6	−2	−1	0	1	3	(2)
−0·8	−1	0	1	1·7	4	(3)
−0·6	1	1·4	1·8	1	3	(4)
	0	1	2	3	5	(5)
0·5	1	0·8	0·6	−0·2		(1)
	−2	−1·6	−1·2	−0·8		(2)
−0·5	−1	−0·8	−0·6	−0·7		(3)
0·5	1	0·8	0·6	−0·8		(4)
−0·5	0	0	−0·6			(1)
−0·25	0	0	−0·3			(3)
	0	0	−1·2			(4)
	0	0				(1)
	0	0				(3)

†39. Complete pivoting can be used, of course, in the solution of linear equations, and indeed in some early accounts it was recommended instead of partial pivoting. Here we notice that the matrix equivalent is very similar to that of partial pivoting, the effect being to multiply the given matrix by $J = J_r J_{r-1} \ldots J_1$, where each J_r is a simple matrix and J^{-1} is easily formed from the multipliers. In our present example we find that the given matrix is expressible as

$$
\overset{C}{\begin{bmatrix} 1 & 1 & 1 & 0\cdot4 & 1 \\ -2 & -1 & 0 & 1 & 3 \\ -1 & 0 & 1 & 1\cdot7 & 4 \\ 1 & 1\cdot4 & 1\cdot8 & 1 & 3 \\ 0 & 1 & 2 & 3 & 5 \end{bmatrix}}
=
\overset{A}{\begin{bmatrix} 1 & -0\cdot5 & 0 & 0\cdot5 & 0\cdot2 \\ 0 & 1 & 0 & 0 & 0\cdot6 \\ 0 & 0\cdot5 & 1 & 0\cdot25 & 0\cdot8 \\ 0 & -0\cdot5 & 0 & 1 & 0\cdot6 \\ 0 & 0 & 0 & 0 & 1 \end{bmatrix}}
\overset{B}{\begin{bmatrix} 0 & 0 & 0 & 0 & 0 \\ -2 & -1\cdot6 & -1\cdot2 & -0\cdot8 & 0 \\ 0 & 0 & 0 & 0 & 0 \\ 0 & 0 & 0 & -1\cdot2 & 0 \\ 0 & 1 & 2 & 3 & 5 \end{bmatrix}}, (93)
$$

the second matrix on the right being the assembly, without interchanges, of the pivotal rows.

Again we can find a product of rectangular matrices by inserting a suitable row-changing orthogonal matrix in the form $C = A I'_r I_r B$, where here

$$
I_r = \begin{bmatrix} 0 & 0 & 0 & 0 & 1 \\ 0 & 1 & 0 & 0 & 0 \\ 0 & 0 & 0 & 1 & 0 \\ 1 & 0 & 0 & 0 & 0 \\ 0 & 0 & 1 & 0 & 0 \end{bmatrix}, \tag{94}
$$

the position of the unit in successive rows denoting, as usual, the order of the pivotal rows in the elimination. We then find

$$
\begin{array}{c}
C \\
\begin{bmatrix}
1 & 1 & 1 & 0\cdot4 & 1 \\
-2 & -1 & 0 & 1 & 3 \\
-1 & 0 & 1 & 1\cdot7 & 4 \\
1 & 1\cdot4 & 1\cdot8 & 1 & 3 \\
0 & 1 & 2 & 3 & 5
\end{bmatrix}
\end{array}
=
\begin{array}{c}
AI'_r \\
\begin{bmatrix}
0\cdot2 & -0\cdot5 & 0\cdot5 \\
0\cdot6 & 1 & 0 \\
0\cdot8 & 0\cdot5 & 0\cdot25 \\
0\cdot6 & -0\cdot5 & 1 \\
1 & 0 & 0
\end{bmatrix}
\end{array}
\begin{array}{c}
I_rB \\
\begin{bmatrix}
0 & 1 & 2 & 3 & 5 \\
-2 & -1\cdot6 & -1\cdot2 & -0\cdot8 & 0 \\
0 & 0 & 0 & -1\cdot2 & 0
\end{bmatrix}
\end{array}.
\tag{95}
$$

We can also carry out the process of § 37 to find the relations between the dependent rows of C and its independent rows. Premultiplication of (95) with the I_r of (94) gives

$$
I_rC = \begin{bmatrix} L_1 \\ \overline{} \\ Q_1 \end{bmatrix}[I_rB] ,
\tag{96}
$$

where

$$
\begin{bmatrix} L_1 \\ \overline{} \\ Q_1 \end{bmatrix} =
\begin{bmatrix}
1 & 0 & 0 \\
0\cdot6 & 1 & 0 \\
0\cdot6 & -0\cdot5 & 1 \\
\hline
0\cdot2 & -0\cdot5 & 0\cdot5 \\
0\cdot8 & 0\cdot5 & 0\cdot25
\end{bmatrix},
\tag{97}
$$

and

$$
Q_1L_1^{-1} = \begin{bmatrix}
0\cdot05 & -0\cdot25 & 0\cdot5 \\
0\cdot275 & 0\cdot625 & 0\cdot25
\end{bmatrix},
\tag{98}
$$

so that

$$
\left.\begin{aligned}
R_4(I_rC) &= 0\cdot05R_1(I_rC)-0\cdot25R_2(I_rC)+0\cdot5R_3(I_rC) \\
R_5(I_rC) &= 0\cdot275R_1(I_rC)+0\cdot625R_2(I_rC)+0\cdot25R_3(I_rC)
\end{aligned}\right\}.
\tag{99}
$$

Compatibility of linear equations

40. By an extension of our methods we can determine whether the equations

$$
Cx = d,
\tag{100}
$$

where here C is $(m \times n)$, x is $(n \times 1)$, and d is $(m \times 1)$, are consistent in the sense that there is some solution, not necessarily unique. The rank of C is still defined as the number of independent rows, and we can discover this in the usual way by performing the elimination process on a row permutation $I_r C$ of C. Moreover we can still write

$$I_r C = AB = \left[\frac{L_1}{Q_1}\right][U_1 \mid Q_2], \tag{101}$$

where, if the rank is p, L_1 is $(p \times p)$, Q_1 is $(m-p \times p)$, U_1 is $(p \times p)$, Q_2 is $(p \times n-p)$.

Then we can partition x into x_1 and x_2, and $I_r d$ into d_1 and d_2, of respective shapes $(p \times 1)$, $(n-p \times 1)$, $(p \times 1)$ and $(m-p \times 1)$, and write

$$\left[\frac{L_1}{Q_1}\right][U_1 \mid Q_2]\left[\frac{x_1}{x_2}\right] = \left[\frac{d_1}{d_2}\right], \tag{102}$$

so that

$$\left.\begin{array}{l} L_1(U_1 x_1 + Q_2 x_2) = d_1 \\ Q_1(U_1 x_1 + Q_2 x_2) = d_2 \end{array}\right\}. \tag{103}$$

From (103) it follows that the consistency condition is given by

$$d_2 = Q_1 L_1^{-1} d_1, \tag{104}$$

in conformity also with (91), and the solution of the equations is

$$x_1 = U_1^{-1}(L_1^{-1} d_1 - Q_2 x_2). \tag{105}$$

We note also, from (102), that we have partitioned the equations $I_r C x = I_r d$ in the form

$$\left[\begin{array}{c|c} P & Q \\ \hline R & S \end{array}\right]\left[\frac{x_1}{x_2}\right] = \left[\frac{d_1}{d_2}\right], \tag{106}$$

where P is a non-singular square matrix of order p, Q has shape $(p \times n-p)$, R has shape $(m-p \times p)$, and S is $(m-p \times n-p)$. We have consistency if

$$d_2 = RP^{-1} d_1, \tag{107}$$

and the solution is

$$x_1 = P^{-1}(d_1 - Q x_2), \qquad x_2 \text{ arbitrary.} \tag{108}$$

41. For example with

$$C = \begin{bmatrix} 2 & -3 \\ 1 & 1 \\ 3 & -2 \end{bmatrix}, \qquad d = \begin{bmatrix} 1 \\ 0 \\ 1 \end{bmatrix}, \tag{109}$$

representing the equations treated by the method of Lanczos in the notes on Chapter 2, and in which $m > n$, we find

$$I_rC = \begin{bmatrix} 3 & -2 \\ 2 & -3 \\ 1 & 1 \end{bmatrix} = \begin{bmatrix} L_1 \\ \hline Q_1 \end{bmatrix} [U_1 \mid Q_2] = \begin{bmatrix} 1 & 0 \\ \frac{2}{3} & 1 \\ \hline \frac{1}{3} & -1 \end{bmatrix} \begin{bmatrix} 3 & -2 \\ 0 & -\frac{5}{3} \end{bmatrix}$$

$$I_rd = \begin{bmatrix} 1 \\ 1 \\ \hline 0 \end{bmatrix} = \begin{bmatrix} d_1 \\ \hline d_2 \end{bmatrix}$$

. (110)

The rank is 2, and in (101) the matrix L_1 is (2×2), Q_1 is (1×2), U_1 is (2×2), Q_2 does not exist, x_1 is (2×1), x_2 does not exist, d_1 is (2×1), $d_2 = (1 \times 1)$, and we find

$$Q_1 L_1^{-1} d_1 = [\tfrac{1}{3} \quad -1] \begin{bmatrix} 1 & 0 \\ -\frac{2}{3} & 1 \end{bmatrix} \begin{bmatrix} 1 \\ 1 \end{bmatrix} = 0 = d_2, \qquad (111)$$

and we have consistency.

Similarly, for another example treated there, with $m < n$ and with

$$C = \begin{bmatrix} 1 & 1 & 1 \\ 2 & 2 & 2 \end{bmatrix}, \qquad d = \begin{bmatrix} 1 \\ 3 \end{bmatrix}, \qquad (112)$$

we find

$$I_rC = \begin{bmatrix} 2 & 2 & 2 \\ 1 & 1 & 1 \end{bmatrix} = \begin{bmatrix} 1 \\ \hline \frac{1}{2} \end{bmatrix} [2 \mid 2 \quad 2]$$

$$I_rd = \begin{bmatrix} 3 \\ \hline 1 \end{bmatrix}$$

. (113)

The rank is 1, and L_1 is (1×1), Q_1 is (1×1), U_1 is (1×1), Q_2 is (1×2), x_1 is (1×1), x_2 is (2×1), d_1 is (1×1) and d_2 is (1×1). We find

$$Q_1 L_1^{-1} d_1 = [\tfrac{1}{2}][1][3] \neq [1], \qquad (114)$$

and there is no solution.

42. It is clear that if as the result of elimination on the equations $Cx = d$ we ever produce a null matrix, square or rectangular, there can be no solution unless we also produce a corresponding null column on the right, a result expressible in the phrase 'the rank of the extended matrix is equal to that of the coefficient matrix'. If there is no row dependence in the case $m \leqslant n$, as for example in the system

$$\left.\begin{aligned} 2x_1 + x_2 + 3x_3 + 5x_4 &= 8 \\ -3x_1 + x_2 - 2x_3 - x_4 &= -3 \end{aligned}\right\}, \tag{115}$$

also treated by Lanczos' method in Chapter 2, there is no problem and there will be a solution for any right-hand side. But if $m > n$ we must have some row dependence or dependencies, since otherwise we have conflicting information.

Note on comparison of methods

43. In deciding which method to use in any particular problem we are concerned only with two requirements, that the method should give as accurate an answer as possible and that the time and labour involved should be a minimum. So far we have assumed that the solutions can be obtained exactly, either as terminating decimal numbers or as the ratios of integers, and in this respect all our methods are equally satisfactory.

This may not apply if the results are not exact, as a consequence of inevitable rounding errors introduced in the course of the computation, and we must note also that our particular computing equipment may not give exact results even for the methods of this chapter. For example the number 0·1 in the decimal scale cannot be represented as a terminating binary fraction, so that a binary machine could not store this number exactly and some rounding error is committed.

44. With regard to time and labour the computing machine plays an even more important part in the evaluation of a method. For example with pen, paper and desk machine we find that both the setting-up of numbers and the recording of intermediate results are relatively slow and error-provoking processes, addition is rapid, and the slower processes of multiplication and division take about the same time as each other. Again, a small change in the size of the problem (order of matrix) is of no great concern to a desk-machine operator, except that a large problem induces fatigue and more blunders, making essential the frequent application of sum checks.

Automatic machines have general characteristics of a different nature. For example recording is rapid and can be done by 'erasure', thereby conserving storage space. The size of the high-speed store is limited, so that a small change in the order of the matrix may necessitate the use of slower backing stores, with consequent increase in computing time. Except with very large problems fatigue is less obvious, and checks are less important within the machine. At the present stage of development, however, the terminal equipment is less reliable, and it pays to check that the given data is properly fed in and that the output punch or printer has not failed.

With regard to the time of numerical operation there is considerable variation among automatic machines. For example the 'fixed-point' machine DEUCE adds in 64 μsec and multiplies in 2 msec, while the 'floating-point' MERCURY adds in 180 μsec and multiplies in the relatively fast time of 300 μsec. The use of 'multi-length' arithmetic also varies in time and convenience.

In assessing the amount of work in methods for automatic machines, therefore, we should compute the total number of operations like addition, multiplication and division which each method involves, together with the amount of storage required. For desk machines we should also consider the number of recordings of intermediate results, and perhaps even the number of 'settings' involved. The question of accuracy of the various methods in general requires some error analysis.

These considerations are deferred until later chapters, so that other allied methods can be evaluated at the same time.

Additional notes and bibliography

§ 5. A check on the back-substitution, which is often recommended, is to repeat this in the sum column $s^{(1)}$. Clearly $s^{(1)} = b + Ae$, where e is the vector all of whose components are unity, so that the solution of $Az = s^{(1)}$ is $z = x + e$, each component of this solution being greater by unity than the computed x.

The amount of work in this check is greater than that of the text, and the latter has the additional advantage that it checks the elimination as well as the back-substitution. In particular it is recommended that this check is performed using column sums obtained from any original data, and the solution as *printed* from the machine, so that possible errors of transmission are also detected.

The check on $s^{(1)}$ might be useful, however, when the calculation is subject to rounding error, and provided that the intermediate check sums are not altered to conform with the row sums of A and b. This process provides a less satisfactory check on the elimination, since the column of values $s^{(1)}$ may begin to deviate by more than the sum of rounding errors from their true values, but the extent to which z_r agrees with $x_r + 1$ will give some indication of the number of reliable figures in the solution.

§ 15. The row-changing matrix I_r of equation (24) is our first example of an *orthogonal* matrix, satisfying the equation $I_r^{-1} = I_r'$. The operation $I_r J^{-1} I_r'$, where J^{-1} is given in equation (22), effectively first interchanges the rows, so that each has then the correct number of zero elements, and then the columns, to put the units back on the diagonal.

§ 27. The normal equations can be obtained more quickly by noticing that $S = (Ax - b)'(Ax - b) = x'A'Ax - x'A'b - b'Ax + b'b$. We can differentiate this expression with respect to each component of the vector x, to obtain immediately the normal equations $A'Ax = A'b$.

§ 39. Complete pivoting was recommended in

Fox, L., Huskey, H. D. and Wilkinson, J. H., 1948, *Notes on the solution of algebraic linear simultaneous equations. Quart. J. Mech.* **1**, 149–173.

This paper describes the methods of Gauss, Jordan and Aitken for general matrices, and other methods for symmetric matrices relevant to Chapters 4 and 5.

Compact Elimination Methods of Doolittle, Crout, Banachiewicz and Cholesky

Introduction

1. IN ALL the elimination methods of the previous chapter our aim was to convert the matrix A, by an ordered sequence of operations, to a form, triangular or diagonal, from which the solution can be obtained with relative ease. In the Gauss process, for example, we *computed* the matrices J_1A, $J_2J_1A,\ldots$, and the vectors J_1b, $J_2J_1b,\ldots$, and *recorded* the elements of all those rows, in all the 'reduced' matrices, which had been changed by the numerical operations. We need, however, only the final upper triangular matrix JA and the corresponding vector Jb in order to complete the solution by back-substitution, and the question arises as to whether we can obtain the triangle and the right-hand vector without calculating and recording the elements of intermediate matrices and vectors.

This is achieved in methods associated with the names of Doolittle, Crout, Banachiewicz and Cholesky, and these methods have close connection with the decomposition of a matrix A into the product LU of lower and upper triangles, already mentioned in § 24 of Chapter 2.

The method of Doolittle

2. Consider an elimination process for the four simultaneous equations relevant to the array

$$\left.\begin{matrix} a_{11} & a_{12} & a_{13} & a_{14} & b_1 \\ a_{21} & a_{22} & a_{23} & a_{24} & b_2 \\ a_{31} & a_{32} & a_{33} & a_{34} & b_3 \\ a_{41} & a_{42} & a_{43} & a_{44} & b_4 \end{matrix}\right\}, \tag{1}$$

which will reduce the problem to the triangular set

$$\left.\begin{matrix} u_{11} & u_{12} & u_{13} & u_{14} & f_1 \\ & u_{22} & u_{23} & u_{24} & f_2 \\ & & u_{33} & u_{34} & f_3 \\ & & & u_{44} & f_4 \end{matrix}\right\}, \tag{2}$$

the rows of (2) being the pivotal rows of the Gauss elimination method applied to (1), with pivots selected in order down the diagonal.

The first row of (2) is clearly identical with that of (1). The second row of (2) is the second (pivotal) row of the first reduced matrix $J_1 A$, $J_1 b$ of the Gauss process, and is obtained by the addition $m_{21} R_1(1) + R_2(1)$, where m_{21} is chosen so that the first element vanishes, so that $m_{21} a_{11} + a_{21} = 0$. These results are contained in the equations

$$\left. \begin{aligned} R_1(2) &= R_1(1) \\ m_{21} a_{11} + a_{21} &= 0 \\ R_2(2) &= m_{21} R_1(1) + R_2(1) \end{aligned} \right\}. \tag{3}$$

Consider now the calculation of the third row of (2). The corresponding row of $(J_1 A, J_1 b)$ would be $m_{31} R_1(1) + R_3(1)$, where $m_{31} a_{11} + a_{31} = 0$, and we add to this a multiple m_{32} of $R_2(2)$ to eliminate the coefficient of x_2. Then

$$\left. \begin{aligned} R_3(2) &= m_{31} R_1(1) + R_3(1) + m_{32} R_2(2) \\ m_{31} a_{11} + a_{31} &= 0 \\ m_{31} a_{12} + a_{32} + m_{32} u_{22} &= 0 \end{aligned} \right\}. \tag{4}$$

Finally, for the last row of (2), we first form implicitly the last row of $(J_1 A, J_1 b)$ in the form $m_{41} R_1(1) + R_4(1)$, where $m_{41} a_{11} + a_{41} = 0$, eliminate from this the coefficient of x_2 by adding a multiple m_{42} of $R_2(2)$, and the coefficient of x_3 by adding a multiple m_{43} of $R_3(2)$, so that we have

$$\left. \begin{aligned} R_4(2) &= m_{41} R_1(1) + R_4(1) + m_{42} R_2(2) + m_{43} R_3(2) \\ m_{41} a_{11} + a_{41} &= 0 \\ m_{41} a_{12} + a_{42} + m_{42} u_{22} &= 0 \\ m_{41} a_{13} + a_{43} + m_{42} u_{23} + m_{43} u_{33} &= 0 \end{aligned} \right\}. \tag{5}$$

3. The multipliers m_{rs} are of course identical in theory with those of the Gauss process, and differ in practice only through possible rounding errors. For example in the last of (5) the quantity $m_{41} a_{13} + a_{43}$ would be recorded, after any necessary rounding, as one of the elements in the first reduced matrix in the Gauss process, and then this quantity plus $m_{42} u_{23}$ would also be recorded and rounded before the final computation of m_{43}. On a desk machine, and on some automatic machines, many of these individual roundings could be avoided.

4. The desk-machine arrangement of the computation might appear as in Table 1.

TABLE 1

			a_{11}	a_{12}	a_{13}	a_{14}	b_1
			a_{21}	a_{22}	a_{23}	a_{24}	b_2
			a_{31}	a_{32}	a_{33}	a_{34}	b_3
			a_{41}	a_{42}	a_{43}	a_{44}	b_4
m_{21}	m_{31}	m_{41}	u_{11}	u_{12}	u_{13}	u_{14}	f_1
	m_{32}	m_{42}		u_{22}	u_{23}	u_{24}	f_2
		m_{43}			u_{33}	u_{34}	f_3
						u_{44}	f_4
			x_1	x_2	x_3	x_4	

In accordance with equations (3), (4) and (5) the order of calculation is:

(i) first row of U and f is the same as that of A and b,

(ii) calculate m_{21} from (3),

(iii) calculate second row of U and f from (3),

(iv) calculate m_{31} and m_{32}, successively, from (4),

(v) calculate third row of U and f from (4),

(vi) calculate m_{41}, m_{42} and m_{43}, successively, from (5),

(vii) calculate fourth row of U and f from (5).

We may note also, in equations (3), (4) and (5), that the quantities in $R_1(1)$ are the same as those in $R_1(2)$, so that the m_{rs} multiply only the u_{rs}, and the positions of the m_{rs} are arranged so that partners in the products are located in the same rows.

The U matrix and the multipliers are obviously those of the Gauss process and the solution is completed by back-substitution. Intermediate row-sum checks, and the final check on the solution, can be included as in previous cases. The method will fail, as in the Gauss process, if any u_{rr} vanishes, since a multiplier cannot then be found.

5. Table 2 shows the computation by this method of the example of Table 1 of Chapter 3.

TABLE 2

			A			b
			4	−9	2	5
			2	−4	6	3
			1	−1	3	4
	m			U		f
−0·5	−0·25		4	−9	2	5
	−2·5			0·5	5	0·5
					−10	1·5
	x'		6·95	2·5	−0·15	

Connexion with decomposition

6. Detailed inspection of equations (3), (4) and (5) shows that the complete calculation of the U matrix and of the multipliers comes from the matrix equation

$$
\begin{matrix} A \end{matrix} \qquad\qquad \begin{matrix} L \end{matrix} \qquad\qquad \begin{matrix} U \end{matrix}
$$

$$
\begin{bmatrix} a_{11} & a_{12} & a_{13} & a_{14} \\ a_{21} & a_{22} & a_{23} & a_{24} \\ a_{31} & a_{32} & a_{33} & a_{34} \\ a_{41} & a_{42} & a_{43} & a_{44} \end{bmatrix} = \begin{bmatrix} 1 & & & \\ -m_{21} & 1 & & \\ -m_{31} & -m_{32} & 1 & \\ -m_{41} & -m_{42} & -m_{43} & 1 \end{bmatrix} \begin{bmatrix} u_{11} & u_{12} & u_{13} & u_{14} \\ & u_{22} & u_{23} & u_{24} \\ & & u_{33} & u_{34} \\ & & & u_{44} \end{bmatrix}. \tag{6}
$$

For example the first of (3) comes from multiplication of the first row of L by successive columns of U, and the equations (5) are equivalent to multiplication of the last row of L by successive columns of U, the results being equated to the relevant terms in A.

Now L is a lower triangular matrix, and we have effectively found all the coefficients in the lower and upper triangles whose product is A. In passing we note a verification of a result in § 14 of Chapter 3, relating the coefficients of this L matrix with the multipliers in the Gauss elimination method and the product of the matrices of type $J_1, J_2, ..., J_{n-1}$ used in that method.

In a similar manner the final vector f, the right-hand side of the triangular array, is obtained from b from the equation

$$
b = Lf. \tag{7}
$$

Then if $Ax = b$, $A = LU$, $b = Lf$, we find

$$
LUx = Lf, \qquad \text{or} \qquad Ux = f, \tag{8}
$$

since L is not singular, so that the required x is obtained by back-substitution from the final triangular array. Alternatively we could defer the calculation of f until L is known, using the equations

$$
Ax = LUx = b, \qquad Ux = f, \qquad Lf = b, \tag{9}
$$

obtaining f from the last of (9) by *forward* substitution.

The method of Crout

7. For desk-machine work Crout records the various matrices in an even more compact form, though with a disadvantage that partners in multiplication are no longer in the same row or column. He has one main additional variation, that the rows of U and f are divided by the diagonal elements of U, before recording, so that his *upper* triangle is a unit triangle.

Corresponding to Table 1 the Crout arrangement has the appearance of Table 3.

TABLE 3

a_{11}	a_{12}	a_{13}	a_{14}	b_1
a_{21}	a_{22}	a_{23}	a_{24}	b_2
a_{31}	a_{32}	a_{33}	a_{34}	b_3
a_{41}	a_{42}	a_{43}	a_{44}	b_4
u_{11}	v_{12}	v_{13}	v_{14}	d_1
n_{21}	u_{22}	v_{23}	v_{24}	d_2
n_{31}	n_{32}	u_{33}	v_{34}	d_3
n_{41}	n_{42}	n_{43}	u_{44}	d_4

The computation is performed by means of the following equations, which may be compared with (3), (4) and (5) for the Doolittle process. For the first row

$$\left. \begin{array}{l} u_{11} = a_{11} \\[4pt] v_{1s} = a_{1s}/u_{11}, \qquad s > 1 \\[4pt] d_1 = b_1/u_{11} \end{array} \right\}, \tag{10}$$

so that each off-diagonal term in the first row is that of the Doolittle method when divided by $u_{11} = a_{11}$. For the second row

$$\left. \begin{array}{l} n_{21} = a_{21} \\[4pt] u_{22} = a_{22} - n_{21}v_{12} \\[4pt] v_{2s} = (a_{2s} - n_{21}v_{1s})/u_{22}, \qquad s > 2 \\[4pt] d_2 = (b_2 - n_{21}d_1)/u_{22} \end{array} \right\}, \tag{11}$$

so that $n_{21} = -a_{11}$ times the corresponding multiplier m_{21} of the Doolittle method, u_{22} is identical with the corresponding Doolittle element, and the other elements in this row are those of Doolittle divided by the diagonal u_{22}. For the third row

$$\left. \begin{array}{l} n_{31} = a_{31} \\[4pt] n_{32} = a_{32} - n_{31}v_{12} \\[4pt] u_{33} = a_{33} - n_{31}v_{13} - n_{32}v_{23} \\[4pt] v_{3s} = (a_{3s} - n_{31}v_{1s} - n_{32}v_{2s})/u_{33}, \qquad s > 3 \\[4pt] d_3 = (b_3 - n_{31}d_1 - n_{32}d_2)/u_{33} \end{array} \right\}, \tag{12}$$

so that $n_{31} = -a_{11}m_{31} = -u_{11}m_{31}$, $n_{32} = -u_{22}m_{32}$, u_{33} is the same as in Doolittle, and other elements in this row are those of Doolittle divided by the diagonal u_{33}.

The continuation is obvious, and the solution is effected by back-substitution using the matrix V and vector d, the diagonal elements of the V matrix having the values unity.

8. Table 4 shows the computation of the previous example, effected by this method.

TABLE 4

4	-9	2	5
2	-4	6	3
1	-1	3	4
4	$-2{\cdot}25$	$0{\cdot}5$	$1{\cdot}25$
2	$0{\cdot}5$	$10{\cdot}0$	$1{\cdot}0$
1	$1{\cdot}25$	$-10{\cdot}0$	$-0{\cdot}15$
x' $6{\cdot}95$	$2{\cdot}5$	$-0{\cdot}15$	

9. The connexion between the methods of Crout and Doolittle has been established, and it is also easy to see that

$$
\begin{bmatrix} a_{11} & a_{12} & a_{13} & a_{14} \\ a_{21} & a_{22} & a_{23} & a_{24} \\ a_{31} & a_{32} & a_{33} & a_{34} \\ a_{41} & a_{42} & a_{43} & a_{44} \end{bmatrix} = \begin{bmatrix} u_{11} & & & \\ n_{21} & u_{22} & & \\ n_{31} & n_{32} & u_{33} & \\ n_{41} & n_{42} & n_{43} & u_{44} \end{bmatrix} \begin{bmatrix} 1 & v_{12} & v_{13} & v_{14} \\ & 1 & v_{23} & v_{24} \\ & & 1 & v_{34} \\ & & & 1 \end{bmatrix}, \quad (13)
$$

giving the decomposition analogue of the Crout method.

Symmetric case

10. If A is symmetric there are simplifications corresponding to those of the Gauss elimination method, with pivots taken down the diagonal, which were mentioned in § 24 of Chapter 3. Only the upper triangular part of A need be recorded, and the multipliers are obtained more easily.

In Table 5, for example, we have the general relations

$$
\left.\begin{aligned} m_{ij} &= -u_{ji}/u_{jj} \\ n_{ij} &= v_{ji}u_{jj} \end{aligned}\right\}, \quad (14)
$$

TABLE 5

	Doolittle					Crout			
a_{11}	a_{12}	a_{13}	a_{14}	b_1		a_{11}	a_{12}	a_{13}	a_{14}	b_1
	a_{22}	a_{23}	a_{24}	b_2			a_{22}	a_{23}	a_{24}	b_2
		a_{33}	a_{34}	b_3				a_{33}	a_{34}	b_3
			a_{44}	b_4					a_{44}	b_4

m_{21}	m_{31}	m_{41}	u_{11}	u_{12}	u_{13}	u_{14}	f_1	u_{11}	v_{12}	v_{13}	v_{14}	d_1
	m_{32}	m_{42}		u_{22}	u_{23}	u_{24}	f_2	n_{21}	u_{22}	v_{23}	v_{24}	d_2
		m_{43}			u_{33}	u_{34}	f_3	n_{31}	n_{32}	u_{33}	v_{34}	d_3
						u_{44}	f_4	n_{41}	n_{42}	n_{43}	u_{44}	d_4

| | | | x_1 | x_2 | x_3 | x_4 | | x_1 | x_2 | x_3 | x_4 |

for the Doolittle and Crout methods respectively. The decomposition expression in (13) can then be written in the form

$$
A = \begin{bmatrix} 1 & & & \\ v_{12} & 1 & & \\ v_{13} & v_{23} & 1 & \\ v_{14} & v_{24} & v_{34} & 1 \end{bmatrix} \begin{bmatrix} u_{11} & & & \\ & u_{22} & & \\ & & u_{33} & \\ & & & u_{44} \end{bmatrix} \begin{bmatrix} 1 & v_{12} & v_{13} & v_{14} \\ & 1 & v_{23} & v_{24} \\ & & 1 & v_{34} \\ & & & 1 \end{bmatrix}, \quad (15)
$$

so that for a symmetric matrix we can write

$$A = LDL', \qquad (16)$$

the unit triangles being each the transpose of the other.

The computation in the Doolittle method effectively associates D with L', so that $A = LU_1$, with L a unit triangle and the rows of U_1 being proportional to the columns of L, while Crout associates D with L, so that $A = L_1U$, with U a unit triangle with rows proportional to the columns of L_1.

11. Table 6 illustrates the computation, by the two methods, for the example of Table 9 in Chapter 3.

TABLE 6

		Doolittle				Crout			
	4	5	2	1		4	5	2	1
		6	−1	2			6	−1	2
			2	0				2	0
−1·25	−0·5	4	5	2	1	4	1·25	0·5	0·25
	−14·0		−0·25	−3·5	0·75	5	−0·25	14·0	−3·0
				50	−11	2	−3·5	50·0	−0·22
		0·26	0·08	−0·22			0·26	0·08	−0·22

The methods of Banachiewicz and Cholesky

12. It has become clear, from consideration of the compact methods of Doolittle and Crout, that the multipliers in all the elimination methods have a significance, both in their nature and in their computation, quite comparable with that of the elements in the final triangle U required in the back-substitution. Indeed they belong to the L matrix in the expression $A = LU$, and the elements of L can be found from this expression, simultaneously with those of U, by applying the rules of matrix multiplication and without any explicit thoughts of 'elimination'. This possibility was discussed in § 24 of Chapter 2.

For desk-machine work we might arrange the computation as shown in Table 7, recording the transpose U' beneath L so that scalar products are calculated more conveniently. Each multiplication $R_s(L)R_t(U')$ provides an equation for the direct determination of one element of U for $t \geqslant s$, or of L for $t < s$. In L we take the diagonal terms to be unity. The product $R_s(L)f'$ also gives, by the same process, successive elements in the vector f given by $Lf = b$, and the final solution is obtained by back-substitution, that is by solving $Ux = f$ as indicated in equation (9). If x is recorded as a column we can use for this purpose the product $C_s(U')x = f_s$ to produce successive elements of x, and the terms to be multiplied are again arranged conveniently. Finally the intermediate sum checks, if required, can be effected by forming the row c', the sums of columns of U' and f', and equating the products $R_s(L)c'$ to the corresponding elements in the column $s(A,b)$ of row sums of A and b. The final check, using the row of column sums of A and b, is performed as in previous methods. Here this row is recorded as a column $c(A,b)$ side by side with x, to facilitate the computation.

Table 7 shows the computation by this method of our standard example.

The choice of units in the diagonal of L is effectively that of Gauss and Doolittle elimination. We could equally well choose U to be a unit triangle, and in this case the diagonal of L has to be calculated and we have the equivalent of Crout.

13. I ascribe to Banachiewicz the method of Table 7. In the case of symmetry we can follow Cholesky, and save much recording, by using the triangular decomposition in the form

$$A = LL', \tag{17}$$

in which L is no longer a unit triangle. This result is equivalent to

TABLE 7

	A		b	$s(A,b)$
4	−9	2	5	2
2	−4	6	3	7
1	−1	3	4	7
	L			
1				
0·5	1			
0·25	2·5	1		
	U'		x	$c(A,b)$
4			6·95	7
−9	0·5		2·5	−14
2	5	−10	−0·15	11

f'	5	0·5	1·5	12
c'	2	6·0	−8·5	

(16) if we write the latter in the form

$$A = (LD^{\frac{1}{2}})(D^{\frac{1}{2}}L'), \qquad (18)$$

where $D^{\frac{1}{2}}$, the 'square root' of D, is a diagonal matrix whose elements are the square roots of those of D.

The formula
$$a_{rs} = R_r(L)C_s(L') \qquad (19)$$
is then equivalent to

$$a_{rs} = R_r(L)R_s(L) = C_r(L')C_s(L'), \qquad (20)$$

so that we can record L or L' only, use the relevant scalar products either by row-by-row or column-by-column multiplication respectively, and obtain successive elements of the triangle, the diagonal elements involving the calculation of a square root. The allied equations

$$Lf = b, \qquad L'x = f, \qquad (21)$$

complete the solution as before, and the positions of f and x can in each case be recorded conveniently.

14. Table 8 shows the two possible arrangements, including all checks, the 'check' values being denoted by dashes.

TABLE 8

	A		b	$s(A,b)$			A		b	$s(A,b)$
a_{11}			b_1	—		a_{11}	a_{12}	a_{13}	b_1	—
a_{12}	a_{22}		b_2	—			a_{22}	a_{23}	b_2	—
a_{13}	a_{23}	a_{33}	b_3	—				a_{33}	b_3	—
	L			$c(A,b)$			L'			c
l_{11}			x_1	—		l_{11}	l_{21}	l_{31}	f_1	—
l_{21}	l_{22}		x_2	—			l_{22}	l_{32}	f_2	—
l_{31}	l_{32}	l_{33}	x_3	—				l_{33}	f_3	—
f_1	f_2	f_3				x_1	x_2	x_3		
c' —	—	—		—	$c'(A,b)$	—	—	—	—	

Table 9 illustrates the solution of a symmetric set of three equations.

TABLE 9

A			b	$s(A,b)$
4			1	12
5	6·5		2	15
2	1·5	6	0	9·5

	L			x	$c(A, b)$
	2·0			−17·25	11
	2·5	0·5		13·0	13
	1·0	−2·0	1·0	2·5	9·5
f'	0·5	1·5	2·5		3
c'	6·0	0·0	3·5		

15. If we try by this method the example of Table 6 we find that, for one diagonal term of L, we have to take the square root of a negative number. This corresponds to the fact that the D matrix in (16) has as elements the pivotal terms of the elimination process, and the second of these is negative, as apparent from the results in Table 6. This situation is inconvenient, especially for automatic computing, but it is not catastrophic. The resulting number is purely imaginary and imaginary elements will then occur in all the other positions in that column of L and the relevant element of f. We shall have to multiply and divide by imaginary quantities, but it is clear that we never have the more serious worry of dealing with complex numbers $a + ib$.

Nevertheless the use of Cholesky's method, at least by automatic computer, is best reserved for those cases in which we are sure of positive pivots, and we have already seen, in §§ 27 and 28 of Chapter 3, that symmetric matrices will have this property provided they are also positive definite.

16. Table 10 shows the Cholesky computation without checks of

TABLE 10

A			b
4			1
5	6		2
2	−1	2	0

	L			x
	2			0·26
	2·5	0·5i		0·08
	1	7·0i	$\sqrt{50}$	−0·22
f'	0·5	−1·5i	$\dfrac{-11}{\sqrt{50}}$	

the 'unsatisfactory' example of Table 6, the complications for desk-machine work being not excessive.

Close examination of the computation shows that the effect of the imaginary quantities can be expressed as follows.

(i) Associate a marker symbol (i say) with the square root of any negative number.

(ii) When dividing by such a marked number carry out the normal (real) arithmetic operation but change the sign of the result and give it the symbol. These results will appear only in the other positions in the corresponding column of L and f'.

(iii) Marked numbers multiply only marked numbers, and the numerical result should have its sign changed, without a symbol.

(iv) The symbol is ignored in the final back-substitution.

These conclusions are apparent when we consider that we could write A in the form LU, in which the diagonal elements of L are correspondingly the same in absolute value as those of U, differing in sign only when the pivot is negative, in which case the other elements in corresponding columns of L and U' will also have the same values but with opposite signs. For example in this case we would find

$$
\begin{array}{cc}
L & U' \\
\begin{bmatrix} 2 & & \\ 2{\cdot}5 & 0{\cdot}5 & \\ 1 & 7{\cdot}0 & \sqrt{50} \end{bmatrix}, &
\begin{bmatrix} 2 & & \\ 2{\cdot}5 & -0{\cdot}5 & \\ 1 & -7{\cdot}0 & \sqrt{50} \end{bmatrix}.
\end{array} \tag{22}
$$

$$f' \quad 0{\cdot}5 \quad 1{\cdot}5 \quad \frac{-11}{\sqrt{50}}$$

The numerical quantities in L are the same as those of Table 10, while the corresponding elements in U' and f, when the diagonal of U is negative, have opposite signs to those of Table 10. We always *divide* by a diagonal term of U', corresponding to (ii), and *multiply* corresponding terms of $[U', f]$ and $[L]$, corresponding to (iii).

17. We note finally that in Table 10 the quantities involving $\sqrt{50}$ would in practice be expressed as truncated decimal or binary numbers, so that, unlike all the other methods used for this example, Cholesky's method will not necessarily give exact results even for this problem.

Inversion. Connexion with Doolittle and Crout

18. If we want the solutions for several right-hand sides, with the same matrix on the left, we can treat all the right-hand sides simultaneously in Gauss elimination and finally perform back-substitution for each of the solutions. In the case of inversion the right-hand sides would be the columns of the unit matrix. Table 11 shows the computation of an inverse by the method of Doolittle. Here the columns involved in successive back-substitutions form the matrix labelled C. Successive back-substitutions yield successive rows of $(A^{-1})'$.

TABLE 11

			A			I	
		4	−9	2	1	0	0
		2	−4	6	0	1	0
		1	−1	3	0	0	1
M			U			C	
−0·5	−0·25	4	−9	2	1	0	0
	−2·5		0·5	5	−0·5	1	0
				−10	1	−2·5	1
			$(A^{-1})'$				
		0·3	0	−0·1			
		−1·25	−0·5	0·25			
		2·3	1·0	−0·1			

†19. Similarly for inversion by Banachiewicz we seek directly that matrix, A^{-1}, which satisfies the equation

$$AA^{-1} = I. \tag{23}$$

If A is decomposed into the product LU we can write

$$LUA^{-1} = I, \tag{24}$$

and, defining the matrix C by the equation

$$LC = I, \tag{25}$$

we have finally the required A^{-1} from

$$UA^{-1} = C. \tag{26}$$

Now it is clear that the matrix C of these equations, equivalent to L^{-1} from (25), is exactly the C of Table 11, so that the two methods are effectively identical. The computation, however, might be arranged as in Table 12 and we use the rules of matrix multiplication, with suitable location of partners in products, to find L^{-1} and A^{-1} from

$$R_r(L^{-1})C_s(L) = C_r(L^{-1})'C_s(L) = (r, s) \text{ element of } I, \tag{27}$$

and $\quad R_r(U)C_s(A^{-1}) = C_r(U')C_s(A^{-1}) = (s, r) \text{ element of } (L^{-1})'. \tag{28}$

TABLE 12

A					
4	−9	2			
2	−4	6			
1	−1	3			

L				$(L^{-1})'$	
1			1	−0·5	1
0·5	1			1	−2·5
0·25	2·5	1			1

U'				A^{-1}	
4			0·3	−1·25	2·3
−9	0·5		0·0	−0·50	1·0
2	5	−10	−0·1	0·25	−0·1

20. The Doolittle process then effectively calculates L, U and $C = L^{-1}$, and in this respect is equivalent to the Banachiewicz method of Table 12. Some alternative methods of this type have no immediate connexion with the compact elimination processes.

We might, for example, write equations (23)–(26) in the alternative forms

$$A^{-1}A = I, \quad A^{-1}LU = I, \quad A^{-1}L = U^{-1}, \qquad (29)$$

compute L and U as before, but invert U instead of L and find A^{-1} from the last of (29), written in the form

$$C_r(A^{-1})'C_s(L) = (r, s) \text{ element of } U^{-1}. \qquad (30)$$

Similarly we find U^{-1} from

$$R_r(U)C_s(U^{-1}) = C_r(U')C_s(U^{-1}) = (r, s) \text{ element of } I. \qquad (31)$$

The computation is shown on the left of Table 13, and there is clearly nothing equivalent in the Doolittle method to the evaluation of U^{-1}.

A little thought, however, shows that if we start with the transpose A' of A and carry out the operation $A' = LU$, but choosing U to be the unit triangle, the L and U we obtain will be exactly the transposed U and L respectively of Table 12, and the remaining work is identical with that of Table 12. Further, recalling the connexion between Crout and Doolittle, we see that the methods of equations (29)–(31) are identical with the Crout method applied to the transpose A' of A. This is shown on the right of Table 13.

21. A second possibility, with no obvious previous analogy, is to write

$$A^{-1} = U^{-1}L^{-1}, \qquad (32)$$

inverting both U and L and computing the product. Table 14 shows a possible arrangement, with row-by-row or column-by-column multiplication in all cases. We shall in fact see in § 27 that this method is closely related to the Aitken elimination process for inversion.

TABLE 13

A					A'			I		
4	−9	2			4	2	1	1	0	0
2	−4	6			−9	−4	−1	0	1	0
1	−1	3			2	6	3	0	0	1
L				(A⁻¹)'						
1			0·3	0	−0·1	4	0·5	0·25	0·25	0	0
0·5	1		−1·25	−0·5	0·25	−9	0·5	2·5	4·5	2	0
0·25	2·5	1	2·3	1·0	−0·1	2	5	−10	2·3	1	−0·1
U'			U⁻¹			A⁻¹					
4			0·25	4·5	2·3	0·3	−1·25	2·3			
−9	0·5		2	1·0		0	−0·5	1·0			
2	5	−10			−0·1	−0·1	0·25	−0·1			

TABLE 14

A			A⁻¹		
4	−9	2	0·3	−1·25	2·3
2	−4	6	0	−0·5	1·0
1	−1	3	−0·1	0·25	−0·1

L			(L⁻¹)'		
1			1	−0·5	1
0·5	1			1	−2·5
0·25	2·5	1			1
U'			U⁻¹		
4			0·25	4·5	2·3
−9	0·5			2	1·0
2	5	−10			−0·1

22. We can even obtain A^{-1} without inverting either L or U, using the equations

$$UA^{-1} = L^{-1}, \quad A^{-1}L = U^{-1}, \tag{33}$$

and concentrating on those row-column products which produce the diagonal terms of L^{-1} (all units if L is selected to be the unit triangle), and the zero elements in L^{-1} and U^{-1}. That is we use

$$R_r(U)C_s(A^{-1}) = (L^{-1})_{rs}, \quad r < s, \tag{34}$$

$$R_r(A^{-1})C_s(L) = (U^{-1})_{rs}, \quad r > s, \tag{35}$$

each equation producing another element of A^{-1}. This method has no explicit 'elimination' analogy.

There is no possible arrangement of the computing sheet which will permit row-by-row or column-by-column multiplication through-out, without the recording of some matrix and also its transpose. The first part of Table 15 shows such an arrangement which records both L and L' and uses (34) and (35) in the forms

$$C_r(U')C_s(A^{-1}) = (L^{-1})_{rs}, = 0 \text{ if } r < s, = 1 \text{ if } r = s, \qquad (36)$$

$$R_r(A^{-1})R_s(L') = (U^{-1})_{rs}, = 0 \text{ if } r > s. \qquad (37)$$

The order of calculation is

(i) last column of A^{-1}, from (36) with $r = 3, 2, 1$;
(ii) remainder of last row of (A^{-1}), from (37) with $s = 2, 1$;
(iii) repetition of (i) and (ii), for other columns and rows.

An alternative arrangement which does not involve copying, which allows convenient multiplication for the determination of A^{-1}, but not for that of L and U, is shown on the right of Table 15. Equations (36) and (37) are replaced by

$$\left. \begin{array}{l} R_r(U)R_s(A^{-1})' = (L^{-1})_{rs}, = 0 \text{ if } r < s, = 1 \text{ if } r = s \\ C_r(A^{-1})'C_s(L) = (U^{-1})_{rs} = 0 \text{ if } r > s \end{array} \right\}, \qquad (38)$$

and the order of calculation becomes

(i) last row of $(A^{-1})'$, from the first of (38) with $r = 3, 2, 1$;
(ii) last column of $(A^{-1})'$, from the second of (38) with $s = 2, 1$;
(iii) repetition for succeeding rows and columns, working backwards.

This method involves the minimum amount of recording.

TABLE 15

A			A			U		
4	-9	2	4	-9	2	4	-9	2
2	-4	6	2	-4	6		0·5	5·0
1	-1	3	1	-1	3			-10

L			L'		
1			1	0·5	0·25
0·5	1			1	0·5
0·25	2·5	1			1

L			(A⁻¹)'		
1			0·3	0	-0·1
0·5	1		-1·25	-0·5	0·25
0·25	2·5	1	2·3	1·0	-0·1

U'			A⁻¹		
4			0·3	-1·25	2·3
-9	0·5		0	-0·5	1·0
2	5·0	-10	-0·1	0·25	-0·1

Inversion. Symmetric case

23. If A is symmetric we again have several possibilities following the computation of L from $A = LL'$. First, we could find L^{-1}, and therefore implicitly $(L')^{-1}$, and compute A^{-1} from $A^{-1}L = (L')^{-1}$. Alternatively we could form A^{-1} from the product $(L')^{-1}L^{-1}$, and

the amount of recording is the same in each case. A possible arrangement for the second method is shown in Table 16, in which we use the equation

$$R_r(L)C_s(L') = C_r(L')C_s(L') = (r, s) \text{ element of } A \qquad (39)$$

to find L',

$$R_r(L)C_s(L^{-1}) = C_r(L')C_s(L^{-1}) = (r, s) \text{ element of } I \qquad (40)$$

to find L^{-1} and

$$(A^{-1})_{rs} = R_r(L^{-1})'C_s(L^{-1}) = C_r(L^{-1})C_s(L^{-1}) \qquad (41)$$

to find the elements of the inverse A^{-1}, which is symmetric if A is symmetric.

TABLE 16

A			A⁻¹		
4	5	2	36·75	−27	−5·5
	6·5	1·5		20	4·0
		6			1·0

L′			L⁻¹		
2·0	2·5	1·0	0·5		
	0·5	−2·0	−2·5	2·0	
		1·0	−5·5	4·0	1·0

24. By analogy with the general case we need not invert L but can find A^{-1} from either of the equations

$$A^{-1}L = (L')^{-1}, \qquad L'A^{-1} = L^{-1}, \qquad (42)$$

using those scalar products $R_r(A^{-1})R_s(L')$, $r \geqslant s$, or $C_r(L)C_s(A^{-1})$, $r \leqslant s$, which give the zero and diagonal terms in $(L')^{-1}$ and L^{-1} respectively, and taking advantage of the symmetry of A^{-1}. The diagonal terms of the inverse triangle, of course, are the reciprocals of those of the triangle. Table 17 illustrates two arrangements, those diagonal terms being recorded in a row beneath L' and L respectively.

TABLE 17

A			A		
4	5	2	4		
	6·5	1·5	5	6·5	
		6	2	1·5	6

L′			L			A⁻¹		
2·0	2·5	1·0	2·0			36·75	−27·0	−5·5
	0·5	−2·0	2·5	0·5		−27·0	20·0	4·0
		1·0	1·0	−2·0	1·0	−5·5	4·0	1·0
0·5	2·0	1·0	0·5	2·0	1·0			

A⁻¹		
36·75	−27·0	−5·5
−27·0	20·0	4·0
−5·5	4·0	1·0

Connexion with Jordan and Aitken

25. The Jordan method reduced the original matrix A to diagonal form by premultiplication with matrices whose product had no special form. There is clearly no possibility of a 'compact' Jordan method, and no connexion between the Jordan method and that of Banachiewicz.

26. The Aitken method, however, discussed in §§ 18–20 in Chapter 3, has certain analogies with the methods of this chapter. We recall that the process started with the array

$$\left[\begin{array}{c|c} A & B \\ \hline C & O \end{array}\right], \tag{43}$$

reduced A to U by elimination and performed the corresponding operations on B, so that we then had the array

$$\left[\begin{array}{c|c} U & F \\ \hline C & O \end{array}\right]. \tag{44}$$

We then added multiples of rows of U to C, to reduce C to the null matrix, and the same operations on F produced the result $-CA^{-1}B$. In particular, if $B = b$ and $C = c'$, a single row, we computed the number $-c'x$ where $Ax = b$, and if $B = C = I$ we produced the negative inverse matrix $-A^{-1}$.

Now it is clear that the first stage corresponds to the equations

$$A = LU, \qquad B = LF, \tag{45}$$

where L is the unit lower triangle and U the upper triangle of our standard matrix decomposition. The matrix U is the assembly of pivotal rows and the matrix L is formed by the multipliers in the elimination process.

The elimination of C is then equivalent to the production of a matrix K such that

$$KU + C = O, \tag{46}$$

and the corresponding operation on F gives $KF = -CU^{-1}F = -CA^{-1}B$. The matrix K was formed of the multipliers in the extended elimination process.

27. All these operations can be conducted by direct application of the rules of matrix multiplication, from the set of equations

$$R_r(L)C_s(U) = (r, s) \text{ element of } A, \text{ giving } L \text{ and } U, \quad (47)$$

$$R_r(L)C_s(F) = (r, s) \text{ element of } B, \text{ giving } F, \quad (48)$$

$$R_r(-K)C_s(U) = (r, s) \text{ element of } C, \text{ giving } -K, \quad (49)$$

$$R_r(-K)C_s(F) = (r, s) \text{ element of } -KF = CA^{-1}B, \quad (50)$$

giving the required result.

An arrangement of this computation for the Aitken example of Table 7 in Chapter 3 is shown in Table 18. The arrangement allows row-by-row multiplication throughout, according to the equations

$$\left.\begin{array}{l} R_r(L)R_s(U') = a_{rs} \\ R_r(L)R_s(F') = b_{rs} \\ R_r(-K)R_s(U') = c_{rs} \\ R_r(-K)R_s(F') = (r, s) \text{ element of result} \end{array}\right\}. \quad (51)$$

TABLE 18

	A	
4	−9	2
2	−4	6
1	−1	3

	L			*B*	
1			3	1	
0·5	1		−1	4	
0·25	2·5	1	2	1	

	U'			*C*	
4			1	2	3
−9	0·5		4	−5	6
2	5	−10			

	F'	
3	−2·5	7·5
1	3·5	−8·0

	−*K*	
0·25	8·5	4·0
1·00	8·0	3·6

	CA⁻¹B
9·5	−2·0
10·0	0·2

We see the saving in recording at all stages of the computation and note the numerical connexion between these results and those of the standard Aitken method.

28. When B and C are each equal to I, so that we are using the method to find the inverse of A, it is clear from (45) that $F = L^{-1}$ and from (46) that $-K = U^{-1}$. We are therefore performing the exact equivalent of the Banachiewicz variation of Table 14, so that the latter is a true compact analogy of the Aitken elimination method for inversion.

Row interchanges

†**29.** In all the methods described so far we have at some stage performed the decomposition $LU = A$, and have ignored the possibility of a break-down of the process due to one of the diagonal elements, usually of U, having the value zero, so that subsequent division by this value is prohibited.

We have already seen that if the last diagonal element vanishes the determinant of the matrix A is zero, the matrix is singular, and we can neither invert A nor solve uniquely linear equations involving A. If an earlier diagonal term vanishes, however, it means only that a leading sub-matrix of A has a zero determinant, and does not necessarily imply singularity of A. For example the matrix

$$
A
$$

$$
\begin{bmatrix}
4 & 3 & 2 & 2 \\
1 & -1 & 4 & 4 \\
2 & -2 & 8 & -6 \\
1 & -1 & -3 & 4
\end{bmatrix}
$$

has the perfectly good inverse

$$
A^{-1}
$$

$$
\frac{1}{14}
\begin{bmatrix}
2 & -2 & 2 & 4 \\
2 & 0 & -2 & -4 \\
0 & 2 & 0 & -2 \\
0 & 2 & -1 & 0
\end{bmatrix},
$$

but the matrix of order three obtained by omitting the last row and column of A is singular, and we find

$$
\begin{matrix}
B & & L & & U
\end{matrix}
$$

$$
\begin{bmatrix} 4 & 3 & 2 \\ 1 & -1 & 4 \\ 2 & -2 & 8 \end{bmatrix} = \begin{bmatrix} 1 & & \\ 0{\cdot}25 & 1 & \\ 0{\cdot}5 & 2 & 1 \end{bmatrix} \begin{bmatrix} 4 & 3 & 2 \\ & -1{\cdot}75 & 3{\cdot}5 \\ & & 0 \end{bmatrix}. \qquad (52)
$$

The bordering of this B by extra rows and columns to produce A does not, of course, affect the first three rows and columns of L and U, but the computation can proceed no further. Moreover the use of a small pivot, a diagonal term of U, might give rise to large errors in the computed results.

These difficulties were avoided in the Gauss elimination method by choosing as pivot the largest element, not necessarily on the diagonal, in the relevant column of the reduced matrix. The corresponding row was effectively interchanged with the leading row, and as we saw the result was equivalent to the decomposition into LU of a row-permutation of the original A.

30. We now consider whether this more valuable Gauss process can be performed in a compact way. Only the decomposition $A = LU$ need be discussed, since the earlier compact methods are, as we have seen, effectively equivalent to this.

An example will probably best illustrate the procedure, and we take the equations

$$
\left. \begin{aligned} 2{\cdot}4x_1 + 6{\cdot}0x_2 - 2{\cdot}7x_3 + 5{\cdot}0x_4 &= 14{\cdot}6 \\ -2{\cdot}1x_1 - 2{\cdot}7x_2 + 5{\cdot}9x_3 - 4{\cdot}0x_4 &= -11{\cdot}4 \\ 3{\cdot}0x_1 + 5{\cdot}0x_2 - 4{\cdot}0x_3 + 6{\cdot}0x_4 &= 14{\cdot}0 \\ 0{\cdot}9x_1 + 1{\cdot}9x_2 + 4{\cdot}7x_3 + 1{\cdot}8x_4 &= -0{\cdot}9 \end{aligned} \right\}. \qquad (53)
$$

We proceed to calculate the elements of L, U, y and x such that

$$
LU = I_r A, \qquad Ly = I_r b, \qquad Ux = y, \qquad (54)
$$

where $I_r A$ is a row permutation of A, $I_r b$ the corresponding interchange in position of the elements of b, and the interchanges ensure that all multipliers in the Gauss process are not greater than unity, corresponding to the choice of 'largest element in column' as pivot. This means that no off-diagonal element in the L matrix will exceed unity and that the diagonal element of the U matrix is as large as possible at each stage.

In the first stage the possible values of u_{11} are the elements of the first column of A. We choose 3·0, which means interchanging rows one and three of (A, b), and it also fixes the first row of (U, y). We then have the arrays

$$
\begin{array}{cc}
I_1 A & I_1 b
\end{array}
\quad L \quad
\begin{array}{c}
U \quad\quad y
\end{array}
$$

$$
\left[\begin{array}{cccc|c}
3\cdot0 & 5\cdot0 & -4\cdot0 & 6\cdot0 & 14\cdot0 \\
-2\cdot1 & -2\cdot7 & 5\cdot9 & -4\cdot0 & -11\cdot4 \\
2\cdot4 & 6\cdot0 & -2\cdot7 & 5\cdot0 & 14\cdot6 \\
0\cdot9 & 1\cdot9 & 4\cdot7 & 1\cdot8 & -0\cdot9
\end{array}\right]
\left[\begin{array}{ccc}
1 & & \\
-0\cdot7 & 1 & \\
0\cdot8 & & 1 \\
0\cdot3 & & & 1
\end{array}\right]
\left[\begin{array}{ccccc}
3\cdot0 & 5\cdot0 & -4\cdot0 & 6\cdot0 & 14\cdot0 \\
& u_{22} & & & \\
& & & & \\
& & & &
\end{array}\right].
\quad (55)
$$

This choice of u_{11} also fixes the *values* of the other elements in $C_1(L)$, but their *positions* are not yet fixed, since we can still interchange any of the last three rows in $(I_1 A, I_1 b)$. The possible values of l_{21} are $-0\cdot7$, $0\cdot8$, $0\cdot3$, and the corresponding values of u_{22} would then be $0\cdot8$, $2\cdot0$, $0\cdot4$. These quantities are the elements in the pivotal column of the first Gauss reduced matrix. We choose the largest, and the corresponding multipliers are the ratios of the other elements to this. These multipliers occupy the second column of L, and we can insert them, after interchanging rows 2 and 3 of $(I_1 A, I_1 b)$ and making the corresponding change in L. The first two rows of L are now fixed, together with those of $(I_2 A, I_2 b)$, so that we can also complete the second row of (U, y). At this stage we then have the arrays

$$
\begin{array}{cc}
I_2 A & I_2 b
\end{array}
\quad L \quad
\begin{array}{c}
U \quad\quad y
\end{array}
$$

$$
\left[\begin{array}{cccc|c}
3\cdot0 & 5\cdot0 & -4\cdot0 & 6\cdot0 & 14\cdot0 \\
2\cdot4 & 6\cdot0 & -2\cdot7 & 5\cdot0 & 14\cdot6 \\
-2\cdot1 & -2\cdot7 & 5\cdot9 & -4\cdot0 & -11\cdot4 \\
0\cdot9 & 1\cdot9 & 4\cdot7 & 1\cdot8 & -0\cdot9
\end{array}\right]
\left[\begin{array}{ccc}
1 & & \\
0\cdot8 & 1 & \\
-0\cdot7 & 0\cdot4 & 1 \\
0\cdot3 & 0\cdot2 & & 1
\end{array}\right]
\left[\begin{array}{ccccc}
3\cdot0 & 5\cdot0 & -4\cdot0 & 6\cdot0 & 14\cdot0 \\
& 2\cdot0 & 0\cdot5 & 0\cdot2 & 3\cdot4 \\
& & u_{33} & & \\
& & & &
\end{array}\right],
$$

$$(56)$$

and there is still the necessity of selecting the final order of the last two rows of $(I_2 A, I_2 b)$ and hence of the corresponding elements in L.

If we use the present order we find the possible values 2·9 and 5·8 for u_{33}, so that we must interchange rows 3 and 4 of $(I_2 A, I_2 b)$, and also of L, and can fill in the last missing element of L since the corresponding multiplier in the Gauss process is $2\cdot9/5\cdot8 = 0\cdot5$.

We can also complete the third and fourth rows of U, and end with the arrays

$$
\begin{array}{cc}
I_3 A & I_3 b
\end{array}
\quad L \quad
\begin{array}{c}
U \quad\quad y
\end{array}
$$

$$
\left[\begin{array}{cccc|c}
3\cdot0 & 5\cdot0 & -4\cdot0 & 6\cdot0 & 14\cdot0 \\
2\cdot4 & 6\cdot0 & -2\cdot7 & 5\cdot0 & 14\cdot6 \\
0\cdot9 & 1\cdot9 & 4\cdot7 & 1\cdot8 & -0\cdot9 \\
-2\cdot1 & -2\cdot7 & 5\cdot9 & -4\cdot0 & -11\cdot4
\end{array}\right]
\left[\begin{array}{cccc}
1 & & & \\
0\cdot8 & 1 & & \\
0\cdot3 & 0\cdot2 & 1 & \\
-0\cdot7 & 0\cdot4 & 0\cdot5 & 1
\end{array}\right]
\left[\begin{array}{ccccc}
3\cdot0 & 5\cdot0 & -4\cdot0 & 6\cdot0 & 14\cdot0 \\
& 2\cdot0 & 0\cdot5 & 0\cdot2 & 3\cdot4 \\
& & 5\cdot8 & -0\cdot04 & -5\cdot78 \\
& & & 0\cdot14 & -0\cdot07
\end{array}\right].
$$

$$(57)$$

The final solution is obtained in the usual way, effectively back-substitution, from $Ux = y$.

Three interchanges were involved, so that the determinant $|A|$ is $(-1)^3$ times the product of the diagonal terms of U.

31. In the case of inversion we merely have to note that any of the methods described, such as the multiplication of inverses $U^{-1}L^{-1}$, will produce the inverse $A^{-1}I_r^{-1}$ of I_rA, so that this result must be post-multiplied by I_r to produce A^{-1}. In the particular example we have

$$I_r = \begin{bmatrix} 0 & 0 & 1 & 0 \\ 1 & 0 & 0 & 0 \\ 0 & 0 & 0 & 1 \\ 0 & 1 & 0 & 0 \end{bmatrix}, \tag{58}$$

corresponding to the new order 3, 1, 4, 2 of pivotal rows, so that we find the true A^{-1} from $U^{-1}L^{-1}$ by writing the *columns* in order 2, 4, 1, 3, the position of the unit in successive *columns* of I_r.

32. In the case of the compact equivalent of Aitken's method, using the equations

$$A = LU, \qquad B = LF, \qquad KU + C = O, \qquad KF = -CA^{-1}B \tag{59}$$

to produce $CA^{-1}B$, the process of interchanges would replace the first of (59) by

$$I_rA = LU, \tag{60}$$

and in the second of (59) we must use

$$I_rB = LF \tag{61}$$

to produce F. Then

$$K = -CU^{-1}, \quad KF = -CU^{-1}L^{-1}I_rB = -CA^{-1}I_r^{-1}I_rB = -CA^{-1}B, \tag{62}$$

which verifies the remarks on the Aitken process, using 'largest pivots', at the end of § 21 in Chapter 3.

†33. There is no possibility of row interchanges which preserve symmetry in the symmetric case to give $I_rA = LL'$ since, as in Gaussian elimination, symmetry is lost unless pivots are taken down the diagonal, which corresponds to the use of the original A with rows in natural order in the calculation of L and L'. If A were positive definite there is, as we have seen, no possibility of breakdown. If A were not known to be positive definite we could treat it by the interchange method as if it were a general asymmetric matrix, producing $I_r A = LU$, and this will often be the best method.

Operations with complex matrices

34. If the elements of A and b are complex the solution of the equations $Ax = b$ will also be complex in general. We could find this, by using complex arithmetic, by any of the methods treated in the last two chapters. Alternatively, if we write

$$A = B + iC, \qquad b = c + id, \qquad x = y + iz, \qquad (63)$$

we can use the real equations

$$\left. \begin{aligned} By - Cz &= c \\ Cy + Bz &= d \end{aligned} \right\}, \qquad (64)$$

expressible in the partitioned matrix form

$$\left[\begin{array}{c|c} B & -C \\ \hline C & B \end{array} \right] \left[\begin{array}{c} y \\ \hline z \end{array} \right] = \left[\begin{array}{c} c \\ \hline d \end{array} \right]. \qquad (65)$$

Here, of course, the order of the matrix in (65) is $2n$, and the amount of arithmetic involved in solution, inversion or determinant evaluation is some eight times that of the real case of order n.

35. We might, however, do better to take advantage of the classical theory that the algebra of complex numbers $\alpha + \beta i$ is the same as that of matrices of the type $\alpha I + \beta J$, where

$$I = \begin{bmatrix} 1 & 0 \\ 0 & 1 \end{bmatrix}, \qquad J = \begin{bmatrix} 0 & 1 \\ -1 & 0 \end{bmatrix}. \qquad (66)$$

For $I^2 = I$, $J^2 = -I$, $IJ = JI = J$ so that, for example,

$$\left. \begin{aligned} (\alpha I + \beta J)^2 &= (\alpha^2 - \beta^2) I + 2\alpha\beta J, \qquad (\alpha + \beta i)^2 = \alpha^2 - \beta^2 + 2\alpha\beta i \\ (\alpha I + \beta J)^{-1} &= (\alpha^2 + \beta^2)^{-1}(\alpha I - \beta J), \qquad (\alpha + \beta i)^{-1} = (\alpha^2 + \beta^2)^{-1}(\alpha - \beta i) \end{aligned} \right\}. \qquad (67)$$

We can therefore replace every element $a_{rs} + \alpha_{rs} i$ of A by the (2×2) matrix

$$\begin{bmatrix} a_{rs} & \alpha_{rs} \\ -\alpha_{rs} & a_{rs} \end{bmatrix}, \qquad (68)$$

carry out real arithmetic on the resulting $(2n \times 2n)$ matrix, $\bar{A}$ say, and interpret the results, which appear in (2×2) partitioned form with matrices like (68), in terms of complex numbers.

36. Consider, for example, the equations

$$(3+5i)x_1+(1+\ i)x_2 = -9+19i \\ (2+4i)x_1-(1+3i)x_2 = -25+17i \Big\}, \qquad (69)$$

which have the solution $x_1 = 2+4i$, $x_2 = 1-4i$. We show in Table 19 the arithmetic of the reduction for both the right-hand sides in (69) and for the corresponding unit matrix for the purpose of inversion. The elimination process is that of § 30 in Chapter 3 appropriate to exact integers, with the divisions by previous pivots, here taken down the diagonal, made without comment.

TABLE 19

($\bar{A}$)				($\bar{b}$)		(I)			
3	5	1	1	−9	19	1	0	0	0
−5	3	−1	1	−19	−9	0	1	0	0
2	4	−1	−3	−25	17	0	0	1	0
−4	2	3	−1	−17	−25	0	0	0	1
	34	2	8	−102	68	5	3	0	0
	2	−5	−11	−57	13	−2	0	3	0
	26	13	1	−87	1	4	0	0	3
		−58	−130	−578	102	−26	−2	34	0
		130	−58	−102	−578	2	−26	0	34
			596	2384	596	96	52	−130	−58

We note the various appearances of (2×2) matrices of the standard form, and the back substitution produces two columns, one for each of the two columns of $\bar{b}$, and four columns for those of the unit matrix, in which this arrangement is predominant. The computed columns, arranged as matrices of shapes (4×2) and (4×4), have the values

$$\begin{bmatrix} 2 & 4 \\ -4 & 2 \\ 1 & -4 \\ 4 & 1 \end{bmatrix}, \quad \tfrac{1}{596}\begin{bmatrix} 46 & -62 & 6 & -34 \\ 62 & 46 & 34 & 6 \\ 52 & -96 & -58 & 130 \\ 96 & 52 & -130 & -58 \end{bmatrix}, \qquad (70)$$

and we deduce the solutions

$$x = \begin{bmatrix} 2+4i \\ 1-4i \end{bmatrix}, \quad A^{-1} = \tfrac{1}{596}\begin{bmatrix} 46-62i & 6-34i \\ 52-96i & -58+130i \end{bmatrix}. \qquad (71)$$

We also note the saving in computing time which this method gives, though we are unaware of any machine programme of this type.

†37. The computation of the determinant $|A|$ needs a little more thought. If we consider the triangular resolution of the A of our example, without interchanging the rows, we find

$$
\overset{A}{\begin{bmatrix} 3+5i & 1+i \\ 2+4i & -(1+3i) \end{bmatrix}} = \overset{L}{\begin{bmatrix} 1 & \\ \frac{1}{34}(26+2i) & 1 \end{bmatrix}} \overset{U}{\begin{bmatrix} 3+5i & 1+i \\ & -\frac{1}{34}(58+130i) \end{bmatrix}}, \quad (72)
$$

and $|A|$ is the product of the diagonal elements of U. In terms of our matrix equivalents of the complex numbers the corresponding decomposition looks like

$$
\overset{\bar{A}}{\begin{bmatrix} 3 & 5 & 1 & 1 \\ -5 & 3 & -1 & 1 \\ 2 & 4 & -1 & -3 \\ -4 & 2 & 3 & -1 \end{bmatrix}} = \overset{L}{\begin{bmatrix} 1 & & & \\ 0 & 1 & & \\ \frac{26}{34} & \frac{2}{34} & 1 & \\ -\frac{2}{34} & \frac{26}{34} & 0 & 1 \end{bmatrix}} \overset{\bar{U}}{\begin{bmatrix} 3 & 5 & 1 & 1 \\ -5 & 3 & -1 & 1 \\ & & -\frac{58}{34} & -\frac{130}{34} \\ & & \frac{130}{34} & -\frac{58}{34} \end{bmatrix}},
$$
$$(73)$$

and the matrix $\bar{U}$ is not quite triangular. The true triangular decomposition is given by

$$
\overset{\bar{A}}{\begin{bmatrix} 3 & 5 & 1 & 1 \\ -5 & 3 & -1 & 1 \\ 2 & 4 & -1 & -3 \\ -4 & 2 & 3 & -1 \end{bmatrix}} = \overset{\bar{\bar{L}}}{\begin{bmatrix} 1 & & & \\ -\frac{5}{3} & 1 & & \\ \frac{2}{3} & \frac{2}{34} & 1 & \\ -\frac{4}{3} & \frac{26}{34} & -\frac{130}{58} & 1 \end{bmatrix}} \overset{\bar{\bar{U}}}{\begin{bmatrix} 3 & 5 & 1 & 1 \\ & \frac{34}{3} & \frac{2}{3} & \frac{8}{3} \\ & & -\frac{58}{34} & -\frac{130}{34} \\ & & & -\frac{596}{58} \end{bmatrix}},
$$
$$(74)$$

and we note that the first and third rows of $\bar{\bar{U}}$ are identical with those of $\bar{U}$.

It can be shown that this is true in general, the odd-numbered rows of $\bar{\bar{U}}$ being the same as those of $\bar{U}$, and this of course is all that is required to find $|A|$. Here, for example,

$$|A| = (3+5i)(-\tfrac{58}{34}-\tfrac{130}{34}i). \quad (75)$$

We can in fact easily express (73) in the form (74) by simple transformations which eliminate the terms below the diagonal in $\bar{U}$ without changing the product $L\bar{U}$. This is performed by simultaneously *adding* a multiple of the first row of $\bar{U}$ to its second row, to effect the elimination, and *subtracting* the same multiple of the second column of L from its first column, and so on for the other eliminations.

It is then also clear that the product of the first two diagonal terms of $\bar{\bar{U}}$ in (74) is the square of the modulus of the first diagonal complex term of U in (72), the product of the third and fourth diagonal term of $\bar{\bar{U}}$ is the square of the modulus of the second complex factor of $\bar{U}$, and so on. In fact

$$\{\mathrm{mod}\ |A|\}^2 = \prod_{r=1}^{2n} \bar{\bar{u}}_{rr}, \tag{76}$$

in this case the number 596, which is also the final pivot of the elimination process of Table 19.

Additional notes and bibliography

§§ 19–22, 29–32. These and other devices and computing arrangements are described in detail in

Fox, L., 1954, Practical solution of linear equations and inversion of matrices. *U.S. Bur. Stand. Appl. Math. Ser.* **39**, 1–54.

They are interesting in respect of the relations between the various methods, and valuable for desk-machine computation and for the insight they reveal into the development of research in the more practical aspects of numerical analysis. We shall see in Chapter 7 that they all involve effectively the same amount of arithmetic, and the preferred methods for electronic computing are those for which storage requirements are least and the accuracy achieved is as good as possible.

The methods of §§ 19–22 are effectively equivalent to elimination with pivots taken down the diagonal, and the possible consequent difficulties are avoided by using the compact row-interchanging processes of §§ 29–32.

§ 33. It is worth noting that in the positive-definite case, with the matrix scaled so that all $|a_{ij}| < 1$, then the matrix L in $A = LL'$ has all its elements less than unity. A proof of this is given in Chapter 7. On the other hand, in the corresponding processes of Gauss, Doolittle or Crout it is not certain, even in this case, that the use of pivots in order down the diagonal would keep the multipliers smaller than unity. It is, however, certain that the largest element of the matrix, and in all reduced matrices, will lie on the diagonal, so that Gauss elimination with *complete* pivoting will both preserve symmetry and also give the advantages of multipliers smaller than unity.

§ 37. This result was given by

Goodwin, E. T., 1950, Note on the evaluation of complex determinants. *Proc. Camb. Phil. Soc.* **46**, 450–2.

He also noted that we can write $|A| = re^{i\theta}$, where r is obtained from equation (76), and

$$\theta = -\sum_{r=1}^{n} \tan^{-1}(\bar{\bar{l}}_{2r,2r-1}),$$

where $\bar{\bar{l}}_{ij}$ is an element of $\bar{\bar{L}}$ in (74). With this result the amount of complex arithmetic required is reduced to a minimum.

5

Orthogonalisation Methods

Introduction

†1. THE methods discussed in this chapter have a common origin, that the required solution x of the linear equations $Ax = b$ is obtained as a linear combination of a special set of vectors whose properties permit easy evaluation of the coefficients of this combination.

Such special vectors might satisfy certain orthogonality properties, of the kind discussed in Chapter 2 with relation to the latent vectors of matrices. For example we might have two sets $x^{(r)}$ and $y^{(r)}$ of *biorthogonal* vectors, satisfying the relations

$$y^{(s)\prime}x^{(r)} = x^{(r)\prime}y^{(s)} = 0, \quad \text{for } r \neq s. \tag{1}$$

Or the vectors $x^{(r)}$ and $y^{(r)}$ may coincide, as in the case of latent vectors of a symmetric matrix, so that (1) becomes

$$x^{(s)\prime}x^{(r)} = x^{(r)\prime}x^{(s)} = 0, \quad \text{for } r \neq s. \tag{2}$$

We shall also use *conjugate* vectors, orthogonal with respect to a matrix A. Corresponding to (1) we would have the conjugacy condition

$$y^{(s)\prime}Ax^{(r)} = x^{(r)\prime}A'y^{(s)} = 0, \quad \text{for } r \neq s, \tag{3}$$

and corresponding to (2) for a symmetric matrix we would have

$$x^{(s)\prime}Ax^{(r)} = x^{(r)\prime}Ax^{(s)} = 0, \quad \text{for } r \neq s. \tag{4}$$

We start from the observation that the solution x can be expressed in the form

$$x = \sum_{r=1}^{n} \alpha_r x^{(r)}, \tag{5}$$

where the $x^{(r)}$ represent an independent set of vectors forming a non-singular matrix X, but are otherwise arbitrary at this stage. For such a set of vectors the coefficients α_r are clearly unique.

In the next few sections we consider methods for determining suitable vectors and the coefficients of their linear combination which gives the solution of $Ax = b$, for both symmetric and unsymmetric matrices.

2. We show that these methods are implicitly equivalent to the transformation of the original equations into other sets whose matrices are of simpler form. In the second part of the chapter we consider two methods which operate explicitly on the original matrix to produce a transformation of the required form.

Symmetric case

†3. Taking first the symmetric case we assume a solution (5) in which the vectors are conjugate, that is they satisfy equation (4). From (5) and the equation $Ax = b$ we find

$$\sum_{r=1}^{n} \alpha_r A x^{(r)} = b, \tag{6}$$

and premultiplication with $x^{(r)'}$ gives from (4) the result

$$\alpha_r = x^{(r)'} b / x^{(r)'} A x^{(r)}. \tag{7}$$

We are left with the problem of finding suitable vectors $x^{(r)}$. The simplest set of independent vectors forms the columns $e^{(r)}$, $r = 1, 2, \ldots, n$ of the unit matrix, and from these we can produce an independent set of conjugate vectors from the formulae

$$\left.\begin{aligned}
x^{(1)} &= e^{(1)} \\
x^{(2)} &= e^{(2)} - \alpha_{12} x^{(1)} \\
x^{(3)} &= e^{(3)} - \alpha_{13} x^{(1)} - \alpha_{23} x^{(2)} \\
\text{etc.}
\end{aligned}\right\}, \tag{8}$$

computing the coefficients α_{rs} so that all the conjugacy conditions (4) are satisfied. The conjugacy of $x^{(1)}$ and $x^{(2)}$ gives

$$\alpha_{12} = x^{(1)'} A e^{(2)} / x^{(1)'} A x^{(1)}. \tag{9}$$

Then premultiplication of the third of (8) by $x^{(1)'} A$ gives the result

$$\alpha_{13} = x^{(1)'} A e^{(3)} / x^{(1)'} A x^{(1)}, \tag{10}$$

since $x^{(1)}$ and $x^{(2)}$ are already conjugate. Similarly

$$\alpha_{23} = x^{(2)'} A e^{(3)} / x^{(2)'} A x^{(2)}, \tag{11}$$

and we can easily deduce the general expression

$$\alpha_{rs} = x^{(r)'} A e^{(s)} / x^{(r)'} A x^{(r)}, \qquad r < s. \tag{12}$$

This expression can be simplified. From (8) we see that $x^{(r)}$ is a linear combination of $e^{(r)}$ and previous vectors $x^{(p)}$, $p = 1, 2, ..., r-1$. Then

$$x^{(r)\prime} A x^{(r)} = x^{(r)\prime} A e^{(r)}, \qquad \alpha_{rs} = x^{(r)\prime} A e^{(s)} / x^{(r)\prime} A e^{(r)}. \qquad (13)$$

Moreover $A e^{(r)}$ is just the rth column of A, so that (12) reduces to

$$\alpha_{rs} = x^{(r)\prime} C_s(A) / x^{(r)\prime} C_r(A). \qquad (14)$$

Similarly (7) simplifies to

$$\alpha_r = x^{(r)\prime} b / x^{(r)\prime} C_r(A), \qquad (15)$$

and the denominators in (14) and (15) are the same.

4. The matrix X, formed from the conjugate vectors $x^{(r)}$ computed from (8), is obviously a unit upper triangle, and the conjugate relationships are included in the equation

$$X'AX = D, \qquad (16)$$

where D is a diagonal matrix with elements $d_r = x^{(r)\prime} C_r(A)$. Equations (8), moreover, can be written in the form

$$XC = I, \qquad (17)$$

where

$$C = \begin{bmatrix} 1 & \alpha_{12} & \alpha_{13} & \cdots \\ & 1 & \alpha_{23} & \cdots \\ & & 1 & \cdots \end{bmatrix}, \qquad (18)$$

a unit upper triangle whose elements are the multipliers α_{rs}. The matrix C is therefore the inverse of X. Finally we see from (7) that the multipliers α_r in (5) form a vector c given by

$$c = D^{-1} X' b, \qquad (19)$$

the quantities $x^{(r)\prime} A x^{(r)}$ being the diagonal elements d_r of D. Then (5) gives

$$x = Xc = X D^{-1} X' b = X(X'AX)^{-1} X' b = A^{-1} b. \qquad (20)$$

It follows that this method is closely related to previous compact elimination methods involving symmetric matrices. In particular the matrix C is the upper triangle of the Crout method for symmetric matrices, and the vector c is the 'right-hand side' of that method prior to the back-substitution. The matrix X' is then effectively the *inverse* of the lower triangle of the Crout method.

5. An example and an arrangement of the computation is shown in Table 1. In this arrangement we record the matrix A and beneath it the right-hand sides written as a row-vector b'. Beneath this comes an array whose elements above the diagonal belong to the matrix C with the units omitted, and below the diagonal we have the matrix X', also with the units omitted. The diagonal terms in this array are the elements of the matrix $D = X'AX$. At the right we have the column-vector c and below we have the row-vector solution x'. The arrangement permits easy row-by-row or column-by-column multiplication in most of the calculation.

TABLE 1

A

	4	5	2	
	5	6	−1	
	2	−1	2	
b'	1	2	0	
				c
	4	1·25	0·5	0·25
	−1·25	−0·25	14·0	−3·0
	17·0	−14·0	50·0	−0·22
x'	0·26	0·08	−0·22	

We can compare these figures with the Crout computation for the same example in Table 6 of Chapter 4, but the relationship disappears if we choose our orthogonal vectors in any other way, which is of course quite permissible. We have in fact chosen the easiest set and any other choice would increase considerably the labour involved.

The method would fail, as before, if any element of D were zero, a situation quite possible unless the matrix A is positive definite. The presence of the small and negative diagonal term $−0·25$ in this computation, for example, is not far from a catastrophic case and proves incidentally that A is not positive definite.

If needed we can easily compute the inverse A^{-1}, preferably from the equation

$$CA^{-1} = D^{-1}X', \qquad (21)$$

in which C, D and X' are already known and recorded. The appropriate computational scheme is that of § 24 and Table 17 in Chapter 4.

Unsymmetric case

6. For the unsymmetric case an analogous process will produce simultaneously the solution of two sets of linear equations, one involving the matrix A and one involving its transpose A'. Here we shall need to use two biconjugate sets of vectors $x^{(r)}$ and $y^{(r)}$ satisfying the conjugacy relations (3). We choose the set $x^{(r)}$ as before, to satisfy equations (8), and the $y^{(r)}$ from similar equations in which α_{rs} is replaced by β_{rs}. The conjugacy conditions then give

$$\alpha_{rs} = \frac{y^{(r)\prime} A e^{(s)}}{y^{(r)\prime} A x^{(r)}} = \frac{y^{(r)\prime} C_s(A)}{y^{(r)\prime} C_r(A)}, \qquad \beta_{rs} = \frac{x^{(r)\prime} A' e^{(s)}}{x^{(r)\prime} A' y^{(r)}} = \frac{x^{(r)\prime} R_s(A)}{x^{(r)\prime} R_r(A)}, \quad (22)$$

where $R_r(A)$ and $C_r(A)$ are respectively the rth row and rth column of A. The coefficients α_r, and β_r if required, in the solutions

$$x = \sum_{r=1}^{n} \alpha_r x^{(r)}, \qquad y = \sum_{r=1}^{n} \beta_r y^{(r)} \qquad (23)$$

of the respective equations

$$A x = b^{(1)}, \qquad A' y = b^{(2)}, \qquad (24)$$

are then given by

$$\alpha_r = \frac{y^{(r)\prime} b^{(1)}}{y^{(r)\prime} A x^{(r)}} = \frac{y^{(r)\prime} b^{(1)}}{y^{(r)\prime} C_r(A)}, \qquad \beta_r = \frac{x^{(r)\prime} b^{(2)}}{x^{(r)\prime} A' y^{(r)}} = \frac{x^{(r)\prime} b^{(2)}}{x^{(r)\prime} R_r(A)}. \quad (25)$$

We note that the denominators of the expressions for α_{rs}, β_{rs}, α_r and β_r are all the same, which simplifies the computation.

7. In matrix terms we see that the assembly of vectors $x^{(r)}$ forms a unit upper triangle X, that of $y^{(r)}$ a unit upper triangle Y, and the conjugacy relations are included in the equation

$$Y' A X = X' A' Y = D, \qquad (26)$$

a diagonal matrix whose elements are the denominators in (22) and (25). The coefficients α_{rs} and β_{rs} form unit upper triangles C_1 and C_2, for which

$$C_1 = X^{-1}, \qquad C_2 = Y^{-1}, \qquad (27)$$

the coefficients α_r and β_r form respective vectors $c^{(1)}$ and $c^{(2)}$ which satisfy

$$c^{(1)} = D^{-1} Y' b^{(1)}, \qquad c^{(2)} = D^{-1} X' b^{(2)}, \qquad (28)$$

and the solutions are given by

$$\left.\begin{array}{l} x = X c^{(1)} = X D^{-1} Y' b^{(1)} = X(Y' A X)^{-1} Y' b^{(1)} = A^{-1} b^{(1)} \\ y = Y c^{(2)} = Y D^{-1} X' b^{(2)} = Y(X' A' Y)^{-1} X' b^{(2)} = (A')^{-1} b^{(2)} \end{array}\right\}. \quad (29)$$

8. An example is shown in Table 2, together with a proposed arrangement for desk computing. The arrays labelled $[C_2' \backslash Y]$ and $[X' \backslash C_1]$ do not include the unit diagonal elements of these four triangular matrices, their places being taken by the (common) diagonal terms of D.

The numbers may be compared with those of Table 4 in Chapter 4 for the Crout method of solving the single set $Ax = b^{(1)}$. We can also see the relations with other methods quite easily from a combination of the various equations of § 7. For example we have

$$Y'AX = D, \qquad A = (Y')^{-1}DX^{-1} = C_2'DC_1, \tag{30}$$

so that C_2' corresponds to the lower triangle, C_1 to the upper triangle, of our general decomposition $A = LDU$. In the computation we have recorded $L\ (= C_2')$, $D\ (= D)$, and $U\ (= C_1)$, together with the inverses $L^{-1}\ (= Y')$ and $U^{-1}\ (= X)$. Then in the solution of $Ax = b^{(1)}$

TABLE 2

		A		$b^{(1)}$			$[C_2' \backslash Y]$		y
	4	−9	2	5		4	−0·5	1·0	0·1
	2	−4	6	3		0·5	0·5	−2·5	−2·75
	1	−1	3	4		0·25	2·5	−10·0	7·1
$b^{(2)'}$	2	3	5		$c^{(2)'}$	0·5	15·0	7·1	

		$[X \backslash C_1]$		$c^{(1)}$
	4	−2·25	0·5	1·25
	2·25	0·5	10·0	1·0
	−23·0	−10·0	−10·0	−0·15
x'	6·95	2·5	−0·15	

we first find the auxiliary vector $z = D^{-1}L^{-1}b$, which is just our $c^{(1)}$, following which we back-substitute in $Ux = z$ which is just $x = Xc^{(1)}$. The solution of $A'y = b^{(2)}$ has an obvious similar analogy.

The inverse of A is given by

$$A^{-1} = XD^{-1}Y', \tag{31}$$

in which we have already recorded X, Y and D. Alternatively we could use

$$C_1A^{-1} = D^{-1}Y', \qquad A^{-1}C_2' = XD^{-1}, \tag{32}$$

analogous to the method of § 22 of Chapter 4.

Again any other choice of conjugate vectors would be possible, and the connexion with other methods would disappear. The vanishing of any diagonal term of D would be catastrophic, and any attempt to produce a process equivalent to 'taking largest element as pivot' would clearly be very involved.

Matrix orthogonalisation

9. Our methods of conjugate vectors effectively carry out a transformation of the original matrix, and the associated equations, into simpler forms for which the solution is more easily calculated. In the symmetric case, for example, we find a set of vectors $x^{(r)}$ forming a matrix X such that $X'AX = D$, and in the unsymmetric case we have two sets, $x^{(r)}$ and $y^{(r)}$, such that $Y'AX = D$. The equations $Ax = b$ are then transformed to

$$X'AX(X^{-1}x) = X'b, \qquad x = XD^{-1}(X'b), \qquad (33)$$

or to

$$Y'AX(X^{-1}x) = Y'b, \qquad x = XD^{-1}(Y'b). \qquad (34)$$

We have effectively transformed A into a diagonal D. A second class of methods transforms A into an *orthogonal* matrix B. Such a matrix we have defined previously to satisfy the condition $BB' = I$. Here we take the less stringent definition

$$BB' = D, \qquad (35)$$

where D is diagonal. If now we can operate on $Ax = b$ to produce $Bx = c$, where B satisfies (35), the solution follows immediately from

$$x = B^{-1}c = B'D^{-1}c, \qquad (36)$$

and again we have produced a simpler set of equations.

10. In the first of these methods we build up the rows of the matrix B from those of A, in the form

$$\left.\begin{aligned}
R_1(B) &= R_1(A) \\
R_2(B) &= R_2(A) - \alpha_{12}R_1(B) \\
R_3(B) &= R_3(A) - \alpha_{13}R_1(B) - \alpha_{23}R_2(B) \\
\text{etc.}
\end{aligned}\right\}, \qquad (37)$$

choosing the coefficients α_{rs} such that the orthogonality conditions

$$R_r(B)R_s(B) = 0, \qquad r \neq s, \qquad (38)$$

are satisfied. The coefficients α_{rs} are clearly given by

$$\alpha_{rs} = \frac{R_r(B)R_s(A)}{R_r(B)R_r(B)}, \qquad r < s, \qquad (39)$$

and the denominators in these expressions are the elements of the diagonal matrix D. In fact the elements α_{rs} form the unit upper triangular matrix C of type (18), such that

$$B'C = A', \qquad C'B = A, \qquad B = (C')^{-1}A, \qquad (40)$$

and the equations $Ax = b$ are transformed to

$$Bx = c, \qquad c = (C')^{-1}b. \qquad (41)$$

We do not evaluate $(C')^{-1}$, but build up c, like B, from the equations

$$\left.\begin{aligned}c_1 &= b_1\\ c_2 &= b_2 - \alpha_{12}c_1\\ c_3 &= b_3 - \alpha_{13}c_1 - \alpha_{23}c_2\\ \text{etc.}\end{aligned}\right\} , \tag{42}$$

analogous to (37).

An example is shown in Table 3, and the numbers are expressed in the form of rational fractions so that there are no rounding errors and the solution is exact. The row d' contains the diagonal elements of D.

TABLE 3

			A			b	
		4	-9	2		5	
		2	-4	6		3	
		1	-1	3		4	

C			B			c	$D^{-1}c$
1	$\dfrac{56}{101}$ $\dfrac{19}{101}$	4	-9	2		5	$\dfrac{5}{101}$
	1 $\dfrac{34}{63}$	$-\dfrac{22}{101}$	$\dfrac{100}{101}$	$\dfrac{494}{101}$		$\dfrac{23}{101}$	$\dfrac{23}{2520}$
	1	$\dfrac{23}{63}$	$\dfrac{10}{63}$	$-\dfrac{1}{63}$		$\dfrac{185}{63}$	$\dfrac{185}{10}$
d'	101		$\dfrac{2520}{101}$	$\dfrac{10}{63}$			
x'	6·95		2·50	$-0·15$			

11. For the inverse we have

$$A^{-1} = B^{-1}(C')^{-1}, \qquad A^{-1}C' = B'D^{-1}. \tag{43}$$

The right-hand side of (43) is easily available, and C' is lower triangular, so that the elements of A^{-1} are obtained in succession in ways discussed in § 20 of Chapter 4.

We also note a connexion between this method and that of conjugate vectors. Since $BB' = D$ we have, from (40),

$$(C')^{-1}AA'C^{-1} = D, \qquad AA' = C'DC, \tag{44}$$

and comparing this with the symmetric conjugate vector method we see that the matrix C in (44) is exactly the matrix of the same name in § 4 when that method is applied to the symmetric matrix AA'.

†12. A second matrix orthogonalisation process comes from the observation that the rows of A^{-1} are orthogonal to the columns of A, and we can compute successive matrices B_1, B_2,..., B_n, starting with an arbitrary B_0, for which the rows of B_r are orthogonal to the first r columns of A, and $B_n = A^{-1}$.

Consider for example a matrix of order three, and start with

$$
\left.
\begin{aligned}
R_1(B_1) &= R_1(B_0) \\
R_2(B_1) &= R_2(B_0) - \alpha_{21} R_1(B_1) \\
R_3(B_1) &= R_3(B_0) - \alpha_{31} R_1(B_1)
\end{aligned}
\right\},
\tag{45}
$$

choosing the coefficients so that $R_s(B_1)C_1(A) = 0$, $s = 2, 3$. This gives

$$
\alpha_{s1} = R_s(B_0)C_1(A)/R_1(B_0)C_1(A), \qquad s = 2, 3,
\tag{46}
$$

since $R_1(B_1) = R_1(B_0)$.

Then we build up B_2 from the equations

$$
\left.
\begin{aligned}
R_1(B_2) &= R_1(B_1) - \alpha_{12} R_2(B_2) \\
R_2(B_2) &= R_2(B_1) \\
R_3(B_2) &= R_3(B_1) - \alpha_{32} R_2(B_2)
\end{aligned}
\right\},
\tag{47}
$$

choosing the coefficients so that $R_s(B_2)C_2(A) = 0$, $s = 1, 3$, which gives

$$
\alpha_{s2} = R_s(B_1)C_2(A)/R_2(B_1)C_2(A),
\tag{48}
$$

since $R_2(B_2) = R_2(B_1)$. The important thing to note is that

$$
R_s(B_2)C_1(A) = 0, \qquad s = 2, 3,
\tag{49}
$$

so that our B_2 has its rows orthogonal to both $C_1(A)$ and $C_2(A)$.

Finally we find B_3 for which

$$
\left.
\begin{aligned}
R_1(B_3) &= R_1(B_2) - \alpha_{13} R_3(B_3) \\
R_2(B_3) &= R_2(B_2) - \alpha_{23} R_3(B_3) \\
R_3(B_3) &= R_3(B_2)
\end{aligned}
\right\},
\tag{50}
$$

choosing the coefficients so that $R_s(B_3)C_3(A) = 0$, $s = 1, 2$, that is

$$
\alpha_{s3} = R_s(B_2)C_3(A)/R_3(B_2)C_3(A),
\tag{51}
$$

and again it is easy to see that we have not destroyed the orthogonality of the rows of B_3 with respect to the first and second columns of A. The resulting matrix B_3 satisfies

$$
B_3 A = D,
\tag{52}
$$

where D is a diagonal matrix whose rth element is

$$R_r(B_{r-1})C_r(A) = R_r(B_r)C_r(A).$$

The inverse A^{-1} is then just $D^{-1}B_3$.

13. The method will fail if any diagonal terms of D, except the last, should happen to vanish, and this depends on the choice of B_0. Consider first the choice $B_0 = I$. We can then show that the method is identical with the Jordan elimination process with pivots taken down the diagonal. For equations (45) are then equivalent to the matrix equation

$$J_1 B_1 = I, \tag{53}$$

where J is the lower triangular matrix

$$J_1 = \begin{bmatrix} 1 & & \\ \alpha_{21} & 1 & \\ \alpha_{31} & 0 & 1 \end{bmatrix}, \quad \text{and} \quad J_1^{-1} = \begin{bmatrix} 1 & & \\ -\alpha_{21} & 1 & \\ -\alpha_{31} & 0 & 1 \end{bmatrix}. \tag{54}$$

Moreover $\alpha_{s1} = a_{s1}/a_{11}$, $s = 2, 3$, where a_{pq} is an element of A, and these are the negatives of the multipliers in the Jordan elimination, so that $J_1^{-1}A = B_1A$ is exactly the first 'reduced' matrix in the Jordan process.

Similarly $J_2 B_2 = B_1$, where

$$J_2 = \begin{bmatrix} 1 & \alpha_{12} & \\ & 1 & \\ & \alpha_{32} & 1 \end{bmatrix}, \quad J_2^{-1} = \begin{bmatrix} 1 & -\alpha_{12} & \\ & 1 & \\ & -\alpha_{32} & 1 \end{bmatrix}, \tag{55}$$

and the $-\alpha_{s2}$, $s = 1, 3$, are the multipliers in the second Jordan elimination, since

$$(B_1 A)_{s,2}/(B_1 A)_{2,2} = R_s(B_1)C_2(A)/R_2(B_1)C_2(A) = \alpha_{s2}. \tag{56}$$

The continuation is obvious and proves the equivalence of this method with Jordan elimination.

14. This choice of B_0 does not guarantee the non-vanishing of a diagonal element of D, but we find that the diagonal elements of D are such that d_1, $d_1 d_2$, $d_1 d_2 d_3, \ldots$, are respectively equal to the determinants of successive leading submatrices of $B_0 A$. If we take $B_0 = A'$ this matrix is positive definite, so that all d_r will be positive.

Additional notes and bibliography

§§ 1, 2. Again the most interesting fact about these methods is that, notwithstanding the different language in which they are expressed, they are effectively equivalent, at least in their simpler applications, to one or other of our various elimination methods, and as such their practical value for actual computation is very small.

§ 3. This method was described by Fox, Huskey and Wilkinson (1948, *loc. cit.* p. 98), under the name 'method of orthogonal vectors'. They remarked that, through rounding errors, the quantities $x^{(s)'}Ax^{(r)}$ will not necessarily vanish exactly, and equation (7) is not then the exact expression for the coefficient α_r. On the other hand we can accept the computed vectors $x^{(r)}$ as exact, provided that we express equation (6) in the form

$$\sum_{r=1}^{n} \alpha_r x^{(s)'}Ax^{(r)} = x^{(s)'}b, \qquad s = 1, 2, \ldots, n,$$

and the elements of the matrix of these equations for α_r, and the right-hand sides, can be computed as exactly as desired, if necessary with a small amount of double-length arithmetic. The matrix of these equations is almost diagonal, so that their accurate solution can be obtained rapidly by iteration, with one of the methods of Chapter 8.

§ 12. This and other similar processes are discussed in

HESTENES, M. R., 1958, Inversion of matrices by biorthogonalisation and related results. *J. Soc. industr. appl. Math.* **6**, 51–90.

6

Condition, Accuracy and Precision

Introduction

1. WE MUST now turn attention to problems of accuracy and precision, and there are two main points to consider. First, if the data of the problem are slightly inaccurate or inexact, what is the 'tolerance' of the results, that is what is the maximum 'error' in the solution? Second, what errors are introduced by the solving process, in virtue of the rounding errors committed at many stages of the work?

The rounding errors, of course, have not occurred in our previous examples, since all numbers have been exact terminating decimals, or recorded as the ratio of two integers or even with a square root sign with deferred explicit evaluation. In practical computation all numbers would be stored to a fixed number of significant figures (floating-point operation), or a fixed number of decimal places, (fixed-point operation) and in high-speed computing the scale of two is generally preferred, so that the phrase 'decimal places' is replaced by 'binary places'. There is the possibility that the computing machine cannot even start with the correct coefficients, since a decimal number like $0 \cdot 1$ is a non-terminating binary fraction, just as $\frac{4}{7}$ is a non-terminating decimal fraction.

2. We find that the effect of such errors varies considerably from problem to problem, and it is clear from simple cases that the magnitude of the coefficients of the inverse matrix is an important factor. For example the difference between the solutions of $Ax = b$ and $Ax = b + \delta b$ is just $A^{-1} \delta b$, and this 'error' may be large if A is nearly singular.

Again the difference δx between the solution of $Ax = b$ and that of

$$(A + \delta A)(x + \delta x) = b + \delta b \tag{1}$$

can be expressed as

$$\delta x = (A + \delta A)^{-1}(\delta b - \delta A x), \tag{2}$$

and again its value depends critically on the behaviour of the inverse matrix.

If small errors in the coefficients, or in the solving process, have little effect on the solution we say that the problem is *well-conditioned*:

if the effect is large we speak of an *ill-conditioned* problem. We might sometimes speak of an ill-conditioned matrix, but this is a meaningless phrase until we have specified what we want to do with it. If we want to solve linear equations the matrix is ill-conditioned if it is nearly singular, that is if small changes in its elements would cause singularity, but such a matrix may occur in a perfectly well-conditioned way in other contexts, such as the determination of latent roots.

Symptoms, causes and effects of ill-conditioning

3. Consider the equations

$$
\left.
\begin{aligned}
0{\cdot}20000x_1+0{\cdot}16667x_2+0{\cdot}14286x_3+0{\cdot}12500x_4 &= 1 \\
0{\cdot}16667x_1+0{\cdot}14286x_2+0{\cdot}12500x_3+0{\cdot}11111x_4 &= 1 \\
0{\cdot}14286x_1+0{\cdot}12500x_2+0{\cdot}11111x_3+0{\cdot}10000x_4 &= 1 \\
0{\cdot}12500x_1+0{\cdot}11111x_2+0{\cdot}10000x_3+0{\cdot}09091x_4 &= 1
\end{aligned}
\right\},
\qquad (3)
$$

which have a symmetric matrix. If we solve this set by the Gauss elimination process, ignoring symmetry, taking pivots in order of size and rounding every calculated number to five decimals, we have the elimination array shown in Table 1.

TABLE 1

	(1)	0·20000	0·16667	0·14286	0·12500	1·00000
−0·83335	(2)	0·16667	0·14286	0·12500	0·11111	1·00000
−0·71430	(3)	0·14286	0·12500	0·11111	0·10000	1·00000
−0·62500	(4)	0·12500	0·11111	0·10000	0·09091	1·00000
−0·57205	(2)		0·00397	0·00595	0·00694	0·16665
−0·85735	(3)		0·00595	0·00907	0·01071	0·28570
	(4)		0·00694	0·01071	0·01278	0·37500
	(2)			−0·00018	−0·00037	−0·04787
−0·61111	(3)			−0·00011	−0·00025	−0·03581
	(3)				−0·00002	−0·00656

The corresponding Banachiewicz method, in which scalar products are produced from the sum of *unrounded* multiplications, divisions are performed *before* the numerator is rounded, and in which the rows are interchanged to give the effect of largest pivot, is shown in Table 2 with the operations on the right-hand side omitted.

4. The problem is notoriously ill-conditioned, and this is made manifest by the steady decrease, in absolute value, of the size of the pivots. In particular the last pivot has almost vanished in the Gauss method and is truly zero, to five decimals, in the triangular decomposition. Back-substitution involves division by these quantities and the results are obviously very unreliable.

TABLE 2

A

0·20000	0·16667	0·14286	0·12500
0·12500	0·11111	0·10000	0·09091
0·16667	0·14286	0·12500	0·11111
0·14286	0·12500	0·11111	0·10000

L

1·00000			
0·62500	1·00000		
0·83335	0·57141	1·00000	
0·71430	0·85701	0·66750	1·00000

U'

0·20000			
0·16667	0·00694		
0·14286	0·01071	−0·00017	
0·12500	0·01278	−0·00036	0·00000

In particular it is easy to see that the addition of a quantity ϵ to the element a_{34} of A in Table 1 would be carried forward directly to the last pivot, which appears in the (3, 4) position, so that in the Gauss method this would become $-0\cdot00002+\epsilon$. If ϵ is the possible 'tolerance' in a_{34}, this means that the final pivot is uncertain to within this amount, even if the figure $-0\cdot00002$ is correct and uncontaminated by rounding errors. In practice ϵ is very likely to have a maximum value of half a unit in the last figure given, so that in this case the last pivot lies between $-0\cdot000015$ and $-0\cdot000025$. The corresponding uncertainty in x_4 is clearly very large, and hardly one significant figure has any real meaning. The general conclusion follows that if we perform the Gauss process, working with a fixed number of decimals, the significance of the results is not greater than the number of significant figures in the final pivot. The significance may be less than this because the final pivot is likely to be incorrect through rounding errors committed in the solving process, and these will have greater relative effect on a very small last pivot.

It is also worth noting that the relative inaccuracies of x_1, x_2, x_3 and x_4 are not in general dependent on the relative sizes of their respective pivots, since it is clear, for example, that we could perform the elimination in reverse order, giving x_4 the largest pivot and x_1 probably the smallest.

5. In Table 1 the pivots lose figures systematically, but it is not uncommon to have a sudden decline, say in just the last pivot. For example the matrix

$$A = \begin{bmatrix} 4 & -4 & 4 & 6 \\ -1 & 6 & 6{\cdot}5 & 1 \\ 2 & -6 & 5 & 5 \\ 3 & -2 & 0 & 2{\cdot}99 \end{bmatrix} \tag{4}$$

would have successive pivots 4, 5, 9 and $-0{\cdot}01$. Again we cannot infer that x_1, x_2 and x_3 are determinable more accurately than x_4. In fact the inverse of this matrix, to one decimal, is

$$A^{-1} = \begin{bmatrix} -84{\cdot}1 & 17{\cdot}8 & 44{\cdot}2 & 88{\cdot}9 \\ 15{\cdot}9 & -3{\cdot}3 & -8{\cdot}5 & -16{\cdot}7 \\ -42{\cdot}3 & 9{\cdot}0 & 22{\cdot}3 & 44{\cdot}4 \\ 95{\cdot}0 & -20{\cdot}0 & -50{\cdot}0 & -100{\cdot}0 \end{bmatrix}, \tag{5}$$

so that the x_2 is least affected by uncertainties in the right-hand sides, whereas x_1 is affected as badly as x_4.

6. Ill-conditioning occurs, of course, when the rows of A are almost linearly dependent. In (4), for example, we find

$$0{\cdot}95R_1 - 0{\cdot}2R_2 - 0{\cdot}5R_3 - R_4 = 0,\ 0,\ 0,\ 0{\cdot}01, \tag{6}$$

where R_s is the sth row of A. This is very nearly the null vector, and it is significant that the coefficients in (6) are closely proportional to the corresponding elements of A^{-1} in successive *columns*.

Special near-dependency occurs if two *rows* are very similar, and in this case we find that the corresponding *columns* of A^{-1} are relatively large. Table 3 shows the results of Gauss elimination, with the unit matrix on the right, in a typical case of two very similar rows.

TABLE 3

A				I			
4	−4	4	6	1	0	0	0
−1	6	6·5	1	0	1	0	0
3	−2	0	3	0	0	1	0
3	−2	0	2·99	0	0	0	1
4	−4	4	6	1			
	5	7·5	2·5	0·25	1		
		−4·5	−2·0	−0·8	−0·2	1	
			−0·01	0	0	−1	1

Back-substitution clearly gives large values only in the last two solutions. The fact that we have not always chosen the 'largest pivot' does not affect this argument.

If more rows of A are very similar we would expect all the corresponding columns of A^{-1} to be large, as illustrated in Table 4.

TABLE 4

	A				I		
4	−4	4	6	1	0	0	0
3	−2	0	3	0	1	0	0
3	−1·98	−0·01	3·01	0	0	1	0
3	−2·01	0	2·99	0	0	0	1
4	−4	4	6	1			
	1	−3	−1·5	−0·75	1		
		0·05	0·04	0·015	−1·02	1	
			−0·001	0·0015	−1·602	0·6	1

Clearly the last three solutions, giving the last three columns of A^{-1}, have much larger elements than the first. The uncertainties in the answers due to small uncertainties in the right-hand sides come from the equation $\delta x = A^{-1}\delta b$, and each component is likely to be affected significantly if the rows of A are almost dependent. If this dependence springs from near similarity of certain rows the components of δb in these rows will have the more important effect.

7. If the rows are nearly dependent the columns must also be nearly dependent. For example in the case of (4) we find

$$-0\cdot89C_1+0\cdot17C_2-0\cdot44C_3+C_4 = 0,\ 0\cdot05,\ 0,\ -0\cdot02, \qquad (7)$$

where C_s is the sth column of A. These coefficients are nearly proportional to the corresponding elements in successive *rows* of A^{-1}.

Some rows of A^{-1} may be substantially larger than others, giving poorer definition in the corresponding components of the solution, if the corresponding *columns* of A are very similar. For example, the approximate inverse of the matrix

$$\begin{bmatrix} 4 & -1 & 3 & 3 \\ -4 & 6 & -2 & -2 \\ 4 & 6\cdot5 & 0 & 0 \\ 6 & 1 & 3 & 2\cdot99 \end{bmatrix} \text{ is } \begin{bmatrix} -0\cdot15 & -0\cdot22 & 0\cdot18 & 0 \\ 0\cdot09 & 0\cdot13 & 0\cdot04 & 0 \\ -88\cdot33 & 17\cdot00 & -44\cdot67 & 100 \\ 88\cdot89 & -16\cdot67 & 44\cdot44 & -100 \end{bmatrix}.$$

We note that the elements x_3 and x_4 are rather poorly determined, but the sum x_3+x_4 will have small errors, comparable with those of

x_1 and x_2. This is a special case of a not uncommon phenomenon, in which some or all components of the solution $Ax = b$ are very unreliable, whereas a linear combination $p = c'x$, say, has a relatively stable value. This will happen when the vector $c'A^{-1}$ has relatively small values, even though the elements of A^{-1} are large, and as we have seen it is indeed possible to choose a suitable c' when A is ill-conditioned, since successive columns of A^{-1} are likely to have the same general behaviour. Corresponding to the matrix in (4) the elements of c' need only satisfy approximately the equation

$$-0 \cdot 89c_1 + 0 \cdot 17c_2 - 0 \cdot 44c_3 + c_4 = 0, \tag{8}$$

with the same coefficients as in (7), to produce this effect. We return to this point in §§ 37–40.

Measure of condition

8. We can use our matrix norms to find some quantitative measure of the degree of ill-conditioning of a matrix A, which in relation to the solution of linear equations depends on the magnitude of the elements of its inverse A^{-1}. In particular we shall assume that the computed inverse, which we have obtained explicitly if we want the inverse and implicitly if we are solving linear equations, is the exact inverse of a perturbation $A + E$ of the original A.

We might then examine the error of our approximate inverse in the following way. We have

$$(A + E) - A = E, \tag{9}$$

so that by pre- and post-multiplication with $(A + E)^{-1}$ and A^{-1} respectively we find

$$A^{-1} - (A + E)^{-1} = (A + E)^{-1}EA^{-1}. \tag{10}$$

Taking norms, using the fact that $(A + E)^{-1} = (I + A^{-1}E)^{-1}A^{-1}$, and taking advantage of known results in the algebra of norms (§§ 44–7, Chapter 2), we get

$$\|A^{-1} - (A + E)^{-1}\| \leqslant \frac{\|A^{-1}\|^2\,\|E\|}{1 - \|A^{-1}\|\,\|E\|}, \text{ if } \|A^{-1}\|\,\|E\| < 1, \tag{11}$$

the latter condition implying only that our perturbation E should have sufficiently small elements compared with those of A^{-1}.

Then if

$$\frac{\|E\|}{\|A\|} = \epsilon, \quad \|A\|\,\|A^{-1}\| = k, \tag{12}$$

we have

$$\frac{\|A^{-1} - (A+E)^{-1}\|}{\|A^{-1}\|} \leqslant \frac{k\epsilon}{1-k\epsilon}, \quad \text{where} \quad k\epsilon < 1, \tag{13}$$

giving a kind of relative error for the computed inverse. The introduction of ϵ is necessary to relate the perturbation E to the original A.

This error is proportional to $k = \|A\| \, \|A^{-1}\|$, and this is called a *condition number* for the matrix. In particular, if A is symmetric so that $A = A'$, our norm $\|A\|_2$ is $|\lambda_1|$ and $\|A^{-1}\|_2 = |\lambda_n^{-1}|$, where λ_1 and λ_n are respectively the numerically largest and smallest latent roots of A. The condition number is then the ratio of the largest latent root of A to the smallest. We note in passing that even the relative error has the factor $\|A^{-1}\|$, and the absolute error has the square of this possibly large quantity as a factor.

9. The result (13) gives a warning about a method which has often been suggested for solving linear equations $Ax = b$. The suggestion is to premultiply this equation by A', and to solve the equations

$$Bx = c, \qquad B = A'A, \qquad b = A'b, \tag{14}$$

for which the matrix is symmetric and positive definite. We can take advantage of these facts, decomposing B into LL' without interchanges and with no worries about zero diagonal elements.

But apart from the fact that the formation of the derived equations costs something like $\frac{1}{2}n^3$ multiplications, somewhat more than the complete solution of the original equations, the new problem is significantly more ill-conditioned than the old. The condition number of $A'A$ can be as large as the square of that of A, and we might lose significant figures in the elimination process at an alarming rate.

For example if we apply this method to the matrix of equation (3) and carry out the first step of the elimination process, we find the results of Table 5, and note that all the significant figures have virtually disappeared in just one step. At the next step we have nothing more than a 'rounded' unit in the fifth decimal.

TABLE 5

	$A'A$			
	0·10381	0·08889	0·07778	0·06917
−0·85628	0·08889	0·07616	0·06667	0·05931
−0·74925	0·07778	0·06667	0·05838	0·05195
−0·66631	0·06917	0·05931	0·05195	0·04624
		0·00005	0·00007	0·00008
		0·00007	0·00010	0·00012
		0·00008	0·00012	0·00015

Exact and approximate data

10. Detailed analysis of the incidence and extent of rounding errors committed in the calculating process is very complicated, and is of no great *practical* value in determining the accuracy of the final results. The real needs are

(i) to obtain as good a result as the given data will permit,
(ii) to know when this result has been obtained,
(iii) to correct an approximate solution if such a correction is meaningful.

If the given data, the coefficients and right-hand sides in the equations, are known exactly, then there is justification for asking for answers to as many significant figures as may be needed in subsequent applications of the results. Such problems I call *mathematical*, to distinguish them from those, here called *physical*, in which the data are subject to uncertainties of measurement or other causes. In physical problems the answers are limited in accuracy, and figures quoted beyond a certain stage will generally be meaningless except in special circumstances such as those surrounding equation (8).

Mathematical problems. Correction to approximate solution

11. We take first the mathematical problem and look for methods of achieving required accuracy. One obvious method is to work throughout with a large number of figures, using double, triple or even multiple-length working in all the arithmetic operations. Examination of the final pivot in the Gauss process, for example, then gives some estimate of the number of significant figures that can properly be quoted in the answers, and we shall see later that this estimate can be made more precise.

This technique is likely to be lengthy, however, since multi-length arithmetic can be very time-consuming and also uneconomic, since the accuracy required might be obtainable with single-length arithmetic.

A possible alternative is to perform the elimination or an equivalent process using single-length arithmetic, and then make subsequent corrections to this first approximation, stopping when the required accuracy has been obtained.

12. If the matrix A and right-hand sides b can be stored exactly in single-length registers the first approximation is inaccurate only through successive rounding errors, and we can regard this solution $x^{(0)}$, an approximation to $A^{-1}b$, as the result $B^{-1}b$ where B^{-1} is an approximation to A^{-1}. We can then calculate $Ax^{(0)}$, which gives the right-hand side satisfied by $x^{(0)}$, and make a correction $\Delta^{(0)}$ for which $A\Delta^{(0)} = b - Ax^{(0)}$. The solution of this set of equations is simplified by the fact that the elimination has already been performed on A and only the same operations on the right-hand sides, followed by back-substitution, are needed. Again, however, we effectively find

$$\Delta^{(0)} = B^{-1}(b - Ax^{(0)}),$$

using an approximation to A^{-1}.

All this is contained in the iterative scheme

$$x^{(r+1)} = x^{(r)} + B^{-1}(b - Ax^{(r)}), \qquad x^{(0)} = B^{-1}b, \qquad (15)$$

and successive applications give

$$\left.\begin{aligned}
x^{(0)} &= B^{-1}b \\
x^{(1)} &= (I+C)B^{-1}b \\
x^{(2)} &= (I+C+C^2)B^{-1}b \\
&\cdots\cdots\cdots\cdots\cdots\cdots \\
x^{(r)} &= (I+C+C^2+\ldots+C^r)B^{-1}b
\end{aligned}\right\}, \qquad (16)$$

where
$$C = I - B^{-1}A. \qquad (17)$$

We saw in our discussion of vector and matrix norms in Chapter 2 that the series $I+C+C^2+\ldots+C^r+\ldots$ converges to $(I-C)^{-1}$ if $\|C\| < 1$, so that $x^{(r)}$ converges to $(I-C)^{-1}B^{-1}b = A^{-1}b$, the required result. We can also estimate the rate of convergence, for from equation (151) of Chapter 2 we have

$$\|(I-C)^{-1} - (I+C+C^2+\ldots+C^r)\| \leqslant \|C\|^{r+1}/(1 - \|C\|). \qquad (18)$$

Then
$$\begin{aligned}
\|x - x^{(r)}\| &= \|\{(I-C)^{-1} - (I+C+C^2+\ \ldots\ +C^r)\}B^{-1}b\| \\
&\leqslant \|\{(I-C)^{-1} - (I+C+C^2+\ \ldots\ +C^r)\}\|\ \|B^{-1}b\| \\
&\leqslant \frac{\|C\|^{r+1}}{1 - \|C\|}\ \|x^{(0)}\|.
\end{aligned} \qquad (19)$$

We also have

$$x - x^{(r)} = C(x - x^{(r-1)}), \qquad \|x - x^{(r)}\| \leqslant \|C\|\ \|x - x^{(r-1)}\|, \qquad (20)$$

so that at worst the error vector is multiplied by a constant, less than unity, at each stage. Such a rate of convergence is called *linear*, and has the effect that the number of accurate significant figures increases by a roughly constant amount at each step of the iteration.

13. We can obtain similar results in terms of the latent roots and vectors of C, and apply them quantitatively in analysing possible situations which might arise in the iterative process. If the matrix has distinct latent roots, arranged in order of modulus

$$|\lambda_1| > |\lambda_2| > \ldots > |\lambda_n|,$$

so that the corresponding vectors $y^{(s)}$ are independent, we can expand an arbitrary vector uniquely in terms of the latent vectors. From (15) and (16) there then follows the equations

$$\left. \begin{aligned} x - x^{(r)} &= e^{(r)} = Ce^{(r-1)} = C^2 e^{(r-2)} = \ldots = C^r e^{(0)} = C^{r+1}x \\ x^{(r+1)} - x^{(r)} &= \Delta^{(r)} = C\Delta^{(r-1)} \\ &= C^2\Delta^{(r-2)} = \ldots = C^r\Delta^{(0)} = C^{r+1}x^{(0)} \\ b - Ax^{(r)} &= R^{(r)} = BCB^{-1}R^{(r-1)} = BC^2B^{-1}R^{(r-2)} = \ldots \\ &= BC^r B^{-1}R^{(0)} = BC^{r+1}x^{(0)} \end{aligned} \right\}, \qquad (21)$$

for the successive behaviour of the error vector, correcting vector, and residual vector respectively. Then if

$$x = \sum_{s=1}^{n} \alpha_s y^{(s)}, \qquad x^{(0)} = \sum_{s=1}^{n} \beta_s y^{(s)}, \qquad (22)$$

we have

$$\left. \begin{aligned} x - x^{(r)} &= \sum_{s=1}^{n} \alpha_s \lambda_s^{r+1} y^{(s)} \rightarrow \alpha_1 \lambda_1^{r+1} y^{(1)} \\ x^{(r+1)} - x^{(r)} &= \sum_{s=1}^{n} \beta_s \lambda_s^{r+1} y^{(s)} \rightarrow \beta_1 \lambda_1^{r+1} y^{(1)} \\ b - Ax^{(r)} &= \sum_{s=1}^{n} \beta_s \lambda_s^{r+1} B y^{(s)} \rightarrow \beta_1 \lambda_1^{r+1} B y^{(1)} \end{aligned} \right\}, \qquad (23)$$

provided that the root λ of largest modulus is real and r is sufficiently large. If this root is complex we must add to each right-hand side of (23) the corresponding complex conjugate. We note also that, if any vector on the left of (23), with $r = 0$, is *deficient* in $y^{(1)}$, so that α_1 or β_1 is small, the approach to the right-hand side may be slow, though a multiple of $y^{(1)}$ will certainly appear as soon as we commit any rounding error, and this will ultimately dominate.

For convergence, of course, it is necessary that $|\lambda_1| < 1$, corresponding to the condition $\|C\| < 1$ on the norm. We can then say, if (23) holds, that ultimately

$$\left.\begin{aligned} e^{(r)} &= \lambda_1 e^{(r-1)} \\ \Delta^{(r)} &= \lambda_1 \Delta^{(r-1)} \\ R^{(r)} &= \lambda_1 R^{(r-1)} \end{aligned}\right\}, \qquad (24)$$

so that the ratios of corresponding components of the error vector, correcting vector, and residual vector are all ultimately constant and equal to λ_1.

We can use these results to accelerate the rate of convergence. For if r is large enough for the truth of (24) we can estimate λ_1 from either of the last two of (24), and substitute in the first of (24) to find x in the form

$$x = \{x^{(r)} - \lambda_1 x^{(r-1)}\}/(1 - \lambda_1) = x^{(r-1)} + (1 - \lambda_1)^{-1} \Delta^{(r-1)}. \qquad (25)$$

Alternatively we can avoid the computation of λ_1 by using the first of (24) for two successive values of r, obtaining for each component of the required vector the *Aitken extrapolation formula*

$$x_i = \frac{x_i^{(r+1)} x_i^{(r-1)} - x_i^{(r)2}}{x_i^{(r+1)} - 2x_i^{(r)} + x_i^{(r-1)}} = x_i^{(r+1)} - \frac{(x_i^{(r+1)} - x_i^{(r)})^2}{x_i^{(r+1)} - 2x_i^{(r)} + x_i^{(r-1)}}. \qquad (26)$$

This is identical with (25) if for each component in (25) the value of λ_1 is estimated separately from the corresponding ratio of $\Delta^{(r)}$ to $\Delta^{(r-1)}$.

The result (25), incidentally, reveals the danger of stopping an iterative process when the current correction vector is smaller than the tolerable error. If λ_1 is very close to unity, a situation common to many iterative processes, the real correction may be very much larger.

14. This acceleration device must be modified if the root λ_1 is complex, since the first of (23), for example, is then

$$x - x^{(r)} \to \alpha_1 \lambda_1^{r+1} y^{(1)} + \bar{\alpha}_1 \bar{\lambda}_1^{r+1} \bar{y}^{(1)}, \qquad (27)$$

and the ratio of corresponding components of successive vectors does not become constant. We can here eliminate all the unknown quantities and obtain x from five successive estimates $x^{(r)}$, rather than three in the real case, finding the formula

$$x_i \begin{vmatrix} 1 & 1 & 1 \\ \Delta_i^{(r)} & \Delta_i^{(r+1)} & \Delta_i^{(r+2)} \\ \Delta_i^{(r+1)} & \Delta_i^{(r+2)} & \Delta_i^{(r+3)} \end{vmatrix} = \begin{vmatrix} x_i^{(r)} & x_i^{(r+1)} & x_i^{(r+2)} \\ x_i^{(r+1)} & x_i^{(r+2)} & x_i^{(r+3)} \\ x_i^{(r+2)} & x_i^{(r+3)} & x_i^{(r+4)} \end{vmatrix}. \qquad (28)$$

This same formula could be used if the two largest roots are real but have comparable modulus, so that there are two significant real terms on the right of (27) even for large r.

15. We consider now some illustrations of these results, taking first the equations

$$\left. \begin{array}{l} 5x_1 + 7x_2 + 6x_3 + 5x_4 = 23 \\ 7x_1 + 10x_2 + 8x_3 + 7x_4 = 32 \\ 6x_1 + 8x_2 + 10x_3 + 9x_4 = 33 \\ 5x_1 + 7x_2 + 9x_3 + 10x_4 = 31 \end{array} \right\}, \qquad (29)$$

and assuming that the elimination process is equivalent to using the approximate and unsymmetric inverse

$$B^{-1} = \begin{bmatrix} 77.888 & -46.929 & -19.429 & 11.443 \\ -46.977 & 28.584 & 11.493 & -6.897 \\ -19.489 & 11.493 & 5.612 & -3.363 \\ 11.495 & -6.897 & -3.368 & 2.218 \end{bmatrix}. \qquad (30)$$

Table 6 shows the first few steps in the iterative process, all computations being rounded, when necessary, to three decimals.

TABLE 6

$x^{(0)}$	3.272	−0.321	0.472	1.295
$R^{(0)}$	−0.420	−0.535	−0.439	−0.311
$\Delta^{(0)}$	−2.635	1.537	0.619	−0.349
$x^{(1)}$	0.637	1.216	1.091	0.946
$R^{(1)}$	0.027	0.031	0.026	0.024
$\Delta^{(1)}$	0.418	−0.249	−0.105	0.062
$x^{(2)}$	1.055	0.967	0.986	1.008

Convergence is rapid to the true result in which every element is exactly unity, but Aitken's formula (26) applied to $x^{(0)}$, $x^{(1)}$ and $x^{(2)}$ gives immediately the good results 0.998, 1.002, 1.001, 0.999. If we use corresponding different estimates of λ_1 for the various components of the vector defined by (25) we of course have exactly the same results.

16. The fact that this acceleration is likely to be useful is perhaps indicated by the fairly constant ratios of the corresponding components of $\Delta^{(1)}$ to $\Delta^{(0)}$, which are $-0\cdot159$, $-0\cdot162$, $-0\cdot170$ and $-0\cdot178$, giving current estimates of λ_1. On the other hand we find that the ratios of the components of the residual vectors $R^{(1)}$ and $R^{(0)}$, though also reasonably constant with values $-0\cdot064$, $-0\cdot058$, $-0\cdot059$ and $-0\cdot077$, are quite different from those of $\Delta^{(1)}$ and $\Delta^{(0)}$.

Analysis shows that the two largest roots of C are approximately $\lambda_1 = -0\cdot1446$ and $\lambda_2 = -0\cdot0414$, and the corresponding vectors are such that, whereas in (22) β_1 and β_2 are of comparable size, the vector $By^{(2)}$ has very much larger components than $By^{(1)}$, and at this small value of r the ratios are more nearly those of the second latent root λ_2. In general, in fact, we have no real reason to expect systematic trend for such a small r, and the Aitken method, if used at any early stage, must be regarded as a process for giving a better starting vector, for further iteration, than for producing immediately a very accurate result.

17. If we repeat the experiment with a B^{-1} obtained by rounding to one decimal that of (30) we find a much slower rate of convergence, some results being shown in Table 7.

TABLE 7

$x^{(0)}$	4·1	−0·2	−1·1	−0·3	$R^{(4)}$	7·095	9·638	12·984	13·252
$R^{(0)}$	12·0	16·2	23·7	24·8	$\Delta^{(4)}$	−0·138	0·059	0·138	0·099
$\Delta^{(0)}$	−2·04	0·75	0·70	0·20	$x^{(5)}$	1·768	0·676	0·209	0·425
$x^{(1)}$	2·06	0·55	−0·40	−0·10	$R^{(5)}$	6·049	8·217	11·069	11·297
					$\Delta^{(5)}$	−0·113	0·047	0·117	0·085
$x^{(3)}$	2·035	0·564	−0·083	0·208	$x^{(6)}$	1·655	0·723	0·326	0·510
$R^{(3)}$	8·335	11·323	15·236	15·544					
$\Delta^{(3)}$	−0·129	0·053	0·154	0·118	$x^{(7)}$	1·559	0·764	0·425	0·582

In this case we find that successive ratios of $R^{(r)}$ settle down to a constant value much faster than those of $\Delta^{(r)}$. For example the ratios of the former for $r = 4, 3$ and $r = 5, 4$ are respectively $0\cdot851$, $0\cdot851$, $0\cdot852$, $0\cdot853$ and $0\cdot853$, $0\cdot853$, $0\cdot853$, $0\cdot853$, whereas for the correcting vector the corresponding figures are $1\cdot07$, $1\cdot11$, $0\cdot90$, $0\cdot84$ and $0\cdot82$, $0\cdot80$, $0\cdot85$, $0\cdot86$. These latter ratios, moreover, are somewhat lacking in significant figures, and with λ_1 close to unity give poor precision in $(1-\lambda_1)^{-1}$.

The Aitken formula (26) applied to $x^{(5)}$, $x^{(6)}$ and $x^{(7)}$, which is equivalent to using the ratios of the correcting vectors for estimating

λ_1, gives for the estimated solution the vector 1·017, 1·044, 0·970 and 0·981, but the use of (25), with $r = 7$ and $\lambda_1 = 0·853$, the obvious choice from the ratios of the residual vectors, gives the much better results 1·002, 1·002, 0·999, 1·000.

18. As final examples we consider the equations

$$\left. \begin{array}{l} 4x_1 - 4x_2 + 4·0x_3 + 6·00x_4 = 10·00 \\ -x_1 + 6x_2 + 6·5x_3 + 1·00x_4 = 12·50 \\ 2x_1 - 6x_2 + 5·0x_3 + 5·00x_4 = 6·00 \\ 3x_1 - 2x_2 + 0·0x_3 + 2·99x_4 = 3·99 \end{array} \right\}, \qquad (31)$$

for which the true answers are again $x_1 = x_2 = x_3 = x_4 = 1$. Some results of using two slightly different approximate inverses B^{-1}, the differences appearing only in the last row, are shown in Table 8, the respective inverses I and II being

$$\text{I} \qquad\qquad\qquad\qquad \text{II}$$

$$\begin{bmatrix} -84·1 & 17·8 & 44·2 & 88·9 \\ 15·9 & -3·3 & -8·5 & -16·7 \\ -42·3 & 9·0 & 22·3 & 44·4 \\ 94·9 & -20·1 & -50·1 & -100·0 \end{bmatrix} \text{ and } \begin{bmatrix} -84·1 & 17·8 & 44·2 & 88·9 \\ 15·9 & -3·3 & -8·5 & -16·7 \\ -42·3 & 9·0 & 22·3 & 44·4 \\ 95·0 & -20·0 & -50·0 & -100·0 \end{bmatrix}. \quad (32)$$

TABLE 8

	I					II			
$x^{(0)}$	1·4110	0·1170	0·4560	−1·8500	$x^{(0)}$	1·4110	0·1170	0·4560	1·0000
$R^{(0)}$	14·1000	12·0950	10·8500	5·5225	$R^{(0)}$	−3·0000	9·2450	−3·4000	−2·9990
$\Delta^{(0)}$	0·0012	−0·1742	−0·4210	−0·8545	$\Delta^{(0)}$	−0·0301	0·7748	1·1294	0·0000
$x^{(1)}$	1·4122	−0·0572	0·0350	−2·7045	$x^{(1)}$	1·3809	0·8918	·1·5854	1·0000
$R^{(1)}$	20·2094	16·7324	16·1799	7·7255	$x^{(2)}$	0·8841	1·2811	1·3344	1·0000
$x^{(2)}$	1·5869	−0·4897	−0·4070	−4·3167	$x^{(3)}$	0·8319	1·0939	0·8519	1·0000
$R^{(2)}$	29·2218	23·9873	23·5065	11·1568	$x^{(4)}$	1·0079	0·9147	0·8464	1·0010
$x^{(3)}$	1·8343	−1·1450	−1·0466	−6·6683	$x^{(5)}$	1·0630	0·9483	1·0237	1·0010

With matrix I the process is clearly diverging, the latent root λ_1 being somewhat larger than unity. There would not, however, seem to be any reason for suspecting that formulae like (25) and (26) would fail completely. Indeed if we apply (26) to $x^{(1)}$, $x^{(2)}$ and $x^{(3)}$ we find the respectable approximation 0·9924, 0·7824, 1·0237 and 0·8108, and from (25) with $r = 3$ and with λ_1 deduced from the ratios of $R^{(2)}$ and $R^{(1)}$ we have the much better results 1·0321, 1·0217, 1·0055, 0·9779.

With matrix II the process is converging, but with no systematic trend, and the largest latent roots of the relevant matrix are in fact complex. In this case we try formula (28), with $r = 1$, and find the rather good solution vector 0·9986, 0·9995, 1·0002 and 1·0000.

19. In practice, of course, rounding errors will occur at various stages which might invalidate, or make less accurate, some of the results. In particular we rarely use explicitly an approximate inverse, and the process of solving $A\Delta^{(r)} = R^{(r)}$ might introduce rounding errors both in the calculation of the right-hand sides and in the operations thereon which correspond to the reduction of A to U, and also in the back-substitution.

There is an obvious limitation on the accuracy obtainable, since if $R^{(r)}$ is calculated to a certain number of figures and rounded with an error ϵ, the error in $\Delta^{(r)}$ is $B^{-1}\epsilon$ and cannot be removed without extra-figure work. For this reason it is preferable to use the methods suggested, and given explicitly in (15), rather than the theoretically equivalent process defined by

$$\left. \begin{array}{ll} x^{(0)} = B^{-1}b, \qquad R^{(0)} = b - Ax^{(0)}, \qquad \Delta^{(0)} = B^{-1}R^{(0)} \\[2mm] R^{(1)} = R^{(0)} - A\Delta^{(0)}, \qquad \Delta^{(1)} = B^{-1}R^{(1)} \\[2mm] x = x^{(0)} + \Delta^{(0)} + \Delta^{(1)} + \ldots \end{array} \right\}. \qquad (33)$$

For if each residual $R^{(s)}$ has an error $\epsilon^{(s)}$ it follows that, from this source alone, x has an error equal to $B^{-1}\sum\limits_{s=1}^{\infty}\epsilon^{(s)}$, which may be several times larger than that of the previous method.

20. In practice, also, we would often need only one or two corrections when the first approximation $x^{(0)}$ is already correct to several figures. Even if the matrix is very ill-conditioned, moreover, giving perhaps a poor $x^{(0)}$, the residuals $R^{(0)}$ are likely to be small, and we can use this fact with advantage to reduce the total amount of arithmetic.

It is a good practice to adjust the coefficients in the equations by multiplying the rows by constants, which has no effect, and multiplying the columns by other constants, which multiplies the unknowns by the reciprocals of those constants, so that the modulus of the largest number in each row and column is less than unity and greater say than 0·5 in the decimal scale. We can then consider working with fixed-point arithmetic to a constant number of decimal places, ignoring for the moment the possibility of numbers exceeding unity at subsequent stages.

If there are p decimal places in the coefficients and we record p decimals in the calculated $x^{(0)}$, the product $Ax^{(0)}$ will have $2p$ decimals. but is likely to agree with b in most of its first p figures. If therefore we form accurately the scalar product and subtract the results from b, the numbers left, the residuals, can be rescaled by multiplying by 10^{p-q}, where q is small, rounded by dropping the last q figures, and used as the new single-length right-hand sides for the determination of the first correction.

21. Consider for example the problem, a slight variant of that of equations (29), represented by the equations

$$
\begin{array}{cccc}
A & & & b \\
0\cdot5001 & 0\cdot7001 & 0\cdot6001 & 0\cdot5001 & 0\cdot2300 \\
0\cdot7001 & 1\cdot0000 & 0\cdot8001 & 0\cdot7001 & 0\cdot3200 \\
0\cdot6001 & 0\cdot8001 & 1\cdot0000 & 0\cdot9001 & 0\cdot3300 \\
0\cdot5001 & 0\cdot7001 & 0\cdot9001 & 1\cdot0000 & 0\cdot3100.
\end{array}
\tag{34}
$$

Using the Cholesky method with $LL' = A$, $Ly = b$, $L'x = y$, and recording everything rigorously to four decimals, we find the lower triangle

$$
L =
\begin{array}{cccc}
0\cdot7072 & & & \\
0\cdot9900 & 0\cdot1411 & & \\
0\cdot8486 & -0\cdot2836 & 0\cdot4466 & \\
0\cdot7072 & -0\cdot0002 & 0\cdot6715 & 0\cdot2213,
\end{array}
\tag{35}
$$

the auxiliary vector

$$
y = 0\cdot3252, \quad -0\cdot0138, \quad 0\cdot1122, \quad 0\cdot0211, \tag{36}
$$

and the first approximation

$$
x^{(0)} = 0\cdot0682, \quad 0\cdot1192, \quad 0\cdot1079, \quad 0\cdot0953. \tag{37}
$$

The residuals $b - Ax^{(0)}$ are then found to be

$$
0\cdot00003094, \ 0\cdot00000286, \ 0\cdot00002173, \ 0\cdot00002047, \tag{38}
$$

so that by taking $0\cdot3094$, $0\cdot0286$, $0\cdot2173$ and $0\cdot2047$ as the right-hand sides we can continue to work with single-length arithmetic, remembering only that the corresponding corrections must be multiplied by 10^{-4}. In fact we find

$$
10^4\Delta^{(0)} = 187\cdot1249, \ -113\cdot2523, \ -46\cdot2571, \ 27\cdot5400, \tag{39}
$$

so that, in spite of its small residuals, $x^{(0)}$ is not a very good solution.

In practice we should continue by forming $x^{(1)}$, rounded to 'single-length' with values

$$x^{(1)} = 0{\cdot}0869,\ 0{\cdot}1079,\ 0{\cdot}1033,\ 0{\cdot}0981, \tag{40}$$

calculating the exact residuals

$$10^4 R^{(1)} = -0{\cdot}4962,\ -0{\cdot}6883,\ -0{\cdot}7929,\ -0{\cdot}7981, \tag{41}$$

and a new correction

$$10^4 \Delta^{(1)} = -0{\cdot}1715,\ -0{\cdot}0722,\ -0{\cdot}1925,\ -0{\cdot}4885. \tag{42}$$

It is interesting to observe in passing that the residuals of $x^{(1)}$, though everywhere larger than those for $x^{(0)}$, give much smaller corrections. We can reasonably assert that the results

$$x^{(1)} = 0{\cdot}0869,\ 0{\cdot}1079,\ 0{\cdot}1033,\ 0{\cdot}0981 \tag{43}$$

are correct in all the figures given.

We can often by this process obtain better than single-length accuracy without using double-length working except in the computation of $Ax^{(r)}$ (In fact this 'double-length' working is no more than the exact accumulation of sums of products of single-length numbers, which on some machines is quite automatic). Here, for example, $\Delta^{(1)}$ has a maximum error of less than five units in its second figure, so that x can be quoted to five decimals, though we may not be able to deduce this fact until a later stage.

22. This example shows yet another possibility of rounding error which might invalidate or at least reduce the value of the acceleration processes of Aitken type. There was no error in $Ax^{(0)}$, but $x^{(1)}$ was not the exact sum of $x^{(0)}$ and $\Delta^{(0)}$.

In general we might have errors of three types. First, the elimination process gives a result differing from the direct use of B^{-1} by a vector α. Second, the accepted residuals R differ from $b - Ax$ by a vector β, and third, the accepted new x differs from $x + \Delta$ by a vector γ. Instead of the equations

$$\left.\begin{aligned} x^{(0)} &= B^{-1}b, & R^{(0)} &= b - Ax^{(0)} \\ \Delta^{(0)} &= B^{-1}R^{(0)}, & x^{(1)} &= x^{(0)} + \Delta^{(0)},\ \dots \end{aligned}\right\} \tag{44}$$

we therefore have

$$\left.\begin{aligned} x^{(0)} &= B^{-1}b + \alpha^{(0)}, & R^{(0)} &= b - Ax^{(0)} + \beta^{(0)} \\ \Delta^{(0)} &= B^{-1}R^{(0)} + \alpha^{(1)}, & x^{(1)} &= x^{(0)} + \Delta^{(0)} + \gamma^{(0)},\ \dots \end{aligned}\right\}, \tag{45}$$

so that instead of (16) we have

$$
\left.
\begin{aligned}
x^{(0)} &= B^{-1}b + \alpha^{(0)} \\
x^{(1)} &= (I+C)B^{-1}b + C\alpha^{(0)} + B^{-1}\beta^{(0)} + \alpha^{(1)} + \gamma^{(0)} \\
x^{(2)} &= (I+C+C^2)B^{-1}b + C^2\alpha^{(0)} + CB^{-1}\beta^{(0)} + C(\alpha^{(1)} + \gamma^{(0)}) + \\
&\qquad\qquad\qquad + B^{-1}\beta^{(1)} + (\alpha^{(2)} + \gamma^{(1)}) \\
x^{(3)} &= (I+C+C^2+C^3)B^{-1}b + C^3\alpha^{(0)} + C^2B^{-1}\beta^{(0)} + C^2(\alpha^{(1)} + \gamma^{(0)}) + \\
&\qquad + CB^{-1}\beta^{(1)} + C(\alpha^{(2)} + \gamma^{(1)}) + B^{-1}\beta^{(2)} + (\alpha^{(3)} + \gamma^{(2)})
\end{aligned}
\right\} \quad (46)
$$

etc.

Corresponding to the first of (21) we now have

$$
x - x^{(r)} = C^{r+1}x + E_r, \tag{47}
$$

where E_r contains all powers of C up to the rth. We cannot therefore assert with conviction that the right-hand side of (47) is dominated by a term with a factor λ_1^{r+1}, on which Aitken's process is based. To this extent all these methods should be regarded as possibilities for accelerating convergence, rather than as formulae for producing accurate results. We also note the error, of magnitude

$$
\epsilon = B^{-1}\beta + (\alpha + \gamma), \tag{48}
$$

which is inherent in this iterative method and which cannot be eliminated.

23. So far we have assumed that the matrix A and right-hand side b are stored exactly in single-length registers. This will not be possible if A and b have many figures, or if the coefficients are expressible only in recurrent binary or decimal form. In this case we could proceed by working mostly with say double-length arithmetic, reserving extra-length work only in the computation of residuals.

Alternatively, and particularly if A is reasonably well-conditioned, we could use approximate B and c, equivalent to A and b to single-length accuracy, to effect approximate solutions, and calculate the residuals by using the true double-length A and b, forming scalar products of double-length numbers a_{rs} and single-length numbers x_r. In this way we again reduce to a minimum the amount of multiple-length arithmetic involved.

The theory is almost identical with that of the previous case, except that B^{-1} is now likely to be a poorer approximation to the true A^{-1}, and in severe cases the process will diverge. In fact if we assume all arithmetic performed exactly, take $B = A + \delta A$ and ignore second-order quantities, the matrix C is approximately $A^{-1}\delta A$, and it is necessary for convergence that its latent roots should be less than unity in absolute value.

24. Consider, in illustration, the equations represented by the array

$$
\begin{array}{ccccc}
& A & & & b \\
0{\cdot}50014999 & 0{\cdot}70005001 & 0{\cdot}60005001 & 0{\cdot}50014999 & 0{\cdot}23004999 \\
0{\cdot}70014999 & 0{\cdot}99995001 & 0{\cdot}80005001 & 0{\cdot}70014999 & 0{\cdot}31995001 \\
0{\cdot}60014999 & 0{\cdot}80005001 & 0{\cdot}99995001 & 0{\cdot}90005001 & 0{\cdot}32995001 \\
0{\cdot}50014999 & 0{\cdot}70005001 & 0{\cdot}90005001 & 0{\cdot}99995001 & 0{\cdot}31004999
\end{array}
\tag{49}
$$

of which (34) are correctly rounded equivalents to four decimals. In relation to § 23 A and b have the coefficients of (49), B and c are those of (34), and we use throughout the L of (35) in the solution of the relevant equations.

The first approximation $x^{(0)}$ is the same as before and recorded in (37). We then find the residuals

$$10^{12} R^{(0)} = 84109364, \; -43950636, \; -15552542, \; 83167458, \tag{50}$$

and use the rounded single-length approximations

$$10^4 R^{(0)} = 0{\cdot}8411, \; -0{\cdot}4395, \; -0{\cdot}1555, \; 0{\cdot}8317, \tag{51}$$

to produce, by single-length work, the rounded first correction

$$\Delta^{(0)} = 0{\cdot}0885, \; -0{\cdot}0534, \; -0{\cdot}0226, \; 0{\cdot}0135, \tag{52}$$

and hence a second approximation

$$x^{(1)} = 0{\cdot}1567, \; 0{\cdot}0658, \; 0{\cdot}0853, \; 0{\cdot}1088. \tag{53}$$

Proceeding in this way we find the successive results

$$
\begin{array}{ccccc}
10^4 R^{(1)} & 0{\cdot}1261 & 0{\cdot}1921 & 0{\cdot}4204 & 0{\cdot}4437 \\
\Delta^{(1)} & -0{\cdot}0021 & 0{\cdot}0012 & 0{\cdot}0006 & -0{\cdot}0003 \\
x^{(2)} & 0{\cdot}1546 & 0{\cdot}0670 & 0{\cdot}0859 & 0{\cdot}1085 \\
10^4 R^{(2)} & 0{\cdot}1288 & 0{\cdot}1960 & 0{\cdot}1234 & 0{\cdot}1458 \\
10^4 \Delta^{(2)} & 0{\cdot}8264 & -0{\cdot}2162 & -0{\cdot}5006 & 0{\cdot}3344
\end{array}
\tag{54}
$$

at which stage we would certainly expect to have an answer correct to single length.

As before we might even get more correct figures in a further application. We start with

$$x^{(3)} = 0.1547,\ 0.0670,\ 0.0858,\ 0.1085, \qquad (55)$$

and find

$$10^4 R^{(3)} = 0.2287,\ 0.2959,\ 0.5232,\ 0.5457, \qquad (56)$$

giving a correction

$$10^4 \Delta^{(3)} = -0.1891,\ -0.2069,\ 0.5033,\ 0.3321. \qquad (57)$$

The results $x_2 + \Delta^{(2)}$, $x_3 + \Delta^{(3)}$ agree to within 16, 9, 4 and 2 units in the seventh decimal in the respective components, so that we know something about the answers beyond single-length precision.

In more favourable cases, with a reasonably well-conditioned matrix, we would find that R vanishes to more than single-length precision so that an extra figure would be available in the current residuals, and hence more accuracy in the current correction Δ.

The rounding errors committed at the various stages, however, may have an even more deleterious effect on acceleration processes of Aitken type.

25. In really severe cases, of course, the matrix B^{-1} will be such a poor approximation to A^{-1} that the method will not converge. For example the coefficients of the matrix in (3) are obtained by rounding to four decimals the fractions $\frac{1}{5}$, $\frac{1}{6}$,..., $\frac{1}{11}$, and it is clear that the approximation so obtained is too crude to give any useful result. In particular the arithmetic of Table 2 produces a situation in which the matrix is effectively singular and the equations cannot be solved. This computation, moreover, is certainly more accurate than that of Table 1, and the fact that the latter would provide some approximate solution gives no grounds for expecting ultimate success.

In this case, in fact, it would be necessary to use more figures in all stages of the work, including the elimination process.

Mathematical problems. Correction to the inverse

26. Methods similar to those for linear equations can be applied to improve the accuracy of an approximate inverse, in the case of mathematical problems in which such a correction is warranted.

If B_0^{-1} is an approximation to A^{-1}, the iterative scheme

$$B_{r+1}^{-1} = B_r^{-1}(2I - A B_r^{-1}), \qquad r = 0, 1,..., \qquad (58)$$

will give a sequence of matrices B_r^{-1} which converges under certain conditions to A^{-1}. We find

$$
\left.\begin{aligned}
B_1^{-1} &= B_0^{-1}(I+C) \\
B_2^{-1} &= B_0^{-1}(I+C+C^2+C^3) \\
B_3^{-1} &= B_0^{-1}(I+C+C^2+C^3+C^4+C^5+C^6+C^7) \\
&\cdots\cdots\cdots\cdots\cdots\cdots\cdots\cdots\cdots\cdots\cdots\cdots\cdots\cdots \\
B_r^{-1} &= B_0^{-1}(I+C+C^2+\dots+C^{2^r-1})
\end{aligned}\right\}, \qquad (59)
$$

where $C = I-AB_0^{-1}$, and the process will converge to $B_0^{-1}(I-C)^{-1}$, which is A^{-1}, if $\|C\| < 1$. This matrix C is not exactly that of the linear equation case, which was $I-B_0^{-1}A$, but in neither case can their norm exceed $\|B_0^{-1}\| \, \|B_0-A\|$, so that the criterion for convergence is the same as before and we can say, not surprisingly, that if B_0^{-1} is suitable for iteration for linear equations it is also suitable for iteration to produce an accurate inverse.

27. The rate of convergence of (59), however, is considerably better than that of (16) for the iteration of linear equations. We have

$$
I-AB_r^{-1} = (I-AB_{r-1}^{-1})^2 = (I-AB_{r-2}^{-1})^4 = \dots = (I-AB_0^{-1})^{2^r}, \quad (60)
$$

so that

$$
A^{-1}-B_r^{-1} = A^{-1}C^{2^r}. \qquad (61)
$$

The rate of convergence is measured by

$$
\begin{aligned}
\|A^{-1}-B_r^{-1}\| &= \|A^{-1}C^{2^r}\| \\
&= \|B_0^{-1}(I-C)^{-1}C^{2^r}\| < \|B_0^{-1}\| \, \|(I-C)^{-1}\| \, \|C\|^{2^r}, \quad (62)
\end{aligned}
$$

and we know that $\|(I-C)^{-1}\| < (1-\|C\|)^{-1}$, so that

$$
\left.\begin{aligned}
\|A^{-1}-B_r^{-1}\| &< \|B_0^{-1}\| \, \|C\|^{2^r}/(1-\|C\|) = k \, \|C\|^{2^r} \\
k &= \|B_0^{-1}\| \, (1-\|C\|)^{-1}
\end{aligned}\right\}. \qquad (63)
$$

Moreover we find from (61) that

$$
\begin{aligned}
(A^{-1}-B_r^{-1}) &= A^{-1}(C^{2^{r-1}})^2 = A^{-1}\{A(A^{-1}-B_{r-1}^{-1})\}^2 \\
&= (A^{-1}-B_{r-1}^{-1})A(A^{-1}-B_{r-1}^{-1}),
\end{aligned}
$$

so that

$$
\|A^{-1}-B_r^{-1}\| < \|A\| \, \|A^{-1}-B_{r-1}^{-1}\|^2. \qquad (64)
$$

This type of convergence is called *quadratic* and is altogether desirable, since the error at each step is a multiple of the square of that of the previous step and the number of correct significant figures is approximately doubled at each step.

This improvement in convergence becomes apparent when we consider the equations, corresponding to (58), which would be used if the methods for linear equations were applied to the successive columns of the unit matrix. Instead of (58) we would have

$$B_{r+1}^{-1} = B_r^{-1} + B_0^{-1}(I - A B_r^{-1}),\qquad(65)$$

and the point is that for inversion we are using, for computing corrections, the best approximation to A^{-1} which is currently available.

28. Again, however, this satisfactory rate of convergence will be reduced if we make rounding errors at any stage. If we write (65) in the form

$$B_{r+1}^{-1} = B_r^{-1} + \Delta B_r^{-1},\qquad \Delta B_r^{-1} = B_r^{-1}(I - A B_r^{-1}),\qquad(66)$$

we might make errors in both matrix products involved in the second of (66) and also in the addition involved in the first, errors which are inescapable if we work with arithmetic of 'fixed' length.

Consider, for example, the inversion of the matrix in (34), starting with the approximation

$$10^{-3}B_0^{-1} = \begin{bmatrix} 0{\cdot}6984 & -0{\cdot}4212 & -0{\cdot}1746 & 0{\cdot}1028 \\ -0{\cdot}4212 & 0{\cdot}2568 & 0{\cdot}1028 & -0{\cdot}0617 \\ -0{\cdot}1746 & 0{\cdot}1028 & 0{\cdot}0512 & -0{\cdot}0307 \\ 0{\cdot}1028 & -0{\cdot}0617 & -0{\cdot}0307 & 0{\cdot}0204 \end{bmatrix},\qquad(67)$$

obtained from the L of (35) and the equation $A^{-1}L = (L')^{-1}$. If we form correctly the double-length scalar products, and scale as much as possible to keep all numbers less than unity, we find

$$10(I - A B_0^{-1}) = \begin{bmatrix} -0{\cdot}2054 & 0{\cdot}2233 & -0{\cdot}2487 & 0{\cdot}0692 \\ -0{\cdot}2266 & 0{\cdot}2801 & -0{\cdot}3459 & 0{\cdot}1075 \\ -0{\cdot}3800 & 0{\cdot}3261 & -0{\cdot}3975 & 0{\cdot}1385 \\ -0{\cdot}3026 & 0{\cdot}2616 & -0{\cdot}3794 & 0{\cdot}1896 \end{bmatrix},\qquad(68)$$

and a correction

$$10^{-3}\Delta B_0^{-1} = 10^{-2}\begin{bmatrix} -0{\cdot}1277 & 0{\cdot}0793 & 0{\cdot}0240 & -0{\cdot}0164 \\ 0{\cdot}0793 & -0{\cdot}0474 & -0{\cdot}0153 & 0{\cdot}0100 \\ 0{\cdot}0240 & -0{\cdot}0153 & -0{\cdot}0084 & 0{\cdot}0024 \\ -0{\cdot}0164 & 0{\cdot}0100 & 0{\cdot}0024 & 0{\cdot}0010 \end{bmatrix},\qquad(69)$$

with three figures discarded. In forming B_1^{-1} two more figures are lost and we take

$$10^{-3}B_1^{-1} = \begin{bmatrix} 0{\cdot}6971 & -0{\cdot}4204 & -0{\cdot}1744 & 0{\cdot}1026 \\ -0{\cdot}4204 & 0{\cdot}2563 & 0{\cdot}1026 & -0{\cdot}0616 \\ -0{\cdot}1744 & 0{\cdot}1026 & 0{\cdot}0511 & -0{\cdot}0307 \\ 0{\cdot}1026 & -0{\cdot}0616 & -0{\cdot}0307 & 0{\cdot}0204 \end{bmatrix}. \qquad (70)$$

This matrix is correct in all four figures given, and in fact at least one of the discarded figures in ΔB_0^{-1} is also correct, so that already we have better than single-length accuracy with a small amount of double-length work.

Repetition of this process, however, starting with the B_1^{-1} of (70), produces no more accurate figures, whereas the retention of all the figures in (69) and (70), and accurate double-length work thereafter, produces results with seven decimal places accurate in $10^{-3}B_2^{-1}$.

Physical problems. Error analysis

29. We turn now to cases in which the data of the problem are not exact, that is when the coefficients or the right-hand sides, or both, have uncertainties whose maximum values may be known. Such a case might arise, for example, in the analysis of a compound structure, which causes some physical effect obtained by measurement, and in which the contributions of the individual components are also found by similar measurement.

Here there is a limit to the number of figures which have any meaning, and our computing method should provide at least an approximation to this number. For example the equations (34) have the four-decimal solution

$$0{\cdot}0869, \ 0{\cdot}1079, \ 0{\cdot}1033, \ 0{\cdot}0981, \qquad (71)$$

while those of (49), which differ by less than half a unit in the fourth decimal of (34), are

$$0{\cdot}1547, \ 0{\cdot}0670, \ 0{\cdot}0859, \ 0{\cdot}1085. \qquad (72)$$

If the coefficients in (34) have uncertainties of as much as half a unit in their last figure it is clear that the solution may have uncertainty even in the first decimal place, and only the single-figure results 0·1, 0·1, 0·1, 0·1 are worth quoting.

†30. Even this statement may not be true because there may be some combination of uncertainties in the coefficients which would give results even farther apart than those in (71) and (72), and in very severe cases small changes in the coefficients, within the tolerance limits, may produce a singular or near-singular matrix. For example the matrix in Table 2 is not singular, but the product LU of the triangular matrices of that table, which would produce a singular matrix, differs from A by at most half a unit in the last figure of A.

The search for the worst combination of uncertainties involves the computation of the inverse, and in solving linear equations we need a simpler and reasonably satisfactory criterion for the number of quotable figures in the calculated solution. If the coefficients vary widely in magnitude there is unlikely to be any satisfactory analysis, but we can produce a theory if we consider working with fixed-point arithmetic, arranging the coefficients and right-hand sides so that all numbers are less than unity in absolute value, and with at least one in every row and column having a non-zero digit after the decimal or binary point. We assume further that the word-length of our computer is p digits, and that the coefficients are known with certainty only in the first q digits. We are therefore keeping $p-q$ 'guarding' figures in the calculation. In the following we assume, for simplicity, that we are using the decimal scale.

We are basing our conclusions on the *smallest* pivot located say in position (i, j), and observe that its value is incorrect for two reasons. First there is the inherent uncertainty of amount 0.5×10^{-q} in a_{ij}, affecting the pivot directly, and second the possibility that the solving process, with all quantities rounded to p decimals, has errors which might affect the first q digits. There is in fact a third source of error, in the combined effect of the other (unknown) uncertainties in the a_{rs}, but this we ignore and say that our conclusion is as optimistic as possible. The conclusion is that if the $p-q$ guarding figures are sufficient to limit the error in the smallest pivot to its inherent uncertainty of 0.5×10^{-q}, then if this pivot has t significant figures the number of meaningful digits in the results does not exceed $t-(p-q)$.

Our first requirement, therefore, is to estimate the number of guarding figures needed to avoid significant errors in the reduction to triangular form. We have still to perform the back-substitution, however, and the number $p-q$ must also be large enough to avoid significant error from this source also.

†31. Consider first the reduction to triangular form. Table 9 shows the 'lengthy' Gauss elimination process for a set of equations arranged so that the largest pivots are on the diagonal, and in which we work to five decimal places.

TABLE 9

m		A		b
	0·19995	0·16667	0·14286	0·50505
−0·71448	0·14286	0·12500	0·11111	0·50505
−0·83356	0·16667	0·14286	0·12500	0·50505
	0·00592	0·00904		0·14420
−0·66385	0·00393	0·00592		0·08406
		−0·00008		−0·01167

If we had kept one guarding figure and worked throughout with six decimals the last pivot would have been −0·000087, so that the solving process in Table 9 has in fact produced an error of one unit in the fifth decimal. The retention of the single guarding figure would verify this fact, and allow us to make statements about the worthwhile nature of our solution in virtue of possible errors of $0·5 \times 10^{-5}$ in the given data. In other words we would have eliminated the errors of the solving process, leaving only the inherent error of the data to contaminate our last pivot.

To find how many guarding figures are necessary for this purpose we follow the method of Wilkinson, who avoids previous ideas of attempting to trace the effects of individual errors, and instead determines what changes would be needed in the original data to give rise *exactly* to the recorded multipliers and upper triangular set. He further determines what extra changes would be necessary in the data so that the computed solution would satisfy exactly the 'perturbed equations'. That is he finds that the computed solution satisfies exactly the perturbed equations

$$(A + \delta A)x = b + \delta b. \tag{73}$$

For each individual problem we can in fact compute δA and δb, but all we need in general are upper bounds to the values of the perturbations. If the largest upper bound is some function $f(n)$ of the order of the matrix, we can assert that the number of guarding figures need be only $\log_{10} f(n)$ or $\log_2 f(n)$, according to the arithmetic scale. Then if p and q, defined in § 30, are such that $p - q > \log f(n)$, so that the perturbation does not affect the q figures known to be accurate, we can forget about the errors in the solving process and state that the inherent errors alone are important, limiting the worth-while accuracy to $t-(p-q)$ significant figures.

†32. To find the function $f(n)$ for the 'lengthy' Gauss elimination process, with interchanges to ensure the use of the 'largest coefficient' as pivot, we assume that the rows are already permuted so that the pivot is on the diagonal, that all the coefficients are initially smaller than unity in magnitude, and that we work in fixed-point decimal arithmetic with a word-length of p digits. For binary arithmetic we replace the number 10, in the obvious contexts, with the number 2. In what follows the superscripts relate in order to the original and reduced sets of coefficients.

Now the first set of recorded multipliers satisfies only approximately the relations

$$m_{21} = -a_{21}^{(1)}/a_{11}^{(1)}, \qquad m_{31} = -a_{31}^{(1)}/a_{11}^{(1)},\dots, \qquad m_{r1} = -a_{r1}^{(1)}/a_{11}^{(1)},\dots \quad (74)$$

because the m_{r1} are rounded to p digits. We can, however, make (74) exact by writing

$$a_{11}^{(1)}m_{r1}^{(1)}+a_{r1}^{(1)}+\epsilon_{r1} = 0, \qquad r = 2, 3,\dots, n, \quad (75)$$

that is the elimination is exact, with the recorded multipliers, if we add ϵ_{r1} to $a_{r1}^{(1)}$, that is 'perturb' the non-pivotal elements in the first column. The amount of this perturbation is at most

$$\epsilon_{r1} = 0.5 \times 10^{-p},$$

because if m_{r1} is correctly rounded to p digits we have

$$\left|\frac{a_{r1}^{(1)}}{a_{11}^{(1)}}+m_{r1}\right| \leqslant 0.5 \times 10^{-p}, \quad (76)$$

and hence

$$|a_{r1}^{(1)}+m_{r1}a_{11}^{(1)}| \leqslant 0.5\,|a_{11}^{(1)}| \times 10^{-p}, \; < 0.5\;10^{-p} \; \text{since } |a_{11}^{(1)}| < 1. \quad (77)$$

It is not in fact necessary to have m_{r1} correct to all p figures if the pivot is small, which is common in ill-conditioned systems. We need only use an m_{r1} which has the same number of correct *decimals* as there are *significant* figures in the pivot. For if the latter number is k, and we take the multiplier correct to k decimals, (76) becomes

$$\left|\frac{a_{r1}^{(1)}}{a_{11}^{(1)}}+m_{r1}\right| \leqslant 0.5 \times 10^{-k}, \quad (78)$$

and since $|a_{11}^{(1)}| < 10^{k-p}$, (77) follows as before. In practice, of course, it is more convenient to keep all p digits in the computer, though not in desk-machine work.

Turning to the other elements of the reduced set, we have the general equations

$$\left.\begin{aligned} a_{rs}^{(2)} &= m_{r1}a_{1s}^{(1)}+a_{rs}^{(1)}+\epsilon_{rs}^{(1)} \\ b_r^{(2)} &= m_{r1}b_1^{(1)}+b_r^{(1)}+\epsilon_r^{(1)} \end{aligned}\right\}, \qquad r,s > 2, \qquad (79)$$

and the quantities $\epsilon_{rs}^{(1)}$, $\epsilon_r^{(1)}$, chosen to make these equations exact, have in general the same properties as ϵ_{r1}. There is here the danger that $a_{rs}^{(2)}$ or $b_r^{(2)}$ may exceed unity, so that the row will need division by a factor of at most two. This possibility we ignore temporarily and observe that in the case of three equations the reduced set

$$\begin{bmatrix} a_{11}^{(1)} & a_{12}^{(1)} & a_{13}^{(1)} & b_1^{(1)} \\ 0 & a_{22}^{(2)} & a_{23}^{(2)} & b_2^{(2)} \\ 0 & a_{32}^{(2)} & a_{33}^{(2)} & b_3^{(2)} \end{bmatrix}, \qquad (80)$$

all with p decimals, would have been produced exactly from the original coefficients increased by amounts

$$\begin{bmatrix} 0 & 0 & 0 & 0 \\ \epsilon_{21}^{(1)} & \epsilon_{22}^{(1)} & \epsilon_{23}^{(1)} & \epsilon_2^{(1)} \\ \epsilon_{31}^{(1)} & \epsilon_{32}^{(1)} & \epsilon_{33}^{(1)} & \epsilon_3^{(1)} \end{bmatrix}, \qquad \epsilon_{rs}^{(1)}, \epsilon_r^{(1)} \leqslant 0{\cdot}5 \times 10^{-p}. \qquad (81)$$

We now discard the first row and column of (80), which is possible since these coefficients remain unchanged in future operations, and apply the same analysis to the remaining equations. The finally reduced set for a problem of order three is

$$\begin{bmatrix} a_{11}^{(1)} & a_{12}^{(1)} & a_{13}^{(1)} & b_1^{(1)} \\ 0 & a_{22}^{(2)} & a_{23}^{(2)} & b_2^{(2)} \\ 0 & 0 & a_{33}^{(3)} & b_3^{(3)} \end{bmatrix}, \qquad (82)$$

and would be produced exactly from original coefficients increased by further amounts

$$\begin{bmatrix} 0 & 0 & 0 & 0 \\ 0 & 0 & 0 & 0 \\ 0 & \epsilon_{32}^{(2)} & \epsilon_{33}^{(2)} & \epsilon_3^{(2)} \end{bmatrix}, \qquad (83)$$

the term $\epsilon_{32}^{(2)}$ being involved in the production of the multiplier to give the required zero coefficient in this term in (82).

The various changes are additive, interfering in no way with each other, and their values are exact numbers with $2p$ decimals. In the case of Table 9 we find, corresponding to (81) and (83), the arrays

$$10^{-10}\begin{bmatrix} 0 & 0 & 0 & 0 \\ 2760 & 23816 & 6128 & -18760 \\ 3220 & -5548 & 23816 & -5220 \end{bmatrix} \tag{84}$$

and

$$10^{-10}\begin{bmatrix} 0 & 0 & 0 & 0 \\ 0 & 0 & 0 & 0 \\ 0 & -80 & 12040 & -28300 \end{bmatrix}, \tag{85}$$

and we deduce that the figures in Table 9 belonging to the final upper triangular set of equations, and also the multipliers, would have been produced exactly from the 'perturbed' original set

$$\begin{bmatrix} 0{\cdot}1999500000 & 0{\cdot}1666700000 & 0{\cdot}1428600000 & 0{\cdot}5050500000 \\ 0{\cdot}1428602760 & 0{\cdot}1250023816 & 0{\cdot}1111106128 & 0{\cdot}5050481240 \\ 0{\cdot}1666703220 & 0{\cdot}1428594372 & 0{\cdot}1250035856 & 0{\cdot}5050466480 \end{bmatrix}. \tag{86}$$

In general, for n equations, the maximum changes, assuming that ϵ_{rs} and ϵ_r have their upper bound and accumulate as much as possible, have the upper bounds

$$0{\cdot}5 \times 10^{-p}\begin{bmatrix} 0 & 0 & 0 & 0 & \ldots & 0 & 0 & 0 \\ 1 & 1 & 1 & 1 & \ldots & 1 & 1 & 1 \\ 1 & 2 & 2 & 2 & \ldots & 2 & 2 & 2 \\ \cdot & \cdot & \cdot & \cdot & \cdot & \cdot & \cdot & \cdot \\ 1 & 2 & 3 & 4 & \ldots & n-1 & n-1 & n-1 \end{bmatrix}, \tag{87}$$

with a maximum value of $0{\cdot}5(n-1)10^{-p}$.

†33. This result does not in fact indicate how the pivot is contaminated by errors in the reduction process, but it proves that the computed upper triangular set is the exact reduction of original coefficients perturbed by at most $\frac{1}{2}(n-1)$ units in the last figure. If we keep $p-q = \log_{10}\frac{1}{2}(n-1)$ guarding figures approximately, it follows that we have eliminated the effect of 'solving' errors, and our

answer is affected only by the inherent uncertainty in A and b. Together with the observed number of significant figures in the smallest pivot and in the right-hand sides we have therefore a good indication of the maximum number of meaningful figures in the answers. It is noteworthy that this number of guarding figures is small, at most one for 10 equations, two for 100 equations, and so on.

In practice these conclusions are affected slightly by the possibility, in fixed-point arithmetic, of coefficients becoming greater than unity in the elimination process. If this happens the corresponding rows are scaled by division by two, the largest possible current increase, and the coefficients of $A^{(1)}$ and b, in the same rows and all other rows subsequently derived from them, have perturbations of twice the standard amount. The effect is small: Wilkinson, for example, reports that total increases by factors of more than four are extremely rare in practical problems.

34. We are now faced with the solution of the upper triangular equations

$$Ux = c, \tag{88}$$

and again it is desirable to estimate the possible errors in the solving process. Again, following Wilkinson, we seek those further changes in the original coefficients which would give rise exactly to the solution obtained by fixed-point arithmetic of fixed word-length. This solution will not satisfy (88) exactly, but will be the exact solution of

$$Ux = c+e, \tag{89}$$

where e is a vector of 'corrections' to the right-hand side of (88). Moreover since each element of x comes from substitution in the corresponding row of U, it follows that if we can accumulate scalar products exactly, and if no element of x exceeds unity, then the upper bound of any element of $|e|$ is also $0 \cdot 5 \times 10^{-p}$.

With regard to the original equations no further perturbation of A is involved, but there is an extra perturbation η of b, given by

$$e = L\eta, \qquad \eta = L^{-1}e, \quad \text{where} \quad LA = U. \tag{90}$$

We know that L^{-1} is a unit lower triangle whose off-diagonal elements are the multipliers, with signs changed, in the Gauss elimination process, so that

$$\left. \begin{aligned} \eta_1 &= e_1 \\ \eta_2 &= -m_{21}e_1 + e_2 \\ \eta_3 &= -m_{31}e_1 - m_{32}e_2 + e_3 \\ &\text{etc.} \end{aligned} \right\}, \tag{91}$$

and since the multipliers are less than unity in absolute value it follows that the largest upper bound of $|\eta_r|$ is $r \times 0.5 \times 10^{-p}$.

Combining this with the perturbation of b involved in the elimination, it follows that, if no element of any reduced set of equations exceeds unity, and no component of the solution exceeds unity, then the solution we have obtained is the exact solution of

$$(A + \delta A)x = b + \delta b,$$

where the upper bounds of δA are given in (87), and the upper bounds of δb are $0.5 \times 10^{-p}(1, 3, 5, \ldots, 2n-1)$.

35. A final practical problem is the possibility of results exceeding unity in the back-substitution. If any element has to be scaled for this reason the corresponding element of c in (88) is similarly affected, and if the largest scaling factor is 10^{-r} the maximum element of $|e|$ in (89) is $0.5 \times 10^{r-p}$, and therefore that of $|\eta|$ in (91) is $0.5n \times 10^{r-p}$ instead of just $0.5n \times 10^{-p}$. It is clear, however, that this possibility does not affect the accuracy achieved to the number of *significant figures* we seek in the solution.

This remark can be verified by considering the residuals of our computed solution, that is the vector

$$R = b - Ax. \tag{92}$$

Since x is the exact solution of

$$(A + \delta A)x = b + \delta b, \tag{93}$$

where $|\delta A|$ and $|\delta b|$ have the quoted upper bounds, and no component of $|x|$ exceeds 10^r, we deduce that

$$|R| = |\delta b| + |\delta A| \, |x|,$$

with upper bound

$$|R|_{\max} = 0.5(n-1)10^{-p} + 0.5n \, 10^{r-p} + 0.5 \, 10^{r-p}(1 + 2 + \ldots + n - 1)$$

$$= 0.5\{(n-1)10^{-p} + \tfrac{1}{2}n(n+1)10^{r-p}\}. \tag{94}$$

We compare this with the upper bound for the residual obtained by substituting into the given equations (assumed known exactly) values obtained by rounding their exact solutions to p decimal figures, so that the computation can be performed. We find

$$|b - Ax| < 0.5n \, 10^{r-p}, \tag{95}$$

and the ratio of (94) to (95) is about $\frac{1}{2}(n+1)$, suggesting, as before, the use of about $\log_{10} n$ extra guarding figures in the components of A and b to eliminate the effects of rounding errors.

In practice, of course, the results are likely to be contaminated by smaller amounts, since the various sources of error are likely to be fairly independent and statistically less important.

36. The errors are also considerably reduced if we use the compact methods of elimination, in which only one rounding error is involved in the evaluation of scalar products and in which interchanges are used to keep multipliers smaller than unity. For in this case the reduction to triangular form is incorporated in the equations

$$\left. \begin{array}{l} LU = A \\ Ly = b \end{array} \right\}, \tag{96}$$

and the method of computation ensures that the computed L, U and y satisfy exactly the relations

$$\left. \begin{array}{l} LU = A + \delta A \\ Ly = b + \delta b \end{array} \right\}, \tag{97}$$

in which all components of δA and δb are in general smaller in magnitude than $0 \cdot 5 \times 10^{-p}$, and in particular slightly larger, if any of the elements of U exceed unity, due to the necessary scaling.

The effect on b of the back-substitution process is the same as before but the 'accuracy' of the residuals is considerably better. For in place of (94) we now have

$$|R|_{\max} = 0 \cdot 5 \ 10^{-p} + 0 \cdot 5n \ 10^{r-p} + 0 \cdot 5 \ 10^{r-p}(1 + 1 + \ldots + 1) \tag{98}$$
$$= 0 \cdot 5 \ 10^{-p} + n \ 10^{r-p},$$

little more than twice that of (95), the maximum residual of the truncated exact solution, and this last result is independent of the order of the matrix.

In this case we hardly need any guarding figures and our estimate of the number of meaningful significant figures in the results is more easily made.

Since our aim is to give an introduction to the numerical processes of linear algebra and not to give an encyclopaedic statement about all possible methods, we shall not pursue further the error analysis of Gauss elimination, compact elimination, or any other of the techniques we have used. We remark, however, that a similar analysis can be performed, with effectively identical results, for floating-point arithmetic, and that the other techniques need hardly detain us since we have observed their close relationships, in various ways, with the elimination methods.

Relative precision of components of solution

37. In one sense, however, we have failed to distinguish between the relative precisions of the various components of the solution vector, since we have judged everything in relation to the number of significant figures left in the final pivot. But we saw in § 7 that it is possible, when some columns of A are very similar, to find the corresponding elements of the solution very poorly determined, while the remaining elements have more worth-while figures because the corresponding rows of the inverse matrix are relatively small.

38. To examine this point we note first that if A is singular then at least one of its latent roots is zero, and if A is not quite singular at least one of its roots might be very small. Now if there is a linear combination of the rows which gives a null vector, that is if

$$r'A = o', \tag{99}$$

where r' is a row, it follows that r is the latent vector of the transpose A' corresponding to the zero latent root. Similarly, if there is a column dependence it can be written in the form

$$Ac = o, \tag{100}$$

where c is a column, and c is then the latent vector of A corresponding to the zero latent root.

We also see, from the discussion of cofactors and the inverse matrix in §§ 25–30 of Chapter 2, that each *column* of the matrix of cofactors satisfies the equation $Ax = 0$, so that each *column* of A^{-1} is the same, apart from a normalizing factor, and each column is the required latent vector. Similarly all *rows* of A^{-1} are the same apart from normalizing factors and each *row* is the required latent vector of the transpose A'.

When A is not quite singular these remarks are 'approximate', in the sense that we shall not get exact proportionality but an approximation thereto. In fact the coefficients in (6) are very near to the latent vector of A' in (4) corresponding to the latent root of smallest absolute value, and those of (7) are almost the corresponding vector of A.

The question of the relative precision of the components of the computed solution of $Ax = b$, when two or more columns of A are similar and the corresponding rows of A^{-1} are correspondingly large, can therefore be settled if we can find the latent vector of A corresponding to its very small root. We shall see, in a later chapter on latent roots, that no special method is needed here. If we perform the elimination process on A to form U, and back-substitute in $Ux = e$, where all the components of e are unity, we shall have a good approximation to the required latent vector, hence to a normalized column of the inverse A^{-1}, and hence, with an examination of the number of significant figures in the smallest pivot, to the number of meaningful significant figures we can quote in each component of the solution.

39. It should be noted that it is important to examine the smallest pivot, rather than the final pivot, because with partial pivoting a small pivot may occur early in the process when there is a relation of near dependence between the columns involved at this stage. For example if we take the matrix of § 7 and write it with columns interchanged in the form

$$\begin{bmatrix} 3 & 3 & 4 & -1 \\ -2 & -2 & -4 & 6 \\ 0 & 0 & 4 & 6 \cdot 5 \\ 3 & 2 \cdot 99 & 6 & 1 \end{bmatrix}, \tag{101}$$

the permuted upper triangle after elimination is

$$\begin{bmatrix} 3 & 3 & 4 & -1 \\ & -0 \cdot 01 & 2 & 2 \\ & & 4 & 6 \cdot 5 \\ & & & 7 \cdot 5 \end{bmatrix}, \tag{102}$$

and the largest pivot is the last. The small pivot in the second column, however, is going to decide the worth-while accuracy of the system and if we now back-substitute in (1, 1, 1, 1) we get the approximate vector $67 \cdot 0$, $-66 \cdot 7$, $0 \cdot 03$, $0 \cdot 13$. If the original coefficients had uncertainties of one unit in the third decimal we could say that x_2 is known certainly to about one figure, x_1 to the same number, and x_3 and x_4 to about three more figures.

40. Finally, we remark that the coefficients c_r in (8) are the components of *any other* latent vector of A', since the vectors of A' are orthogonal to those of A, and (8) is the expression of orthogonality.

Additional notes and bibliography

§ 30. The scaling of rows and columns so that each contains at least one element say in the range $\frac{1}{2} < |a_{rs}| < 1$ has been called 'equilibrating'. The process is unfortunately not unique and the 'best' method is not at all well-defined, particularly when the components of the solution have very different orders of magnitude and when we are interested in their relative rather than their absolute accuracy, suggesting the use of floating-point arithmetic.

Consider, for example, a system like

$$\left.\begin{array}{l} 10^9x_1 + 10^9x_2 + 10^9x_3 = 10^9 \\ 10^9x_1 + \quad x_2 + \quad x_3 = 1 \\ 10^9x_1 - \quad x_2 + \quad x_3 = 3 \end{array}\right\},$$

of which the solution is $x_1 = 0$, $x_2 = -1$, $x_3 = 2$. If we perform no scaling and use either partial or complete pivoting in floating-point arithmetic we should choose as pivot the underlined coefficient. In the 'reduced' equations the coefficients of x_2 and x_3 in the second and third members of the original set would play very little part in a machine with a word-length of nine decimal digits, and the results are correspondingly inaccurate.

We could 'equilibrate' by dividing each row by 10^9, giving the equations

$$\left.\begin{array}{l} x_1 + \quad x_2 + \quad x_3 = 1 \\ x_1 + 10^{-9}x_2 + 10^{-9}x_3 = 10^{-9} \\ x_1 - 10^{-9}x_2 + 10^{-9}x_3 = 3 \cdot 10^{-9} \end{array}\right\},$$

which have at least one element in each row and column lying between $0{\cdot}1$ and $1{\cdot}0$ inclusively. The choice of the leading element as pivot, however, produces exactly the same effect as before, so that this scaling has been of no assistance.

Alternatively we could divide the first row by 10^9 and then divide the first column by 10^9, with $y_1 = 10^9x_1$, to give the set

$$\left.\begin{array}{l} 10^{-9}y_1 + x_2 + x_3 = 1 \\ \underline{y_1} + x_2 + x_3 = 1 \\ y_1 - x_2 + x_3 = 3 \end{array}\right\},$$

and the pivotal choice of the underlined number gives a perfectly satisfactory reduced set. While this choice of equilibration might occur readily in this case to an experienced worker, it is clear that the choice might be more difficult in more complicated systems, and especially when the coefficients are computed inside the machine. Further research is needed here.

§§ 31–36. Previous ideas of error analysis have attempted to trace the effects of each individual rounding error, thereby giving an estimate of the accuracy of the computed solution. The classical paper of this type is that of

VON NEUMANN, J. and GOLDSTINE, H. H., 1947, Numerical inverting of matrices of high order. *Bull. Amer. math. Soc.* **53**, 1021–99.

The idea of finding what perturbations in the original data would give rise exactly to the computed solution, or to various numbers involved in the solution, has been used, for example in

TURING, A. M., 1948, Rounding-off errors in matrix processes. *Quart. J. Mech.* **1**, 287–308,

as an aid to the analysis of the error in the computed results.

More details and extensions of the perturbation analysis appear in

WILKINSON, J. H., 1960, Rounding errors in algebraic processes. *Information Processing*, 44–53. UNESCO, Paris; Butterworth, London.

He also gives various general analyses and results in

WILKINSON, J. H., 1961, Error analysis of direct methods of matrix inversion. *J. Ass. comp. Mach.* **8**, 281–330.

This paper includes both fixed-point and floating-point arithmetic, and a necessary preliminary to the latter appears in

WILKINSON, J. H., 1960, Error analysis of floating-point computation. *Numerische Mathematik* **2**, 319–340.

All these works give, for the relative error in the computed inverse of A, results of the form

$$\frac{\|E\|}{\|A^{-1}\|} < f(n) \, \|A^{-1}\|,$$

where $f(n)$ is some small power of n. We note that the results have limited practical value, since the formula contains $\|A^{-1}\|$ whose computation is not trivial, and since the upper bound for the error is hardly ever attained. In fact these bounds are in usual practice a great overestimate, since the individual rounding errors are independent and their statistical effect very much smaller.

§ 32. The restriction that the multipliers should not exceed unity in modulus is more than a convenience. Not only does it give us the certainty of being able to perform the computation and the possibility of producing a useful error analysis, but it also helps materially in obtaining an accurate approximate solution. The use of a zero pivot is obviously impossible, but a small pivot can produce significant errors, even in floating-point arithmetic.

Consider, for example, the system

$$\left.\begin{aligned} 0{\cdot}0003x_1 + 0{\cdot}1000x_2 + 0{\cdot}8000x_3 &= 1 \\ 0{\cdot}8900x_1 + 0{\cdot}2100x_2 - 0{\cdot}3800x_3 &= 0 \\ 0{\cdot}1300x_1 - 0{\cdot}3700x_2 + 0{\cdot}2900x_3 &= 1 \end{aligned}\right\},$$

and suppose we work in four-decimal floating-point arithmetic, defining a number as $10^p \cdot q$, where p is an integer and q is in the range $0{\cdot}1 \leqslant |q| < 1$.

Choosing the first diagonal element as pivot, the multipliers are

$$-\frac{0{\cdot}8900}{0{\cdot}0003} = -0{\cdot}2967 \times 10^4, \qquad -\frac{0{\cdot}1300}{0{\cdot}0003} = -0{\cdot}4333 \times 10^3,$$

correctly rounded in our floating-point convention. Then in the first reduction the new (2, 2) element is

$$\begin{aligned} 0{\cdot}2100 - 0{\cdot}1000(0{\cdot}2967 \times 10^4) &= 0{\cdot}2100 - 0{\cdot}2967 \times 10^3 \\ &= 0{\cdot}0002 \times 10^3 - 0{\cdot}2967 \times 10^3 = -0{\cdot}2965 \times 10^3, \end{aligned}$$

and we see that the rounding is such that the result would be the same for any *original* (2, 2) element in the range 0·1501 to 0·2499.

The reduced equations, obtained in this way, form the array

$$-0{\cdot}2965(3) \qquad -0{\cdot}2374(4) \qquad -0{\cdot}2967(4)$$
$$-0{\cdot}4370(2) \qquad -0{\cdot}3463(3) \qquad -0{\cdot}4323(3),$$

the number in brackets being the power of ten multiplying the fractional part. Continuing, we find a multiplier $-0{\cdot}1474(0)$, and a considerable cancellation, the final array having the values

$$0{\cdot}3628(1) \qquad 0{\cdot}5036(1).$$

Back-substitution then gives the result

$$x_1 = 0{\cdot}1000(1), \qquad x_2 = -0{\cdot}1107(1), \qquad x_3 = 0{\cdot}1388(1),$$

correct to only one figure compared with the accurate solution

$$x_1 = 0{\cdot}9039(0), \qquad x_2 = -0{\cdot}1280(1), \qquad x_3 = 0{\cdot}1410(1).$$

The first reduced set is of course very ill-conditioned, because each equation differs only slightly from a rather large multiple of the first equation of the original set. Moreover the last two or three figures of the coefficients of x_2 and x_3 in the second and third of the original equations are irretrievably lost—they are used in no part of the computation.

§ 33. The perturbations caused by increases in the elements of the reduced matrices can in certain almost pathological cases get very large. The ratio of maximum elements of the successive reduced matrices to that of A is a factor in the analysis of Wilkinson (1961 *loc. cit.* p. 170). The computed results may be correspondingly poor, even in very well-conditioned systems, as the following example illustrates.

Consider a set of equations represented by the array

$$
A \qquad\qquad b
$$

$$
\begin{bmatrix}
1 & 0 & 0 & 0 & 1 \\
1 & 1 & 0 & 0 & -1 \\
-1 & 1 & 1 & 0 & 1 \\
1 & -1 & 1 & 1 & -1 \\
-1 & 1 & -1 & 1 & 1
\end{bmatrix}
\begin{bmatrix}
1 \\
-1 \\
1 \\
-1 \\
1
\end{bmatrix}.
$$

With partial pivoting we produce the upper triangular set

$$
\begin{bmatrix}
1 & 0 & 0 & 0 & 1 \\
 & 1 & 0 & 0 & -2 \\
 & & 1 & 0 & 4 \\
 & & & 1 & -8 \\
 & & & & 16
\end{bmatrix}
\begin{bmatrix}
1 \\
-2 \\
4 \\
-8 \\
16
\end{bmatrix},
$$

with solution (0, 0, 0, 0, 1).

Now suppose that the last element in the matrix were 0·5 instead of 1·0. The last pivot would be $15\frac{1}{2}$, and the solution $(-\frac{1}{31}, \frac{2}{31}, -\frac{4}{31}, \frac{8}{31}, \frac{32}{31})$, considerably changed from the original. But we may not be able to find this solution because of the finite word-length of our machine. For example with a five-digit binary register the number $15\frac{1}{2}$ might have to be rounded, to 16 or 15 depending on the machine, and in the former case the computed solutions are identical. The last pivot in the offending case has of course a small relative error, and the last component is in fact very nearly correct and certainly good to the number of significant figures contained in it. But in previous rows the large off-diagonal terms in the U matrix multiply its error and transmit it with extra force to the remaining components.

This difficulty is avoided here by complete pivoting, which tends to keep all the numbers reasonably small. Corresponding to the ordering 1, 5, 2, 3, 4 of the columns of A the reduced upper triangle for the original problem looks like

$$\begin{bmatrix} 1 & 1 & 0 & 0 & 0 \\ & -2 & 1 & 0 & 0 \\ & & 2 & 1 & 0 \\ & & & 2 & 1 \\ & & & & 2 \end{bmatrix},$$

and the change to 0·5 in the final element would come in with full weight.

Corresponding to this we might find that, in physical problems, the elements of some matrices may have a combination of rounding errors for which our estimation of meaningful figures, based on the number of significant figures in the smallest pivot, is somewhat over-optimistic.

Consider, for example, a physical system with a matrix such as

$$\begin{bmatrix} 1 & 0 & 0 & 0 & 0 & 0\cdot499 \\ -1 & 1 & 0 & 0 & 0 & -0\cdot501 \\ -1 & -1 & 1 & 0 & 0 & -0\cdot501 \\ -1 & -1 & -1 & 1 & 0 & -0\cdot501 \\ -1 & -1 & -1 & -1 & 1 & -0\cdot501 \\ -1 & -1 & -1 & -1 & -1 & -0\cdot501 \end{bmatrix},$$

and the knowledge that the elements of the last column are uncertain to within $\pm0\cdot001$. The computed upper triangular matrix looks like

$$\begin{bmatrix} 1 & 0 & 0 & 0 & 0 & 0\cdot499 \\ & 1 & 0 & 0 & 0 & -0\cdot002 \\ & & 1 & 0 & 0 & -0\cdot004 \\ & & & 1 & 0 & -0\cdot008 \\ & & & & 1 & -0\cdot016 \\ & & & & & -0\cdot032 \end{bmatrix},$$

and complete pivoting is here identical with partial pivoting.

The last pivot is small, but a similar matrix of order 9 would have a last pivot of significant size $-0\cdot256$, and the matrix, according to our previous ideas, would appear to be rather well-conditioned. Neglecting the last figure, which has inherent error, we should say that there are two significant figures in this pivot, and that therefore two significant figures is the maximum which we can meaningfully quote in the solution.

This is an overestimate because, within our 'tolerance' limits, the original matrix for which the last column is $(0.5, -0.5, -0.5, \ldots)$ is in fact singular and no figures in the solution are meaningful. In practice we would expect the error to be distributed in a more random fashion, so that our conclusion would not be so intolerable, but this situation should nevertheless be considered, and we can get an idea of a suitable method by combining some of our results of error analysis.

Suppose that the equation we try to solve is $Ax = b$, whereas this is only one of the sets of equations included in $(A + \delta A)(x + \delta x) = b + \delta b$, where δA and δb are the tolerances in the data and δx the corresponding uncertainty in the solution. We seek an estimate of δx, but can hardly associate this with A^{-1} or $(A + \delta A)^{-1}$ since we cannot find these exactly and since their differences might be large. Indeed there may be a certain δA for which $(A + \delta A)^{-1}$ is singular.

We know that our solving process, by Gauss or compact elimination, gives matrices L and U and a computed solution x, all known exactly, and for which $LU = A + \delta_1 A$, $LUx = b + \delta_1 b$, where $\delta_1 A$ and $\delta_1 b$ are calculable and in any case have known upper bounds. We then have

$$LUx = (A + \delta_1 A)x = b + \delta_1 b,$$

and we want to solve

$$(A + \delta A)(x + \delta x) = b + \delta b.$$

Substituting for A and LUx we find

$$LU\delta x + \delta_2 A(x + \delta x) = \delta_2 b,$$

where

$$|\delta_2 A| = |\delta_1 A| + |\delta A|, \qquad |\delta_2 b| = |\delta_1 b| + |\delta b|.$$

Then, writing $LU = C$, we obtain

$$\delta x = -C^{-1}\delta_2 Ax - C^{-1}\delta_2 A \delta x + C^{-1}\delta_2 b,$$

and, taking norms, we find

$$\|\delta x\| < \|C^{-1}\| \, \|\delta_2 A\| \, \|x\| + \|C^{-1}\| \, \|\delta_2 A\| \, \|\delta x\| + \|C^{-1}\| \, \|\delta_2 b\|,$$

or

$$\|\delta x\| < \frac{\|C^{-1}\| \{ \|\delta_2 A\| \, \|x\| + \|\delta_2 b\| \}}{1 - \|C^{-1}\| \, \|\delta_2 A\|}.$$

By taking advantage of the theories about what problem we have solved exactly we have expressed the result in terms of a quantity like $\|C^{-1}\|$, which is calculable exactly, and other quantities with known upper bound. In our example we can compute L and U exactly, so that $\delta_2 A$, $\delta_2 b = \delta A$, δb, the original tolerances. We find

$$
L = \begin{bmatrix}
1 & & & & & \\
-1 & 1 & & & & \\
-1 & -1 & 1 & & & \\
-1 & -1 & -1 & 1 & & \\
-1 & -1 & -1 & -1 & 1 & \\
-1 & -1 & -1 & -1 & -1 & 1
\end{bmatrix}
\quad \text{and } L^{-1} =
\begin{bmatrix}
1 & & & & & \\
1 & 1 & & & & \\
2 & 1 & 1 & & & \\
4 & 2 & 1 & 1 & & \\
8 & 4 & 2 & 1 & 1 & \\
16 & 8 & 4 & 2 & 1 & 1
\end{bmatrix}.
$$

We have already recorded U, and we can invert it and multiply by L^{-1} to find, sufficiently accurately for our purpose,

$$
C^{-1} = \begin{bmatrix}
250 & 125 & 62 & 31 & 16 & 16 \\
0 & 0{\cdot}5 & -0{\cdot}25 & -0{\cdot}125 & -0{\cdot}0625 & -0{\cdot}0625 \\
0 & 0 & 0{\cdot}5 & -0{\cdot}25 & -0{\cdot}125 & -0{\cdot}125 \\
0 & 0 & 0 & 0{\cdot}5 & -0{\cdot}25 & -0{\cdot}25 \\
0 & 0 & 0 & 0 & 0{\cdot}5 & -0{\cdot}5 \\
-500 & -250 & -125 & -62{\cdot}5 & -31{\cdot}2 & -31{\cdot}2
\end{bmatrix}.
$$

Then $\|C^{-1}\|_\infty$, coming from the last row, is about 1,000, and the formula for $\|\delta x\|$ tells us that any $\delta_2 A$ for which $\|\delta_2 A\|$ is approaching $0{\cdot}001$, as in our example, represents a serious situation. If $\|\delta_2 A\| \ll 0{\cdot}001$ we have a good chance of getting a small $\|\delta x\|$, that is our computed result is not seriously affected by possible uncertainties in A and b.

Another 'difficult' case (communicated privately by Professor W. Kahan) is exemplified by a matrix such as

$$
A = \begin{bmatrix}
1 & -1 & -1 & -1 \\
 & 1 & -1 & -1 \\
 & & 1 & -1 \\
 & & & 1
\end{bmatrix},
$$

which is already in upper triangular form. The first row of its inverse is (1, 1, 2, 4), and for a matrix of order n the last element is 2^{n-2}. For large n small variations in the right-hand sides can therefore cause large changes in the solution of corresponding linear equations, and the problem can be ill-conditioned. Notice that the pivots are perfectly 'well-behaved'.

Examples of these various kinds, though perhaps not frequent in practice, certainly occur and reveal that, while the incidence of small pivots indicates ill-conditioning, the absence of this phenomenon does not guarantee that the problem is well-conditioned. In the last resort some knowledge of the inverse matrix is necessary for this purpose.

Finally, we remark that the rank of a 'physical' matrix strictly has no meaning, but that we may be able to assert that a matrix differing from the original by a small and calculable amount has a certain rank which we calculate by the methods of Chapter 3.

7

Comparison of Methods. Measure of Work

Introduction

1. WE TURN now to a consideration of the relative efficiency of the various methods we have discussed for solving linear equations, inverting matrices, and evaluating matrix expressions. The question of accuracy has already received attention. For example we know that the compact elimination process, with interchanges, is the most efficient of our processes for those machines which can accumulate scalar products accurately, whereas if each separate product is rounded, before the subsequent addition, then the lengthy Gauss elimination, with interchanges, is exactly the same as the compact method in respect of the arithmetic involved. Many of the other methods we saw have close relations with the various elimination techniques.

They may not, however, be exactly equivalent in terms of storage requirements, and some methods may involve more arithmetic operations than others. In this chapter we shall examine some of our processes from the point of view of storage and arithmetical requirements.

Gauss elimination

2. We consider the lengthy Gauss elimination process for the solution of a set of n equations $Ax = b$. First of all we take the forward elimination, in which A is reduced to U assuming that pivots are taken down the diagonal, and in which the same operation is performed on the right-hand side b. The reduction from p to $p-1$ equations involves the following steps.

(i) We form the reciprocal of the pivot.

(ii) To obtain the multipliers, $p-1$ numbers, we multiply the rest of the pivotal column by this reciprocal, involving $p-1$ multiplications in all.

(iii) To obtain the new set of $p-1$ equations each element involves one multiplication and one addition, giving in all $(p-1)$ multiplications and additions on the left, and $p-1$ multiplications and additions on the right. If there are k right-hand sides there are $k(p-1)$ multiplications and additions on the right.

For the full forward elimination, with k right-hand sides, we there-fore form n reciprocals, including that of the final pivot, and

$$\left. \begin{array}{c} \sum_{1}^{n}\{(p-1)+(p-1)^2+k(p-1)\} \text{ multiplications} \\ \sum_{1}^{n}\{(p-1)^2+k(p-1)\} \text{ additions} \end{array} \right\}. \quad (1)$$

Using the results

$$\sum_{1}^{n}p = \tfrac{1}{2}n(n+1), \qquad \sum_{1}^{n}p^2 = \tfrac{1}{6}n(n+1)(2n+1), \quad (2)$$

we easily deduce the totals

$$\left. \begin{array}{c} n\{(\tfrac{1}{3}n^2-\tfrac{1}{3})+\tfrac{1}{2}k(n-1)\} \text{ multiplications} \\ n\{\tfrac{1}{3}n^2-\tfrac{1}{2}n+\tfrac{1}{6}+\tfrac{1}{2}k(n-1)\} \text{ additions} \end{array} \right\}, \quad (3)$$

reducing for $k = 1$, one right-hand side, to

$$n(\tfrac{1}{3}n^2+\tfrac{1}{2}n-\tfrac{5}{6}) \text{ multiplications}, \qquad n(\tfrac{1}{3}n^2-\tfrac{1}{3}) \text{ additions}. \quad (4)$$

In the back-substitution, for each right-hand side the computation of x_n involves one multiplication by the reciprocal of the final pivot, for x_{n-1} we have one addition and two multiplications, down to x_1 which needs $n-1$ additions and n multiplications. The totals are

$$\left. \begin{array}{c} k(1+2+...+n) = \tfrac{1}{2}kn(n+1) \text{ multiplications} \\ k(1+2+...+n-1) = \tfrac{1}{2}kn(n-1) \text{ additions} \end{array} \right\}. \quad (5)$$

The grand totals for the whole process are therefore

$$\left. \begin{array}{c} n \text{ reciprocals}, \ n(\tfrac{1}{3}n^2-\tfrac{1}{3}+nk) \text{ multiplications} \\ n\{\tfrac{1}{3}n^2-\tfrac{1}{2}n+\tfrac{1}{6}+k(n-1)\} \text{ additions} \end{array} \right\}, \quad (6)$$

reducing for $k = 1$ to

$$\left. \begin{array}{c} n \text{ reciprocals}, \ n(\tfrac{1}{3}n^2+n-\tfrac{1}{3}) \text{ multiplications} \\ n(\tfrac{1}{3}n^2+\tfrac{1}{2}n-\tfrac{5}{6}) \text{ additions} \end{array} \right\}. \quad (7)$$

3. With regard to the storage it is clear that we do not need to keep separately all the coefficients of the 'reduced' equations. We can overwrite a previous equation by its new coefficients derived from the elimination process, we can record the multipliers in spaces left by the coefficients which have been eliminated, and in place of the pivots we can store their reciprocals, for use in the formation of the multi-pliers and subsequently in the back-substitution. Proceeding in this way we find the final array appearing almost exactly in the form similar to the results of compact triangular decomposition, looking like

$$\begin{bmatrix} u_{11}^{-1} & u_{12} & u_{13} & \cdots & u_{1n} \\ m_{21} & u_{22}^{-1} & u_{23} & \cdots & u_{2n} \\ m_{31} & m_{32} & u_{33}^{-1} & \cdots & u_{3n} \\ \cdot & \cdot & \cdot & \cdot & \cdot \\ m_{n1} & m_{n2} & m_{n3} & \cdots & u_{nn}^{-1} \end{bmatrix} \begin{bmatrix} c_1 \\ c_2 \\ c_3 \\ \cdots \\ c_n \end{bmatrix} \tag{8}$$

with one right-hand side, and with k columns of type c for k right-hand sides. Back-substitution follows from here. We therefore need $n(n+k)$ locations for the elements in (8) and an extra kn places for the solution vectors.

The array (8) could in fact overwrite the original version of the equations, but we may want to retain a copy of the latter, for example for finding the residuals and correcting a first approximation, and we may also want to store a row of column sums for final checking.

If we use interchanges, to get the largest coefficient as pivot, we clearly perform exactly the same amount of arithmetic and need exactly the same amount of space, but if we want to make corrections, or to evaluate the determinant, we have also to store the necessary information to tell us what interchanges were involved.

4. For the problem of inversion, in which there are n right-hand sides representing the columns of the unit matrix, the replacement of k by n in (6) clearly gives an overestimate, since the original right-hand sides are special. In fact if we ignore multiplications by zero or unity, and observe that the successive forms of the right-hand matrix, for a set of order four, are

$$\begin{bmatrix} 1 & 0 & 0 & 0 \\ 0 & 1 & 0 & 0 \\ 0 & 0 & 1 & 0 \\ 0 & 0 & 0 & 1 \end{bmatrix}, \begin{bmatrix} 1 & 0 & 0 & 0 \\ \times & 1 & 0 & 0 \\ \times & 0 & 1 & 0 \\ \times & 0 & 0 & 1 \end{bmatrix}, \begin{bmatrix} 1 & 0 & 0 & 0 \\ \times & 1 & 0 & 0 \\ \times & \times & 1 & 0 \\ \times & \times & 0 & 1 \end{bmatrix}, \begin{bmatrix} 1 & 0 & 0 & 0 \\ \times & 1 & 0 & 0 \\ \times & \times & 1 & 0 \\ \times & \times & \times & 1 \end{bmatrix}, \tag{9}$$

where the crosses represent non-zero numbers, we observe that the elimination process on the right involves

$$0(n-1)+1(n-2)+2(n-3)+\ldots \text{ multiplications,} \tag{10}$$

and the same number of additions not counting additions to zero. The corresponding work on the left, with $k = 0$ in the previous analysis, is n reciprocals and

$$
\left.
\begin{array}{l}
\{(n-1)+(n-1)^2\}+\{(n-2)+(n-2)^2\}+\ldots \text{ multiplications} \\
(n-1)^2 + \qquad\qquad (n-2)^2 +\ldots \text{ additions}
\end{array}
\right\}. \quad (11)
$$

The combination of (10) and (11) gives

$$
\left.
\begin{array}{l}
n(n-1)+n(n-2)+\ldots = \tfrac{1}{2}n^2(n-1) \text{ multiplications} \\
(n-1)(n-1)+(n-1)(n-2)+\ldots = \tfrac{1}{2}n(n-1)^2 \text{ additions}
\end{array}
\right\}. \quad (12)
$$

In the back-substitution each column except the last involves $\tfrac{1}{2}n(n+1)$ multiplications, the last being reduced by unity since we already have the reciprocal of the pivot and its multiplier is unity, so that the total of multiplications from this source is $\tfrac{1}{2}n^2(n+1)-1$. For the additions we would have $\tfrac{1}{2}n(n-1)$ in each column if all the elements on the right were arbitrary, but the resulting total of $\tfrac{1}{2}n^2(n-1)$ must be reduced by $\tfrac{1}{2}n(n-1)$, the number of zeros on the right, giving a total of $\tfrac{1}{2}n^3-n^2+\tfrac{1}{2}n$. The resulting grand totals for inversion are

$$
n \text{ reciprocals, } n^3-1 \text{ multiplications, } n^3-2n^2+n \text{ additions.} \quad (13)
$$

5. For the storage we have the array on the left of (8), and on the right a unit lower triangle, the last array of (9). We can, however, overwrite the latter with the elements of the inverse, so that only n^2 elements are needed for the inverse and n^2 for the arithmetic which gave rise thereto and which we may want to keep for subsequent further use.

With interchanges the amount of work and of storage is unchanged, but the final matrix on the right is now a row permutation of a lower triangle, and careful programming is needed to remember just where the zeros and units are located. If the pivotal rows are in order 3, 1, 4, 2 for example, the arrays (9) become

$$
\begin{bmatrix} 0 & 0 & 1 & 0 \\ 1 & 0 & 0 & 0 \\ 0 & 0 & 0 & 1 \\ 0 & 1 & 0 & 0 \end{bmatrix},
\begin{bmatrix} 0 & 0 & 1 & 0 \\ 1 & 0 & \times & 0 \\ 0 & 0 & \times & 1 \\ 0 & 1 & \times & 0 \end{bmatrix},
\begin{bmatrix} 0 & 0 & 1 & 0 \\ 1 & 0 & \times & 0 \\ \times & 0 & \times & 1 \\ \times & 1 & \times & 0 \end{bmatrix},
\begin{bmatrix} 0 & 0 & 1 & 0 \\ 1 & 0 & \times & 0 \\ \times & 0 & \times & 1 \\ \times & 1 & \times & \times \end{bmatrix}, \quad (14)
$$

and as we saw in Chapter 3 the units in the final array are in the same positions, respectively 3, 1, 4, 2, of successive rows.

Jordan elimination

6. In the Jordan process we work always on $(n-1)$ rows, since previous pivotal rows are included in the elimination, but the number of columns is reduced by one at each stage. There is also one more step than in the Gauss method, since the very last pivotal row is added suitably to other rows. The elimination then involves n reciprocals and

$$\left. \begin{array}{c} \{(n-1)+(n-1)^2+(n-1)k\}+\{(n-1)+(n-1)(n-2)+(n-1)k\}+ \\ +\dots \text{ multiplications} \\[4pt] \{(n-1)^2+(n-1)k\}+ \qquad \{(n-1)(n-2)+(n-1)k\}+ \\ +\dots \text{ additions} \end{array} \right\}, \tag{15}$$

where the first term in braces refers to the multipliers and where there are n braces in all. This gives n reciprocals and

$$\left. \begin{array}{c} n(n-1)(1+k)+(n-1)\{(n-1)+(n-2)+\dots+1\} \\ = \tfrac{1}{2}n(n^2-1)+kn(n-1) \text{ multiplications} \\[4pt] n(n-1)k+(n-1)\{(n-1)+(n-2)+\dots+1\} \\ = \tfrac{1}{2}n(n-1)^2+kn(n-1) \text{ additions} \end{array} \right\}. \tag{16}$$

The matrix on the left is now diagonal and we complete the solution with an extra kn multiplications, giving grand totals of

n reciprocals, $\tfrac{1}{2}n(n^2-1)+kn^2$ multiplications,

reducing for $k=1$ to $\qquad \tfrac{1}{2}n(n-1)^2+kn(n-1)$ additions, (17)

n reciprocals, $\tfrac{1}{2}n^3+n^2-\tfrac{1}{2}n$ multiplications, $\tfrac{1}{2}n^3-\tfrac{1}{2}n$ additions. (18)

7. This, we note, is substantially greater, in the ratio three to two, than the work of the Gauss elimination for any sufficiently large n. The Jordan process, however, 'recovers' when applied to inversion. Starting with the unit matrix on the right and with pivots taken down the diagonal, the elimination on the right gives for a matrix of order four the successive arrays

$$\begin{bmatrix} 1 & 0 & 0 & 0 \\ 0 & 1 & 0 & 0 \\ 0 & 0 & 1 & 0 \\ 0 & 0 & 0 & 1 \end{bmatrix}, \begin{bmatrix} 1 & 0 & 0 & 0 \\ \times & 1 & 0 & 0 \\ \times & 0 & 1 & 0 \\ \times & 0 & 0 & 1 \end{bmatrix}, \begin{bmatrix} \times & \times & 0 & 0 \\ \times & 1 & 0 & 0 \\ \times & \times & 1 & 0 \\ \times & \times & 0 & 1 \end{bmatrix}, \begin{bmatrix} \times & \times & \times & 0 \\ \times & \times & \times & 0 \\ \times & \times & 1 & 0 \\ \times & \times & \times & 1 \end{bmatrix}, \begin{bmatrix} \times & \times & \times & \times \\ \times & \times & \times & \times \\ \times & \times & \times & \times \\ \times & \times & \times & 1 \end{bmatrix}, \tag{19}$$

with again one more step than in the Gauss process. Ignoring multiplications by unity or zero, and additions to zero, we have for the work in (19) the amounts

$$0(n-1)+1(n-1)+2(n-1)+\ldots+(n-1)(n-1)$$
$$= \tfrac{1}{2}n(n-1)^2 \text{ multiplications and additions.} \quad (20)$$

To complete the inversion we form an extra n^2-1 multiplications (one of the n^2 elements coming from the reciprocal of the final pivot, already known) and no additions, and the grand totals of these numbers, taking (17) with $k=0$, are

$$n \text{ reciprocals, } n^3-1 \text{ multiplications, } n^3-2n^2+n \text{ additions,} \quad (21)$$

exactly the same as in the Gauss process.

Matrix decomposition

8. Turning now to the 'compact' Gauss elimination with interchanges, equivalent to the decomposition $I_rA = LU$, $Ly = I_rb$, $Ux = y$, it is easy to see that the amount of arithmetic is again independent of the interchanges, and we can perform the analysis with the assumption that A is already in the required favourable ordering. Again, however, we have some programming to do to remember the interchanges and we perhaps need a little extra space to carry them out.

We consider first the arithmetic involved in the decomposition typified by

$$
\begin{array}{ccc}
A & L & U
\end{array}
$$

$$
\begin{bmatrix}
a_{11} & a_{12} & a_{13} & a_{14} \\
a_{21} & a_{22} & a_{23} & a_{24} \\
a_{31} & a_{32} & a_{33} & a_{34} \\
a_{41} & a_{42} & a_{43} & a_{44}
\end{bmatrix}
=
\begin{bmatrix}
1 & & & \\
l_{21} & 1 & & \\
l_{31} & l_{32} & 1 & \\
l_{41} & l_{42} & l_{43} & 1
\end{bmatrix}
\begin{bmatrix}
u_{11} & u_{12} & u_{13} & u_{14} \\
 & u_{22} & u_{23} & u_{24} \\
 & & u_{33} & u_{34} \\
 & & & u_{44}
\end{bmatrix}. \quad (22)
$$

The elements of the first row of U are those of the first row of A and involve no arithmetic.

The required $n-1$ elements of the first column of L are obtained from equations like
$$l_{r1}u_{11} = a_{r1}, \quad (23)$$
so that we have one reciprocation (of u_{11}), $n-1$ multiplications and no additions. The $(n-1)$ elements of the second row of U come from equations like
$$l_{r1}u_{1s}+u_{2s} = a_{2s}, \quad (24)$$

and each involves one multiplication and one addition. The $n-2$ elements of the second column of L come from equations like

$$l_{r1}u_{12}+l_{r2}u_{22} = a_{r2}, \tag{25}$$

and each involves two multiplications and one addition, with one reciprocal for the whole column.

Proceeding in this way we find, for the elements of L, n reciprocals and

$$\sum_{r=1}^{n} r(n-r) \text{ multiplications}, \qquad \sum_{r=1}^{n} (r-1)(n-r) \text{ additions}, \tag{26}$$

and for the elements of U we have

$$\sum_{r=1}^{n} r(n-r) \text{ multiplications}, \qquad \sum_{r=1}^{n} r(n-r) \text{ additions}. \tag{27}$$

Performing the summations we find totals, for the decomposition, of

n reciprocals, $\frac{1}{3}n(n^2-1)$ multiplications,

$$\frac{1}{3}n^3 - \frac{1}{2}n^2 + \frac{1}{6}n \text{ additions}. \tag{28}$$

In the production of y from $Ly = b$ we perform

$$(0+1+\ldots+n-1) = \tfrac{1}{2}n(n-1) \text{ multiplications and additions}, \tag{29}$$

and from $Ux = y$ we find x in a further

$$\left. \begin{array}{l} (1+2+\ldots+n) = \tfrac{1}{2}n(n+1) \text{ multiplications} \\ (0+1+\ldots+n-1) = \tfrac{1}{2}n(n-1) \text{ additions} \end{array} \right\}. \tag{30}$$

The grand totals for the solution of the equations are therefore

$$\left. \begin{array}{l} n \text{ reciprocals, } \tfrac{1}{3}n^3 + n^2 - \tfrac{1}{3}n \text{ multiplications} \\ \tfrac{1}{3}n^3 + \tfrac{1}{2}n^2 - \tfrac{5}{6}n \text{ additions} \end{array} \right\}, \tag{31}$$

exactly the same as those of the 'lengthy' Gauss process.

9. For inversion we have various possibilities following the determination of L and U. First we can invert both L and U and find $A^{-1} = U^{-1}L^{-1}$ by direct multiplication, a process which we saw in § 28 of Chapter 4 is equivalent to the Aitken method for inversion. After our previous analysis we need not pursue the arithmetic in detail, and with the usual neglect of multiplications by 1 or 0 and additions to 0 the following results are easily obtained.

(i) The inverse L^{-1} of the unit lower triangle L involves

$$\tfrac{1}{6}n^3 - \tfrac{1}{2}n^2 + \tfrac{1}{3}n$$

multiplications and additions.

(ii) The inverse U^{-1} of the upper triangle U, with the reciprocals of its diagonal elements already known and stored, involves $\frac{1}{6}n^3 + \frac{1}{2}n^2 - \frac{2}{3}n$ multiplications and $\frac{1}{6}n^3 - \frac{1}{2}n^2 + \frac{1}{3}n$ additions.

(iii) The multiplication $U^{-1}L^{-1}$ (with L^{-1} a unit triangle) involves $\frac{1}{3}n^3 - \frac{1}{3}n$ multiplications and $\frac{1}{3}n^3 - \frac{1}{2}n^2 + \frac{1}{6}n$ additions.

We deduce the grand totals, for inversion by this method, of

$$n \text{ reciprocals, } n^3 - n \text{ multiplications, and } n^3 - 2n^2 + n \text{ additions.} \quad (32)$$

The inversion of U and the computation of A^{-1} from $A^{-1}L = U^{-1}$ involves $n^3 - n$ multiplications, and the opposite process of computing L^{-1} and using $UA^{-1} = L^{-1}$ needs $n^3 - 1$ multiplications. If we invert neither L nor U, but use jointly $A^{-1}L = U^{-1}$, $UA^{-1} = L^{-1}$ and account for the zeros of U^{-1} and L^{-1} and the units in the diagonal of L^{-1}, we again need $n^3 - 1$ multiplications, and the number of additions is in all cases the same as before.

10. The rather trivial difference between the amounts $n^3 - n$ and $n^3 - 1$ of necessary multiplications perhaps needs investigating. It appears that the explicit evaluation of U^{-1}, or something equivalent, gives rise to the smaller number, and it would seem that in this case we take more advantage of the known reciprocals of the diagonal terms of U. We can see the nature of this by considering the back-substitution process, following Gauss elimination, to produce the inverse. We are effectively solving $UA^{-1} = L$, where L has the form of the last array of (9), and this we found to involve $\frac{1}{2}n^2(n+1) - 1$ multiplications. But if we divide the rows before back-substitution by the diagonal terms of U we have the arrays typified by

$$\begin{bmatrix} 1 & \times & \times & \times \\ & 1 & \times & \times \\ & & 1 & \times \\ & & & 1 \end{bmatrix}, \quad \begin{bmatrix} \times & 0 & 0 & 0 \\ \times & \times & 0 & 0 \\ \times & \times & \times & 0 \\ \times & \times & \times & \times \end{bmatrix}, \quad (33)$$

and this process involves $n(n-1)$ multiplications, the diagonal terms on the right being the known reciprocals of those of the original U. Back-substitution in (33) now involves $\frac{1}{2}n^2(n-1)$ multiplications, so that the total is $\frac{1}{2}n^3 + \frac{1}{2}n^2 - n$. The forward elimination, we saw in (12), needed $\frac{1}{2}n^3 - \frac{1}{2}n^2$ multiplications, and the total is the expected $n^3 - n$.

There is, of course, no virtue in this process since it involves extra rounding error which we would prefer not to have, and the gain of $n-1$ multiplications, compared with n^3, is a relatively poor reward.

Aitken elimination

11. We should also examine the Aitken process to see whether we can compute $CA^{-1}B$, where C is $(p \times n)$, A is $(n \times n)$, B is $(n \times m)$, with less work than, for example, the computation of $A^{-1}B$ by Gauss elimination followed by multiplication with C.

The 'brute-force' method of computing A^{-1} and then performing two matrix multiplications needs $n(n^2+mn+pm-1)$ multiplications. The evaluation of $A^{-1}B$ by the Gauss process for equations with m right-hand sides needs $n(\frac{1}{3}n^2-\frac{1}{3}+nm)$ multiplications, as we saw in (6), and the multiplication by C a further npm multiplications, giving the total multiplication count of $n(\frac{1}{3}n^2-\frac{1}{3}+nm+pm)$.

The Aitken process involves

$$\sum_{r=0}^{n-1}\{(r+p)+(pr+r^2)+(mr+mp)\} \text{ multiplications,} \tag{34}$$

the first term representing the multipliers, the second the elements on the left and the third the elements on the right. The total is

$$\tfrac{1}{3}n^3-\tfrac{1}{3}n+\tfrac{1}{2}mn^2+\tfrac{1}{2}pn^2+mnp-\tfrac{1}{2}mn+\tfrac{1}{2}pn, \tag{35}$$

smaller than that of the Gauss method by an amount $\frac{1}{2}n(n+1)(m-p)$ multiplications. The gain is zero if $m=p$, and particularly for the evaluation of a linear expression with $m=p=1$. It is positive if $m>p$ and negative if $m<p$. In the latter case we could use the method of § 23 of Chapter 3, in which we compute the transpose $(CA^{-1}B)'=B'(A^{-1})'C'$, so that in the multiplication count m and p are interchanged. But see also exercise 2, p. 316.

Other elimination methods. Symmetry

12. We need hardly consider the compact elimination methods of Crout, Banachiewicz, and Doolittle, since they have the closest possible connexions with triangular decomposition. The Cholesky method, however, for symmetric positive-definite matrices, forms n square roots, and its amount of work depends on the relative time of multiplication and computing square roots, which is usually a function of the particular machine involved.

We might compute, and store in the diagonal positions of the L matrix, the reciprocal of the square root of the true number, and indeed this computation, with an iterative scheme for $a^{-\frac{1}{2}}$ contained in the formula

$$x_{r+1} = \tfrac{1}{2}x_r(3 - ax_r^2), \tag{36}$$

is no slower than the evaluation and reciprocation of $a^{\frac{1}{2}}$. The total arithmetic for the computation of L in $A = LL'$ is then

$$n \text{ root reciprocals,} \quad \left. \begin{matrix} \tfrac{1}{6}n^3 + \tfrac{1}{2}n^2 - \tfrac{2}{3}n \text{ multiplications} \\ \tfrac{1}{6}n^3 - \tfrac{1}{6}n \text{ additions} \end{matrix} \right\}. \tag{37}$$

For solving linear equations from the formulae $Ly = b$, $L'x = y$, each takes $\tfrac{1}{2}n(n+1)$ multiplications and $\tfrac{1}{2}n(n-1)$ additions, so that the full total is

$$n \text{ root reciprocals,} \quad \left. \begin{matrix} \tfrac{1}{6}n^3 + \tfrac{3}{2}n^2 + \tfrac{1}{3}n \text{ multiplications} \\ \tfrac{1}{6}n^3 + n^2 - \tfrac{7}{6}n \text{ additions} \end{matrix} \right\}. \tag{38}$$

This is slightly more than the elimination process, with pivots taken down the diagonal, which involves

$$n \text{ divisions,} \quad \left. \begin{matrix} \tfrac{1}{6}n^3 + \tfrac{3}{2}n^2 - \tfrac{2}{3}n \text{ multiplications} \\ \tfrac{1}{6}n^3 + n^2 - \tfrac{7}{6}n \text{ additions} \end{matrix} \right\}, \tag{39}$$

the extra n multiplications having an obvious source.

For storage we keep only the lower triangle L, with the reciprocals of its pivotal elements in the diagonal positions, and the vectors y and x. There are no interchanges and no corresponding programming problems.

13. In the case of inversion we can either invert the triangle and form $(L')^{-1}L^{-1}$, or use $L'A^{-1} = L^{-1}$ and avoid the explicit inversion of L^{-1}. For the first method the inversion involves $\tfrac{1}{6}n^3 + \tfrac{1}{2}n^2 - \tfrac{2}{3}n$ multiplications, and the product needs a further $\tfrac{1}{6}n^3 + \tfrac{1}{2}n^2 + \tfrac{1}{3}n$ for the upper triangular part of A^{-1}. The grand total of multiplications is $\tfrac{1}{3}n^3 + \tfrac{3}{2}n^2 - n$. For the second the computation of the required upper half of A^{-1} involves $\tfrac{1}{3}n^3 + \tfrac{1}{2}n^2 + \tfrac{1}{6}n$, giving a grand total of $\tfrac{1}{3}n^3 + n^2 - \tfrac{1}{2}n$, smaller by an amount $\tfrac{1}{2}n(n-1)$.

Gauss elimination, with pivots down the diagonal and assuming symmetry of A^{-1}, need take no more than $\tfrac{1}{3}n^3 + \tfrac{1}{2}n^2 - 1$ multiplications, still smaller by a further amount of order $\tfrac{1}{2}n(n-1)$, and with no square roots to be computed. Certain tricks, of course, involving the symmetry of A^{-1}, are used in the back-substitution.

It is clear that in the compact methods we do more arithmetic since we have no unit elements in the triangles, and we do not retain explicitly the square of the diagonal elements of L after taking the root reciprocal. The compact methods, of course, have other advantages which we summarize in the evaluation in the next few sections.

Evaluation and comparison

†14. We can now make the following statements about our various elimination methods.

(i) There is no point in ever evaluating an inverse A^{-1} for the purpose of solving sets of equations of the form $AX = B$. Elimination and back-substitution, or its compact equivalents, are always faster, and the inverse should be obtained only if it is needed explicitly. This is true also for the evaluation of the more complicated matrix expression $CA^{-1}B$.

(ii) The Jordan method is considerably slower than any other for solving equations. The programming is easier since no back-substitution is involved, but for a fixed-point machine the necessity for scaling exists in the elimination, since we cannot guarantee that all the multipliers are smaller than unity in absolute value. Moreover we have no simple error analysis such as that of Chapter 6 which determines what equations we have actually solved, and the efficiency of a method depends considerably on our ability to provide such an analysis.

(iii) For solving a general set of equations the compact elimination with interchanges, starting with the decomposition $I_r A = LU$, is superior to all other methods when scalar products can be accumulated accurately, since the 'perturbed' equations have smaller upper bounds and neither the arithmetic nor the storage is greater than that of 'lengthy elimination' with interchanges.

For inversion by the compact method we apparently save a little by inverting U and using $A^{-1}L = U^{-1}$, but in comparison with the solution of linear equations, for which we have a perturbation error analysis, we do better to invert L and use $UA^{-1} = L^{-1}$.

(iv) For the symmetric case we recommend the compact elimination process with interchanges, taking no account of symmetry, unless the symmetric matrix is also positive definite. In the latter case the decomposition $A = LL'$ has significant advantages. The first is that, if every $|a_{rs}| < 1$, then every $|l_{rs}| < 1$, and there are no scaling problems at this stage. Moreover the computed LL' is then $A + \delta A$, where $|(\delta A)_{rs}|$ does not exceed a single rounding error when scalar products are accumulated accurately.

It is worth emphasising that this is not true for the Gauss process with pivots taken down the diagonal, for the multipliers might still exceed unity. These multipliers are *not* the elements of L but are related to them. For example in the first elimination we have

$$m_{r1} = -\frac{a_{r1}}{a_{11}}, \qquad l_{r1} = \frac{a_{r1}}{(a_{11})^{\frac{1}{2}}}, \tag{40}$$

of which the former may exceed unity while the latter is always smaller than unity if the matrix is positive definite. We have, however, already remarked in the notes on Chapter 4 that *complete* pivoting will have pivots on the diagonal, will preserve symmetry, and will have all multipliers less than unity in absolute value.

Conversely we should also note that the element which determines the number of significant figures which we can usefully quote in the solution of the physical problem is *not* l_{nn}, the last element of the triangle in $A = LL'$, but l_{nn}^2, since the reciprocal of l_{nn}^2 and not that of l_{nn} is an element of A^{-1}.

(v) Finally, there would appear to be no advantage in the use of any of the orthogonalisation methods of Chapter 5, in view of their close relationships with the various elimination methods.

Additional notes

§ 14(iii). Both the Gauss elimination and the triangular decomposition, as we have described them with partial pivoting or its equivalent, effectively need to have the complete matrix and right-hand sides in the high-speed store of the machine. With a very large matrix this may not be possible, and the following variant of the elimination method minimises the storage required.

We can describe the method simultaneously with a numerical example, and consider the equations represented by the array

			A		b
(1)	1	-1	3	5	8
(2)	2	-4	6	-4	0
(3)	4	-9	2	2	-1
(4)	1	-0.25	2.5	-1.5	1.75.

The first two rows are put into the store, and the pivot is selected as the larger of the two elements in the first column, the rows being interchanged if necessary. The smaller element is then eliminated in the usual way, and in our example we would then have the array

(2) $\qquad$ $\underline{2}$ $\quad$ -4 $\quad$ 6 $\quad$ -4 $\qquad$ 0

(1) $\qquad\qquad\qquad$ 1 $\quad$ 0 $\quad$ 7 $\qquad$ 8,

the multiplier, not shown, being -0.5. The leading (2×2) matrix of A has been reduced, with interchange if necessary, into upper triangular form.

Next we feed in the third row, and compare its first element with the original pivot, interchanging these rows according to the magnitudes of these elements. Again we perform the relevant elimination in the first column and find the array

(3) $\qquad$ $\underline{4}$ $\quad$ -9 $\quad$ 2 $\quad$ 2 $\qquad$ -1

(1) $\qquad\qquad$ 1 $\quad$ 0 $\quad$ 7 $\qquad$ 8

(2) $\qquad\qquad$ 0.5 $\quad$ 5 $\quad$ -5 $\qquad$ 0.5.

Before introducing the next equation we perform a further elimination to reduce the leading matrix of order three to upper triangular form, choosing as pivot the largest element in the second column corresponding to the rows which have zeros in the first column. Here no interchange is necessary, and we produce

(3) $\qquad$ $\underline{4}$ $\quad$ -9 $\quad$ 2 $\quad$ 2 $\qquad$ -1

(1) $\qquad\qquad$ $\underline{1}$ $\quad$ 0 $\quad$ 7 $\qquad$ 8

(2) $\qquad\qquad\qquad$ 5 $\quad$ -8.5 $\qquad$ -3.5.

At this stage we feed in the fourth row, compare its leading element with our first pivot, interchange if necessary and eliminate. Here no interchange is necessary and we find

(3) $\qquad$ $\underline{4}$ $\quad$ -9 $\quad$ 2 $\quad$ 2 $\qquad$ -1

(1) $\qquad\qquad$ 1 $\quad$ 0 $\quad$ 7 $\qquad$ 8

(2) $\qquad\qquad\qquad$ 5 $\quad$ -8.5 $\qquad$ -3.5

(4) $\qquad\qquad$ 2 $\quad$ 2 $\quad$ -2 $\qquad$ 2 .

We now seek the largest element in the relevant part of the second column, interchange and eliminate, and produce

(3) $\qquad$ 4 $\quad$ -9 $\quad$ 2 $\quad$ 2 $\qquad$ -1

(4) $\qquad\qquad$ 2 $\quad$ 2 $\quad$ -2 $\qquad$ 2

(2) $\qquad\qquad\qquad$ 5 $\quad$ -8.5 $\qquad$ -3.5

(1) $\qquad\qquad\qquad$ -1 $\quad$ 8 $\qquad$ 7 .

Finally we carry out the last computation, no interchange being necessary, to produce the required upper triangular set

$$
\begin{array}{ccccc}
4 & -9 & 2 & 2 & -1 \\
 & 2 & 2 & -2 & 2 \\
 & & 5 & -8.5 & -3.5 \\
 & & & 6.3 & 6.3,
\end{array}
$$

from which we can find the solution by back-substitution.

At the expense of throwing away the various multipliers we have reduced the storage requirement to something little more than half that of our standard method. We have also retained the advantages of keeping the multipliers smaller than unity, but have lost the ability for accurate accumulation of scalar products.

It is fairly obvious, though perhaps worth remarking, that although all the multipliers are less than unity the final set is not that which would be produced by the standard Gauss process. The determinant, however, is still the product of the diagonal elements of the final set with suitable sign adjustment for the number of interchanges performed.

§ 14(iv). The result of (iv), that if A is symmetric and positive definite and every $|a_{rs}| < 1$, then also $A = LL'$ with every $|l_{rs}| < 1$, follows easily with the use of partitioned matrices given at the end of Chapter 2 for the proof of the more general decomposition $A = LU$.

Suppose that we can perform the decomposition for the leading submatrix A_p of order p, and consider the bordering with an extra row and column. Then consider the matrix equation $A_{p+1} = L_{p+1} L'_{p+1}$ partitioned in the form

$$
\left[\begin{array}{c|c}
A_p & c \\
\hline
c' & a_{p+1,p+1}
\end{array}\right]
=
\left[\begin{array}{c|c}
L_p & o \\
\hline
l'_{p+1} & l_{p+1,p+1}
\end{array}\right]
\left[\begin{array}{c|c}
L'_p & l_{p+1} \\
\hline
o' & l_{p+1,p+1}
\end{array}\right],
$$

where c is a column, l'_{p+1} a row, and $a_{p+1,p+1}$ and $l_{p+1,p+1}$ are respectively the last diagonal terms of A_{p+1} and L_{p+1}. We have $A_p = L_p L'_p$, which is the previous decomposition. Then

$$
c = L_p l_{p+1}, \qquad c' = l'_{p+1} L'_p,
$$

which form just one statement, and

$$
a_{p+1,p+1} = l'_{p+1} l_{p+1} + l^2_{p+1,p+1}.
$$

If A is positive definite the determinants of all its leading sub-matrices, as we saw in §§ 27–28 of Chapter 3, are also positive, so that the matrix L_p is not singular and has real coefficients. Its inverse therefore exists, and we can determine $l_{p+1} = L_p^{-1} c$, giving the new elements, other than the diagonal term, in the next triangular matrix. Moreover we can then compute $l^2_{p+1,p+1}$, and this is positive since $l^2_{p+1,p+1}$ is a factor of the determinant of A_{p+1} which is positive. The equation defining $l^2_{p+1,p+1}$ also states that the sum of the squares of the last row of L_{p+1} is equal to $a_{p+1,p+1}$, and if the latter is less than unity then so is every element of this row and, by previous results, of all other rows.

8

Iterative and Gradient Methods

Introduction

1. WE TURN now to a consideration of iterative methods for solving linear equations. Some iteration has already been used in Chapter 6 to correct approximate solutions of mathematical problems, but there the first approximation was computed by a direct method, and except in very ill-conditioned cases had already many accurate figures. In the iterative methods of this chapter we start effectively with an arbitrary first approximation and improve it successively, stopping when the required precision is reached.

We consider also various gradient methods, whose common basis is the examination of certain functions whose minima are given by the solutions of the linear equations concerned. Some of these methods are iterative, in the sense that we truncate an infinite sequence of operations when we have the required precision, while others are finite and therefore properly classifiable as direct methods.

†2. Iterative methods are used mainly in those problems for which convergence is known to be rapid, so that the solution is obtained with much less work than that of a direct method, and for matrices of large order but with many zero coefficients, the so-called 'sparse' matrices, for which elimination methods would be relatively very laborious and need much storage.

As an example of the first class we consider the equations

$$\begin{bmatrix} 10^6 & 1 & 1 \\ 1 & 10^6 & 1 \\ 1 & 1 & 10^6 \end{bmatrix} \begin{bmatrix} x_1 \\ x_2 \\ x_3 \end{bmatrix} = \begin{bmatrix} b_1 \\ b_2 \\ b_3 \end{bmatrix}, \tag{1}$$

of which the solution is clearly not very different from a first approximation $x^{(0)} = 10^{-6}b$, and which we might expect to improve quickly with an iterative scheme such as

$$\left. \begin{aligned} 10^6 x_1^{(r+1)} &= b_1 - x_2^{(r)} - x_3^{(r)} \\ 10^6 x_2^{(r+1)} &= b_2 - x_1^{(r)} - x_3^{(r)} \\ 10^6 x_3^{(r+1)} &= b_3 - x_1^{(r)} - x_2^{(r)} \end{aligned} \right\}. \tag{2}$$

An important example of the second class comes from finite-difference methods for solving elliptic partial differential equations. The resulting matrix may have a form like

$$
\begin{bmatrix}
\times & \times & & & \times & & & & & & \\
\times & \times & \times & & & \times & & & & & \\
& \times & \times & \times & & & \times & & & & \\
& & \times & \times & & & & \times & & & \\
\times & & & & \times & \times & & \times & & & \\
& \times & & & \times & \times & \times & & \times & & \\
& & \times & & & \times & \times & \times & & \times & \\
& & & \times & & & \times & \times & & & \times \\
& & & & \times & & & & \times & \times & \\
& & & & & \times & & & \times & \times & \times \\
& & & & & & \times & & & \times & \times & \times \\
& & & & & & & \times & & & \times & \times \\
\end{bmatrix}, \qquad (3)
$$

and be of very large order, and it is clear that any elimination method would rapidly 'fill up' the places where coefficients are initially zero in the region enclosed within the sloping lines bounding the array. In fact such problems have given rise to a considerable and rapidly-growing literature, though this is mainly outside the scope of our present investigation.

General nature of iteration

3. The general nature of iterative methods is contained in the equation

$$x^{(r+1)} = a^{(r)} + f_r(x^{(r)}), \qquad (4)$$

where x is a single number, a function, a vector or a matrix according to context, $a^{(r)}$ is independent of x but not necessarily of r, and f_r is some function or operation which may also depend on r, the index number of the iterative operation. If a and f are independent of r the iteration is said to be *stationary*.

In Chapter 6 we noted two stationary iterative processes, one for correcting approximate solutions of linear equations according to the scheme

$$x^{(r+1)} = x^{(r)} + B^{-1}(b - Ax^{(r)}), \qquad (5)$$

where B^{-1} is an approximation to A^{-1}, and one for correcting an approximate inverse according to the scheme

$$B_{r+1}^{-1} = B_r^{-1}(2I - AB_r^{-1}). \qquad (6)$$

We are of course interested in the existence of convergence and in the rate of convergence. For (5) and (6) we found that existence depended on the size of the norms of certain matrices, and with our assumptions about the distinct nature of the latent roots of these matrices we could make both the existence of convergence and its rate depend on the magnitude of the largest latent root. For (5) the convergence is linear and for (6) quadratic, and we could sometimes use the Aitken extrapolation process in (5) to accelerate the linear rate.

4. For solving linear equations it is easy to see that the most general stationary iterative scheme can be written in the form

$$x^{(r+1)} = Cx^{(r)} + c, \tag{7}$$

where C is a matrix and c a vector. If x is ultimately to satisfy $Ax = b$ the scheme is valuable only if the 'fixed point' of the transformation is $x = A^{-1}b$. Then

$$x = Cx + c, \qquad c = (I - C)A^{-1}b, \tag{8}$$

and subtraction of the first of (8) from (7) gives

$$(x^{(r+1)} - x) = C(x^{(r)} - x). \tag{9}$$

We write $e^{(r)} = x - x^{(r)}$, the 'error vector' at stage r, and then

$$e^{(r+1)} = Ce^{(r)}, \tag{10}$$

which gives directly the relation between successive error vectors.

A sufficient condition for convergence is $\|C\| < 1$, and it is generally also necessary and sufficient if we replace $\|C\|$ by the latent root of largest modulus of C. The various iterative methods make certain choices of C and therefore of c in (7), and we now examine some possibilities.

Jacobi and Gauss-Seidel iteration

5. The iterative scheme in (2) is generally attributed to Jacobi. If we denote by D, L and U the matrices representing respectively the diagonal, lower-triangular and upper-triangular parts of A, with L and U having zero elements on the diagonal so that $A = L + D + U$, it can easily be verified that the *Jacobi* scheme of which (2) is typical is represented by the matrix expression

$$Dx^{(r+1)} = b - (L + U)x^{(r)}. \tag{11}$$

In terms of (7) we have

$$C = -D^{-1}(L + U), \qquad c = D^{-1}b, \tag{12}$$

and we can verify that the second of (8) is satisfied.

6. We certainly have convergence if $\|D^{-1}(L+U)\| < 1$ and, if we take our first matrix norm, this implies that in each row of the matrix A the sum of the moduli of the off-diagonal terms should be smaller than that of the diagonal term. For example, in the equations

$$\begin{bmatrix} 3 & 2 & 0 \\ 0 & 2 & 1 \\ 1 & 0 & 2 \end{bmatrix} \begin{bmatrix} x_1 \\ x_2 \\ x_3 \end{bmatrix} = \begin{bmatrix} 5 \\ 3 \\ 3 \end{bmatrix}, \tag{13}$$

the first few successive approximations to the solution vector are $(\frac{5}{3}, \frac{3}{2}, \frac{3}{2})$, $(\frac{2}{3}, \frac{3}{4}, \frac{2}{3})$, $(\frac{7}{6}, \frac{7}{6}, \frac{7}{6})$, and $(\frac{8}{9}, \frac{11}{12}, \frac{11}{12})$, and we are converging rapidly to the solution $(1, 1, 1)$. The successive error vectors are $(-\frac{2}{3}, -\frac{1}{2}, -\frac{1}{2})$, $(\frac{1}{3}, \frac{1}{4}, \frac{1}{3})$, $(-\frac{1}{6}, -\frac{1}{6}, -\frac{1}{6})$, and $(\frac{1}{9}, \frac{1}{12}, \frac{1}{12})$. The 'iteration' matrix is

$$-D^{-1}(L+U) = -\begin{bmatrix} 0 & \frac{2}{3} & 0 \\ 0 & 0 & \frac{1}{2} \\ \frac{1}{2} & 0 & 0 \end{bmatrix}, \tag{14}$$

its 'row' (and column) norm is $\frac{2}{3}$, and its largest latent root is about $-0\cdot55$, the ultimate ratio of corresponding components of successive error vectors.

For the system

$$\begin{bmatrix} 3 & 2 & 1 \\ 0 & 2 & 3 \\ 2 & 0 & 2 \end{bmatrix} \begin{bmatrix} x_1 \\ x_2 \\ x_3 \end{bmatrix} = \begin{bmatrix} 6 \\ 5 \\ 4 \end{bmatrix} \tag{15}$$

we get successive estimates $(2, \frac{5}{2}, 2)$, $(-\frac{1}{3}, -\frac{1}{2}, 0)$, $(\frac{7}{3}, \frac{5}{2}, \frac{7}{3})$,..., and there is no convergence. In this case the iteration matrix has one root at about $-1\cdot1$ and a complex pair with modulus $0\cdot96$.

We should note, however, that the norm condition

$$\|D^{-1}(L+U)\| < 1$$

is sufficient but not necessary. For the system in which

$$A = \begin{bmatrix} 3 & 2 & 2 \\ 0 & 2 & 1 \\ 1 & 0 & 2 \end{bmatrix}, \qquad D^{-1}(L+U) = \begin{bmatrix} 0 & \frac{2}{3} & \frac{2}{3} \\ 0 & 0 & \frac{1}{2} \\ \frac{1}{2} & 0 & 0 \end{bmatrix}, \tag{16}$$

we have a norm larger than unity but a maximum latent root of modulus about $0\cdot75$, and we have convergence.

7. The second iterative scheme, the *Gauss-Seidel* iteration, would be represented for equations (1) by the formulae

$$
\left.
\begin{aligned}
10^6 x_1^{(r+1)} &= b_1 - x_2^{(r)} - x_3^{(r)} \\
10^6 x_2^{(r+1)} &= b_2 - x_1^{(r+1)} - x_3^{(r)} \\
10^6 x_3^{(r+1)} &= b_3 - x_1^{(r+1)} - x_2^{(r+1)}
\end{aligned}
\right\}
\tag{17}
$$

in which at each stage we use the last available previous estimate of any unknown. The general matrix expression, with L, D and U having their previous definitions, is

$$(D+L)x^{(r+1)} = b - Ux^{(r)}, \tag{18}$$

so that corresponding to (7) we have

$$C = -(D+L)^{-1}U, \qquad c = (D+L)^{-1}b,$$

and the second of (8) is satisfied.

8. Again we have convergence, sufficiently if $\|(D+L)^{-1}U\| < 1$, and both sufficiently and necessarily if the largest latent root of $(D+L)^{-1}U$ is less than unity in modulus. An important case in which this is certainly true is that for which the matrix A is both symmetric and positive definite.

To prove this we take λ and y as a root and corresponding vector of the matrix $(D+L)^{-1}U = (D+L)^{-1}L'$, since A is symmetric. Then

$$(D+L)^{-1}L'y = \lambda y, \qquad L'y = \lambda(D+L)y, \tag{19}$$

the matrix $D+L$ being non-singular since no term of D is zero in virtue of the positive-definite nature of A.

Now we cannot assume that either y or λ is real, since the symmetry of A does not ensure that of our iteration matrix. We assume a complex y with conjugate $\bar{y}$, and multiply on the left of the second of (19) by $\bar{y}'$, obtaining

$$\bar{y}'L'y = \lambda\bar{y}'(D+L)y = p+iq. \tag{20}$$

Then
$$\bar{y}'Ly = \overline{y'L'\bar{y}} = \overline{\bar{y}'L'y} = p-iq. \tag{21}$$

Then also
$$\bar{y}'Ay = \bar{y}'(L+D+L')y = 2p+\bar{y}'Dy, \tag{22}$$

and is positive since A is positive definite. Hence, from (20),

$$\lambda = \frac{p+iq}{\bar{y}'Dy+p-iq}, \qquad |\lambda|^2 = \frac{p^2+q^2}{(\bar{y}'Dy+p)^2+q^2}. \tag{23}$$

But $\bar{y}'Dy > 0$ since D is positive definite, and if $p > 0$ it is clear from the second of (23) that $|\lambda| < 1$. If $p < 0$ we have $\bar{y}'Dy > 2|p|$ from (22), so that again $|\lambda| < 1$ and convergence of the iteration is assured.

9. There are few other results, for the convergence of Jacobi or Gauss-Seidel, which are of other than theoretical value, and the dominance of the diagonal term of A is obviously our most important practical criterion. We might expect that the Gauss-Seidel method, which uses the latest information at each stage, would converge whenever the Jacobi iteration converges, and at a faster rate, but this is not true in general. For example the matrix

$$A = \begin{bmatrix} 1 & 0 & 1 \\ -1 & 1 & 0 \\ 1 & 2 & -3 \end{bmatrix} \tag{24}$$

gives rise to the respective iteration matrices

$$\begin{bmatrix} 0 & 0 & 1 \\ -1 & 0 & 0 \\ -\frac{1}{3} & -\frac{2}{3} & 0 \end{bmatrix}, \quad \begin{bmatrix} 0 & 0 & 1 \\ 0 & 0 & 1 \\ 0 & 0 & 1 \end{bmatrix}, \tag{25}$$

for Jacobi and Gauss-Seidel, and their latent roots are respectively

$$\begin{aligned} \lambda_1 &= 0{\cdot}748, & \lambda_{2,3} &= -0{\cdot}374 \pm i\,0{\cdot}868 \\ \lambda_1 &= 1, & \lambda_2 &= 0, & \lambda_3 &= 0 \end{aligned} \Bigg\}, \tag{26}$$

so that the Jacobi process converges and the Gauss-Seidel diverges. The convergence of the former, of course, is very slow and spasmodic, in virtue of the complex roots with moduli $0{\cdot}945$. In § 11 we show an example of the converse possibility.

Acceleration of convergence

10. If we denote by C the matrix of the iteration, we have seen that the error vector $e^{(r)} = x - x^{(r)}$ satisfies the equation

$$e^{(r+1)} = Ce^{(r)}. \tag{27}$$

Moreover if C has distinct roots λ_r and therefore a full set of independent vectors $y^{(r)}$ we can write

$$e^{(0)} = \sum_{s=1}^{n} \alpha_s y^{(s)}, \tag{28}$$

so that

$$e^{(r)} = C^r e^{(0)} = \sum_{s=1}^{n} \alpha_s \lambda_s^r y^{(s)}. \tag{29}$$

If the roots are also different in modulus, if in fact they are all real and there are no equal and opposite pairs, then for large enough r we have

$$e^{(r)} = \alpha_1 \lambda_1^r y^{(1)}, \tag{30}$$

where λ_1 is the root of largest modulus. We then have linear convergence and can apply the Aitken process to accelerate convergence.

11. The more common difficulty arises with complex roots, for then there are two terms on the right of (30) and the unrefined Aitken method cannot be used. Even if A is symmetric there is no guarantee that C has real roots, and Aitken therefore proposed a 'double-sweep' process for the Gauss-Seidel method for symmetric positive-definite systems, which converges and for which the iteration matrix has real roots. In this process we compute the first approximations to $x_1, x_2,\ldots,$ x_n as before, but then come backwards in order $x_{n-1}, x_{n-2},\ldots, x_2, x_1,$ using the latest information in the usual way.

For example in the system

$$\left. \begin{array}{l} x_1 + \tfrac{1}{2}x_2 + \tfrac{1}{2}x_3 = 2 \\[4pt] \tfrac{1}{2}x_1 + x_2 + \tfrac{1}{2}x_3 = 2 \\[4pt] \tfrac{1}{2}x_1 + \tfrac{1}{2}x_2 + x_3 = 2 \end{array} \right\} \tag{31}$$

the Jacobi method will diverge, since one of the roots of the iteration matrix is $\lambda = -1$. The Gauss-Seidel method will converge, since the matrix is symmetric and positive definite, but the iteration matrix has one root zero and two complex roots, so that the unrefined Aitken acceleration process cannot be used.

We can clearly start any process with the best information available, and to show the effect at an early stage we start with the approximation $x_1 = x_2 = x_3 = 0.8$, with results shown in Table 1, the figures on the right being rounded, where necessary, to four decimals.

TABLE 1

	Jacobi			Gauss-Seidel		
$r = 0$	0·8	0·8	0·8	0·8	0·8	0·8
1	1·2	1·2	1·2	1·2	1·0	0·9
2	0·8	0·8	0·8	1·05	1·025	0·9625
3	1·2	1·2	1·2	1·0062	1·0156	0·9891
4	0·8	0·8	0·8	0·9976	1·0067	0·9978
5	1·2	1·2	1·2	0·9977	1·0022	1·0000
6	0·8	0·8	0·8	0·9989	1·0006	1·0003

The Jacobi figures are self-explanatory, and in the Gauss-Seidel case we have fairly quick convergence but with no systematic trend. An attempt to apply the Aitken method to the last three vectors gives 0·9976, 0·9997, 1·0003, with no improvement.

If, however, we perform the double-sweep Aitken process we find the successive approximations

$$\left.\begin{aligned}
x^{(0)} &= 0\text{·}8 \qquad 0\text{·}8 \qquad 0\text{·}8 \\
x^{(1)} &= 1\text{·}075 \qquad 0\text{·}95 \qquad 0\text{·}9 \\
x^{(2)} &= 1\text{·}0297 \quad 0\text{·}9844 \quad 0\text{·}9562 \\
x^{(3)} &= 1\text{·}0120 \quad 0\text{·}9943 \quad 0\text{·}9816
\end{aligned}\right\}. \tag{32}$$

There is here a systematic behaviour, and the Aitken acceleration applied to $x^{(1)}$, $x^{(2)}$ and $x^{(3)}$ gives the significantly better results 1·0006, 0·9983, 1·0025. We perform roughly twice as much work, of course, for each step in the double sweep as in the single sweep.

12. To prove convergence of this method, and that the largest latent root of the iteration matrix is real, we note that the double-sweep is represented by the pair of equations

$$\left.\begin{aligned}
(D+L)x^{(r+\frac{1}{2})} &= b - L'x^{(r)} \\
(D+L')x^{(r+1)} &= b - Lx^{(r+\frac{1}{2})}
\end{aligned}\right\}. \tag{33}$$

The complete iteration is incorporated in the scheme

$$\begin{aligned}
(D+L')x^{(r+1)} &= b - L(D+L)^{-1}(b - L'x^{(r)}) \\
&= L(D+L)^{-1}L'x^{(r)} + \{I - L(D+L)^{-1}\}b. \tag{34}
\end{aligned}$$

Then

$$x^{(r+1)} = (D+L')^{-1}L(D+L)^{-1}L'x^{(r)} + (D+L')^{-1}\{I - L(D+L)^{-1}\}b, \tag{35}$$

and we note another particular choice of C and c in (7) and that (8) is satisfied.

The iteration matrix is

$$C = (D+L')^{-1}L(D+L)^{-1}L' = (D+L')^{-1}(LD^{-1})(I+LD^{-1})^{-1}L', \tag{36}$$

and we wish to show that its largest latent root is real and less than unity in modulus. The important step in the proof is the observation that LD^{-1} is a lower triangular matrix with a null diagonal, and simple experiments reveal that successive powers of this matrix introduce extra sloping lines of zeros, and in fact $(LD^{-1})^n$ is the null matrix. We can therefore write

$$(I+LD^{-1})^{-1} = I-(LD^{-1})+(LD^{-1})^2 - \ldots +(-1)^{n-1}(LD^{-1})^{n-1}, \quad (37)$$

and the series terminates. We can therefore interchange the two middle terms in (36) and write

$$C = (D+L')^{-1}(I+LD^{-1})^{-1}LD^{-1}L'. \quad (38)$$

Alternatively we have

$$LD^{-1}(I+LD^{-1}) = (I+LD^{-1})LD^{-1}, \quad (39)$$

and pre- and post-multiplication by $(I+LD^{-1})^{-1}$, which obviously exists here, produces the required result

$$(I+LD^{-1})^{-1}LD^{-1} = LD^{-1}(I+LD^{-1})^{-1}. \quad (40)$$

Then

$$Cy = (D + L')^{-1}(I + LD^{-1})^{-1}LD^{-1}L'y, \quad (41)$$

and we can express $Cy = \lambda y$ in the form

$$LD^{-1}L'y = (I+LD^{-1})(D+L')\lambda y = (A+LD^{-1}L')\lambda y. \quad (42)$$

Hence

$$\lambda = \frac{\bar{y}'LD^{-1}L'y}{\bar{y}'Ay+\bar{y}'LD^{-1}L'y}, \quad (43)$$

in which every term is real, $\bar{y}'Ay$ is positive and $\bar{y}'LD^{-1}L'y$ is non-negative, so that λ is real and positive in the range $0 < \lambda < 1$.

We cannot, of course, assert anything particular about the rate of convergence of the double-sweep method in relation to that of the Gauss-Seidel process, but we do know that the theoretical justification for the Aitken acceleration is established.

13. The Aitken method, however, is rather laborious, and we may have to wait some time before the ratio of successive error vectors settles down to its ultimately constant value. Another class of acceleration devices attempts to find iteration matrices, of type C in (7), whose largest latent root is smaller than those of the Jacobi, Gauss-Seidel, or Aitken schemes. Some of these we proceed to discuss, but with the qualification that their main application is in the treatment of sparse matrices of large order.

In the Jacobi method the iterative sequence (11) can be written in the form

$$D(x^{(r+1)} -x^{(r)}) = b-(L+D+U)x^{(r)} = b-Ax^{(r)}, \quad (44)$$

showing that the *change* we make in any component is equal to the corresponding component of the residual vector of the previous approximation divided by the diagonal element. Now we might wish to anticipate future behaviour by making a somewhat larger or somewhat smaller current change, that is to multiply this change by a factor w. This factor need not be the same for each component and need not be independent of r, in which case we might replace (44) by

$$D(x^{(r+1)}-x^{(r)}) = C_r(b-Ax^{(r)}), \qquad (45)$$

where C_r is a diagonal matrix with coefficients w_r. For a stationary iteration, however, w is independent of r, and it simplifies the analysis if we take C_r to be the matrix wI.

14. From (45) we then produce the *extrapolated* Jacobi scheme

$$x^{(r+1)} = (I-wD^{-1}A)x^{(r)}+wD^{-1}b, \qquad (46)$$

and the matrix C of (7) is here $(I-wD^{-1}A)$. It is clearly not impossible that we should be able to find a w for which the latent root of largest modulus of $(I-wD^{-1}A)$ is smaller than that of

$$-D^{-1}(L+U) = I-D^{-1}A.$$

In fact if a root of the former is μ_r, and of the latter μ'_r, we have the simple relations

$$w^{-1}(1-\mu) = 1-\mu', \qquad \mu = 1-w+w\mu'. \qquad (47)$$

In particular if A is symmetric and positive definite we can express it in the form
$$A = D^{\frac{1}{2}}BD^{\frac{1}{2}}, \qquad (48)$$

where B is also symmetric and positive definite with units on the diagonal. The equations $Ax = b$ then become

$$By = D^{-\frac{1}{2}}b, \qquad y = D^{\frac{1}{2}}x, \qquad (49)$$

and the iteration matrix for the Jacobi scheme is now just $(I-B)$, that for the extrapolated scheme $(I-wB)$.

We can now relate the convergence of the Jacobi scheme to the latent roots λ of B, and in fact we must have $|1-\lambda| < 1$ for convergence. For the extrapolated scheme the roots are $\mu = 1-\lambda w$, and we need $|1-\lambda w| < 1$ for convergence. Since the roots of B are real and positive we have convergence of the unrefined scheme if $\lambda_1 < 2$, where λ_1 is the largest root, but the extrapolated process converges in any case if w is chosen so that $\lambda_1 w < 2$. The best choice of w is that for which $1-\lambda_n w = -(1-\lambda_1 w)$, for then the largest values of μ have equal and opposite sign. We have

$$w = \frac{2}{\lambda_1 + \lambda_n}, \tag{50}$$

and

$$\mu = |1-\lambda w| = \left|1 - \frac{2\lambda}{\lambda_1 + \lambda_n}\right| \leqslant \left|\frac{\lambda_1 - \lambda_n}{\lambda_1 + \lambda_n}\right| = \left|\frac{\frac{\lambda_1}{\lambda_n} - 1}{\frac{\lambda_1}{\lambda_n} + 1}\right| < 1. \tag{51}$$

We note also that the extrapolation will be most effective when B has all its roots very close together so that λ_1/λ_n is near to unity. Such a matrix would be very ill-conditioned from the point of view of determination of latent vectors, but very well-conditioned for solving linear equations, and confirms a remark in Chapter 6 about the 'condition number' of a matrix.

15. For the Gauss-Seidel process, similarly, the change in the approximation at any stage can be expressed in the form

$$D(x^{(r+1)} - x^{(r)}) = b - Ux^{(r)} - Dx^{(r)} - Lx^{(r+1)}$$
$$= b - Ax^{(r)} - L(x^{(r+1)} - x^{(r)}), \tag{52}$$

and if we apply the factor w to this 'displacement' we have the *extrapolated* Gauss-Seidel scheme

$$(w^{-1}D + L)x^{(r+1)} = b - Ux^{(r)} - (1-w^{-1})Dx^{(r)}, \tag{53}$$

and we are interested in the latent roots of the matrix

$$C = -(w^{-1}D + L)^{-1}\{(1-w^{-1})D + U\}. \tag{54}$$

We cannot here pursue this in detail, and indeed for general matrices the problem of determining a suitable w to give a small maximum latent root is formidable. The theory, however, has been developed with considerable skill and striking success in the solution of certain problems involving large sparse matrices. For some matrices of this kind we can prove that the maximum root of the unrefined Gauss-Seidel iteration, with $w = 1$, is exactly the square of that of the unrefined Jacobi iteration, and it can be shown that if this root is less than unity, giving convergence, the optimum value for w in (54), and the corresponding largest root μ of C, are given by

$$w = \frac{2}{1 + (1-\lambda_1^2)^{\frac{1}{2}}}, \quad \mu = \frac{1 - (1-\lambda_1^2)^{\frac{1}{2}}}{1 + (1-\lambda_1^2)^{\frac{1}{2}}} = w - 1, \tag{55}$$

where λ_1 is the largest root of the Jacobi iteration.

16. The relevant property for matrices of this kind is that they can be expressed in the triple-diagonal partitioned form

$$
A = \begin{bmatrix} D_1 & U_1 & O & \cdots \\ L_1 & D_2 & U_2 & \cdots \\ O & L_2 & D_3 & \cdots \end{bmatrix}, \tag{56}
$$

where the D_r are diagonal matrices, not necessarily of constant order, and L_r and U_r are matrices of corresponding shape. For a matrix of order three, for example, we might have

$$
A = \begin{bmatrix} 4 & 2 & -1 \\ 3 & 5 & 0 \\ -1 & 0 & 2 \end{bmatrix}, \tag{57}
$$

for which convergence is assured by our condition of diagonal dominance. The roots of the Jacobi iteration are those of $-D^{-1}(L+U)$, which gives for λ the determinantal equation

$$
\begin{vmatrix} \lambda & \tfrac{1}{2} & -\tfrac{1}{4} \\ \tfrac{3}{5} & \lambda & 0 \\ -\tfrac{1}{2} & 0 & \lambda \end{vmatrix} = 0, \tag{58}
$$

with roots $\lambda_1 = (0{\cdot}425)^{\frac{1}{2}}$, $\lambda_2 = -(0{\cdot}425)^{\frac{1}{2}}$, $\lambda_3 = 0$. For such matrices non-zero roots always occur in equal magnitudes and opposite signs. The roots of the Gauss-Seidel iteration are those of $-(D+L)^{-1}U$, which gives for λ the determinantal equation

$$
\begin{vmatrix} \lambda & \tfrac{1}{2} & -\tfrac{1}{4} \\ 0 & \lambda-\tfrac{3}{10} & \tfrac{3}{20} \\ 0 & \tfrac{1}{4} & \lambda-\tfrac{1}{8} \end{vmatrix} = 0, \tag{59}
$$

with roots $\lambda_1 = 0{\cdot}425$, $\lambda_2 = 0$, $\lambda_3 = 0$.

The rate of convergence of the Gauss-Seidel iteration is therefore considerably better than that of Jacobi, but we can improve on this still further with the extrapolated Gauss-Seidel, for which (55) gives the approximate values $w = 1{\cdot}138$, $\mu = 0{\cdot}138$, and a much faster rate of convergence. The other roots of the extrapolated iteration are $-0{\cdot}138$ and $0{\cdot}138$.

In practice, of course, the determination of the best w is a lengthy operation, so that this method is best reserved for special problems for which this determination, though not trivial, is perhaps a relatively small part of the whole process.

17. Other possible acceleration devices include the choice of *different* parameters w_r at the successive steps, giving a non-stationary iteration. For example in the extrapolated Jacobi scheme for the matrix B in (49), with unit elements on the diagonal, we might take different values of w_r, so that after k steps we are interested in the behaviour of the latent roots of the matrix

$$C = \prod_{r=1}^{k} (I - w_r B) = P_k(B), \qquad (60)$$

where

$$P_k(t) = \prod_{r=1}^{k} (1 - w_r t). \qquad (61)$$

If B has latent roots λ_i, those of $P_k(B)$ are $P_k(\lambda_i)$, and we would like to choose the parameters w_r so that $P_k(t)$ is as small as possible over the whole range of the latent roots λ_i. Now $P_k(t)$ is a polynomial of degree k, and of all polynomials of the same degree, with leading coefficient unity, that which has the smallest maximum variation in a given range is the Chebyshev polynomial. Suitably normalized in the range $(-1, 1)$ this is defined by

$$T_k(t) = \cos(k \cos^{-1} t), \qquad (62)$$

and has successive maxima and minima of ± 1. If we know that the roots λ_i are in the range $a < \lambda < b$ we have the corresponding polynomial

$$P_k(t) = T_k\!\left(\frac{b+a-2t}{b-a}\right) \bigg/ T_k\!\left(\frac{b+a}{b-a}\right), \qquad (63)$$

and the parameters w_r are chosen to be the reciprocals of the zeros of this polynomial.

There are, of course, some complications. For example we have to fix in advance the selected value of k, repeating the process if necessary for one or more runs of k steps, and we must also have reasonably close estimates for a and b. This method, again, has had success with large sparse matrices and has considerable literature in this and other contexts.

Labour and accuracy

18. We cannot of course make any worth-while statement about the amount of work involved in these iterative methods. Much depends on the form of the matrix, the rate of convergence, the ability to find optimum or even good accelerating factors, and so on. We can, however, see easily that one complete cycle of any method, for a general matrix, involves n^2 multiplications, though the production of the relevant equations of type (53) is a relatively painless operation if w is known. There is therefore no gain over the direct elimination methods unless the number of cycles is less than about $\frac{1}{3}n$, and for general matrices, of a size which will not involve too great problems of storage, the direct methods must have preference.

For large sparse matrices, of course, the number of multiplications may be very considerably smaller than n^2, and with the large amount of storage needed for elimination methods our iterative devices can be more valuable.

19. With regard to accuracy we have an apparent advantage over the elimination methods in that the original matrix A, or something easily and accurately derived from A, is used at every stage, so that rounding errors need not accumulate. While this is true in well-conditioned systems it has no particular advantage, and it may fail for very ill-conditioned systems in which the largest latent root λ of the iteration matrix is very near to unity. For then, in the equation

$$x - x^{(r+1)} = \lambda_1(x - x^{(r)}), \tag{64}$$

which ultimately connects successive error vectors, the change $x^{(r+1)} - x^{(r)}$ is very small. It may even be too small to be noticed by our machine, and we make no further progress.

For example the system

$$\left. \begin{array}{l} x_1 + 0\cdot70710x_2 = 0\cdot29290 \\ 0\cdot70710x_1 + 0\cdot50000x_2 = 0\cdot20711 \end{array} \right\}, \tag{65}$$

for which the matrix is symmetric and positive definite so that the Gauss-Seidel process has theoretical convergence, has the solution $x_1 = 0\cdot26267...$, $x_2 = 0\cdot04275...$. If we work with five-figure arithmetic, however, and round all products to this precision, the first step of the Gauss-Seidel process gives $x_1 = 0\cdot29290$, $x_2 = 0\cdot00000$, and the process stops. We have converged to a solution, though not the true solution. But our solution, in the spirit of the error analysis of the elimination process, is the exact solution of a slightly perturbed

set of equations
$$x_1 + 0 \cdot 70710x_2 = 0 \cdot 29290$$
$$0 \cdot 70710x_1 + 0 \cdot 50000x_2 = 0 \cdot 20711 - 0 \cdot 00000041 \qquad (66)$$

In other words our rounding errors mean that we have not used, in the iteration process, the exact coefficients of the original system, and the perturbations are of the same order as those of the elimination process.

20. We should make two further remarks. First, ill-conditioned systems, as we have just illustrated, will have a very poor rate of convergence, and this is analogous to the loss of significant figures in the pivots of the elimination methods.

This becomes obvious when we examine the determinantal equations for the latent roots of the iterative processes. For the Jacobi and Gauss-Seidel methods the respective equations are

$$
\begin{vmatrix}
\lambda a_{11} & a_{12} & a_{13} & \cdots & a_{1n} \\
a_{21} & \lambda a_{22} & a_{23} & \cdots & a_{2n} \\
& & \cdot \\
& & \cdot \\
& & \cdot \\
a_{n1} & a_{n2} & a_{n3} & \cdots & \lambda a_{nn}
\end{vmatrix} = 0,
\qquad
\begin{vmatrix}
\lambda a_{11} & a_{12} & a_{13} & \cdots & a_{1n} \\
\lambda a_{21} & \lambda a_{22} & a_{23} & \cdots & a_{2n} \\
\cdot \\
\cdot \\
\lambda a_{n1} & \lambda a_{n2} & \lambda a_{n3} & \cdots & \lambda a_{nn}
\end{vmatrix} = 0, \qquad (67)
$$

and it is clear that if $|A|$ is nearly zero, corresponding to near-singularity of A, then $\lambda = 1$ is nearly a root of both equations (67).

Consistent ordering

21. Second, all our schemes are dependent on the ordering of the equations, and rates of convergence may be improved by suitable permutations of the rows. For example a set of equations whose matrix is the unit matrix will give convergence for both Jacobi and Gauss-Seidel, but any permutation of the rows will make both D and $D+L$ singular, so that no process will converge.

On the other hand the ordering of both rows and columns, while not affecting the Jacobi method, will certainly affect the rate of convergence of the Gauss-Seidel method. For a set of three equations, for example, the Jacobi method uses the scheme defined by

$$
\begin{aligned}
a_{11}x_1^{(r+1)} &= b_1 - a_{12}x_2^{(r)} - a_{13}x_3^{(r)} \\
a_{22}x_2^{(r+1)} &= b_2 - a_{21}x_1^{(r)} - a_{23}x_3^{(r)} \\
a_{33}x_3^{(r+1)} &= b_3 - a_{31}x_1^{(r)} - a_{32}x_2^{(r)}
\end{aligned}
\right\}, \qquad (68)
$$

and the arithmetic is the same, for the same starting approximation, in whichever *order* we compute the new x_1, x_2 and x_3. The corresponding Gauss-Seidel equations are

$$\left.\begin{aligned}
a_{11}x_1^{(r+1)} &= b_1 - a_{12}x_2^{(r)} - a_{13}x_3^{(r)} \\
a_{22}x_2^{(r+1)} &= b_2 - a_{21}x_1^{(r+1)} - a_{23}x_3^{(r)} \\
a_{33}x_3^{(r+1)} &= b_3 - a_{31}x_1^{(r+1)} - a_{32}x_2^{(r+1)}
\end{aligned}\right\}, \tag{69}$$

but we could also use other schemes, such as

$$\left.\begin{aligned}
a_{22}x_2^{(r+1)} &= b_2 - a_{21}x_1^{(r)} - a_{23}x_3^{(r)} \\
a_{33}x_3^{(r+1)} &= b_3 - a_{31}x_1^{(r)} - a_{32}x_2^{(r+1)} \\
a_{11}x_1^{(r+1)} &= b_1 - a_{12}x_2^{(r+1)} - a_{13}x_3^{(r+1)}
\end{aligned}\right\}, \tag{70}$$

which effectively uses rows and columns in order 2, 3, 1, the equations becoming $I_r A I_r'(I_r x) = I_r b$, with an obvious I_r. The arithmetic here is not identical and the rate of convergence will be different. In (70), for example, we use the *old* value of x_1 to compute a new x_2, while in (69) we have the *new* x_1 on the right of the equation for the new x_2.

Now there are some other orderings which are *consistent* with those of the original equations. For example, with respect to (57), one Gauss-Seidel scheme is given by

$$\left.\begin{aligned}
4x_1^{(r+1)} &= b_1 - 2x_2^{(r)} + x_3^{(r)} \\
5x_2^{(r+1)} &= b_2 - 3x_1^{(r+1)} - 0x_3^{(r)} \\
2x_3^{(r+1)} &= b_3 + x_1^{(r+1)} - 0x_2^{(r+1)}
\end{aligned}\right\}, \tag{71}$$

but the ordering 1, 3, 2, represented here by

$$\begin{bmatrix} 4 & -1 & 2 \\ -1 & 2 & 0 \\ 3 & 0 & 5 \end{bmatrix} \begin{bmatrix} x_1 \\ x_3 \\ x_2 \end{bmatrix} = \begin{bmatrix} b_1 \\ b_3 \\ b_2 \end{bmatrix}, \tag{72}$$

gives rise, in virtue of the zero coefficients, to exactly the same arithmetic. The ordering of (72) is consistent with that of (57), and of course has exactly the same rate of convergence.

22. It is less obvious that the ordering

$$\begin{bmatrix} 5 & 0 & 3 \\ 0 & 2 & -1 \\ 2 & -1 & 4 \end{bmatrix} \begin{bmatrix} x_2 \\ x_3 \\ x_1 \end{bmatrix} = \begin{bmatrix} b_2 \\ b_3 \\ b_1 \end{bmatrix}, \tag{73}$$

which does not have the same arithmetic, should also have the same ultimate rate of convergence. This is so here, however, and for all matrices of type (56), and the latent roots of the iteration matrices corresponding to (57) and (73) are identical. The latent vectors, however, are not the same, so that the initial rate of convergence may be different.

Gradient methods

†23. We consider now a class of methods, called *gradient methods*, whose origin is the minimization of a certain quadratic form of the components $x_1, ..., x_n$ of the required solution of $Ax = b$. The vector $x = A^{-1}b$, for example, minimizes the sum of squares of the components of the residual vector $r = b - Ax$, given by

$$r'r = (b' - x'A')(b - Ax), \qquad (74)$$

and instead of writing down the resulting normal equations and solving them, as in Chapter 3, we might proceed by successive approximation, making such successive changes in the components of a starting approximation $x^{(0)}$ that the quadratic form is steadily reduced to a minimum. We can represent this process by the non-stationary iterative scheme

$$x^{(r+1)} = x^{(r)} + \alpha_r w^{(r)}, \qquad (75)$$

where α_r is a scalar and $w^{(r)}$ a vector as yet undefined. Various methods come from different choices of α and w.

Consider first the possibility of choosing α, for given w, so that $s_r = r^{(r)'}r^{(r)}$ is reduced in the change from $x^{(r)}$ to $x^{(r+1)}$. A little matrix manipulation gives

$$s_{r+1} - s_r = \alpha_r^2 u^{(r)'}u^{(r)} - 2\alpha_r u^{(r)'}r^{(r)}, \qquad u^{(r)} = Aw^{(r)}, \qquad (76)$$

and $s_{r+1} < s_r$ if $2\alpha_r u^{(r)'}r^{(r)} > \alpha_r^2 u^{(r)'}u^{(r)}$, that is if

$$\alpha_r < 2u^{(r)'}r^{(r)}/u^{(r)'}u^{(r)}. \qquad (77)$$

We make the most substantial reduction, that is we make s_{r+1} a minimum with respect to α, with the choice

$$\alpha_r = \frac{u^{(r)'}r^{(r)}}{u^{(r)'}u^{(r)}}. \qquad (78)$$

24. Then comes the question of choosing suitable vectors $w^{(r)}$. An obvious simple choice is $w^{(r)} = e^{(k)}$, where $e^{(k)}$ is the kth column of the unit matrix, so that just one component of $x^{(r)}$ is changed at this

stage. Then $u^{(r)} = Ae^{(k)}$, and is the kth column of A, denoted previously by $C_k(A)$. Dropping the parenthesized A for convenience, we have from (78) the best current α_r given by

$$\alpha_r = \frac{C_k' r^{(r)}}{C_k' C_k}. \tag{79}$$

The change in the kth component is therefore a weighted linear combination of the current residuals, the weights being proportional to the corresponding elements of the kth column.

This process is very similar to the Jacobi method, in which each component is changed by the amount of the current residual divided by the diagonal element. Formula (79), which of course involves rather more work, gives a 'better' current change, and a convergent process. The convergence may be slow, as for example in the system (15) for which unrefined Jacobi diverges. Starting with the approximation $x^{(0)} = (2, \frac{5}{2}, 2)$, the first Jacobi approximation, and taking k in cyclic order 1, 2, 3, we find successive vectors, rounded to one decimal, given by (2·0, 2·5, 2·0), (−0·2, 2·5, 2·0), (−0·2, 0·9, 2·0), (−0·2, 0·9, 1·7), (0·7, 0·9, 1·7), (0·7, 0·5, 1·7), (0·7, 0·5, 1·4), (1·0, 0·5, 1·4), (1·0, 0·6, 1·4), (1·0, 0·6, 1·2), and convergence is sure but slow. The residual vector for the first approximation is (−7, −6, −4), and for the last is (0·6, 0·2, −0·4).

We might improve convergence by taking k in some other order, such as that for which the current change in $r'r$ would be maximized, but the amount of work involved is clearly considerable.

25. A second choice for w is that for which the quadratic form s_r changes most rapidly, moving towards its minimum along the line of 'steepest descent'. The components of w are then the derivatives of the quadratic form $s_r = r^{(r)'} r^{(r)}$ with respect to the components of the current approximation $x^{(r)}$. This gives simply $w^{(r)} = A' r^{(r)}$, and the amount of 'travel' in this direction is that which makes s_{r+1} a minimum with respect to α, and produces the value (78) for α_r. The computational cycle is then

$$\left. \begin{array}{ll} r^{(r)} = b - Ax^{(r)}, \quad w^{(r)} = A' r^{(r)}, \quad u^{(r)} = Aw^{(r)} \\ \alpha_r = u^{(r)'} r^{(r)} / u^{(r)'} u^{(r)}, \quad x^{(r+1)} = x^{(r)} + \alpha_r w^{(r)} \end{array} \right\}, \tag{80}$$

and we note in passing that the new residual vector is expressible in the form

$$\begin{aligned} r^{(r+1)} = b - Ax^{(r+1)} &= r^{(r)} - A(x^{(r+1)} - x^{(r)}) \\ &= r^{(r)} - \alpha_r Aw^{(r)} = r^{(r)} - \alpha_r u_r. \end{aligned} \tag{81}$$

The computation is rather lengthy but the rate of convergence is generally better. For the problem of equations (15) we find, with the same starting approximation as before, the successive rounded vectors $(2 \cdot 0, 2 \cdot 5, 2 \cdot 0)$, $(0 \cdot 9, 1 \cdot 5, 0 \cdot 7)$, $(1 \cdot 0, 1 \cdot 2, 0 \cdot 9)$, and at this stage we have a respectable result.

26. There is of course a connexion between the Jacobi iteration and this 'gradient' method or 'method of steepest descents'. From (80), for example, we find

$$x^{(r+1)} = x^{(r)} + \alpha_r w^{(r)} = x^{(r)} + \alpha_r A'(b - Ax^{(r)})$$

$$= (I - \alpha_r A'A)x^{(r)} + \alpha_r A'b, \tag{82}$$

and this is of the form (46), the extrapolated Jacobi scheme, with D^{-1} replaced by A' and α_r allowed to vary in a non-stationary iteration.

Symmetric positive-definite case

27. For symmetric positive-definite matrices it seems unnecessarily complicated to work with $A'A = A^2$, and we consider the quadratic form

$$s = x'Ax - 2x'b = (x - A^{-1}b)'A(x - A^{-1}b) - b'A^{-1}b$$

$$= r'A^{-1}r - b'A^{-1}b, \tag{83}$$

which takes its minimum value $-b'A^{-1}b$ when $x = A^{-1}b$ and r is the null vector.

Proceeding as before we find that the change to $x^{(r+1)} = x^{(r)} + \alpha_r w^{(r)}$ produces a change in s_r of amount $\alpha_r^2 w^{(r)\prime} A w^{(r)} - 2\alpha_r w^{(r)\prime} r^{(r)}$. This change is negative if $\alpha_r < 2w^{(r)\prime} r^{(r)} / w^{(r)\prime} A w^{(r)}$, and s_{r+1} is a minimum if

$$\alpha_r = w^{(r)\prime} r^{(r)} / w^{(r)\prime} A w^{(r)}, \tag{84}$$

the result analogous to (78).

With the choice $w^{(r)} = e^{(k)}$ we find $\alpha_r = r_k^{(r)}/a_{kk}$, so that this choice gives a method identical with that of Gauss-Seidel if we take k in the order $1, 2, ..., n$, and then start again from the beginning.

If $w^{(r)}$ is taken in the direction of steepest descent we find $w^{(r)} = r^{(r)}$, the current residual vector, and the iteration formula is

$$x^{(r+1)} = x^{(r)} + \alpha_r(b - Ax^{(r)}), \tag{85}$$

which for a constant α is exactly the extrapolated Jacobi method for an A with unit diagonal elements.

The optimum α_r is given by

$$\alpha_r = r^{(r)\prime}r^{(r)}/r^{(r)\prime}Ar^{(r)}, \qquad (86)$$

which we note is the reciprocal of the Rayleigh quotient for the residual vector. The resulting iteration has generally a useful rate of convergence. For the equations of (31), for example, if we start with the first Gauss-Seidel approximation $x^{(0)} = (2\cdot0, 1\cdot0, 0\cdot5)$ the residual vector $r^{(0)}$ is $(-0\cdot75, -0\cdot25, 0\cdot00)$, and we find $\alpha_0 = \frac{10}{13}$ and

$$x^{(1)} = (1\cdot4, 0\cdot8, 0\cdot5),$$

rounded to one decimal. Then $r^{(1)} = (-0\cdot05, 0\cdot25, 0\cdot40)$ and

$$x^{(2)} = (1\cdot4, 1\cdot0, 0\cdot8).$$

The next two approximate solutions are $(1\cdot2, 0\cdot9, 0\cdot8)$ and $(1\cdot2, 1\cdot0, 0\cdot9)$ to this precision. If we start with $x^{(0)} = (0\cdot8, 0\cdot8, 0\cdot8)$, as in § 11, we produce the correct result in one step.

A finite iterative process

28. Finally, for the symmetric case, we describe briefly the method of 'conjugate gradients'. Here we choose the α_r at each stage to make s_{r+1} a minimum, so that α_r is given by (84). For $w^{(r)}$ we take a combination of the current residual and the previous vector $w^{(r-1)}$, so that

$$w^{(r)} = r^{(r)} + \beta_{r-1}w^{(r-1)}, \qquad (87)$$

and we can choose successive constants β_{r-1} so that our process terminates in exactly n steps, giving the best features of both direct and iterative methods. To this end we make $w^{(r)}$ 'conjugate', that is orthogonal with respect to the matrix A, to the previous $w^{(r-1)}$, that is we make

$$w^{(r)\prime}Aw^{(r-1)} = 0, \qquad (88)$$

and the corresponding value of β is obtained from (87) as

$$\beta_{r-1} = -r^{(r)\prime}Aw^{(r-1)}/w^{(r-1)\prime}Aw^{(r-1)}. \qquad (89)$$

The full computational cycle is given by the formulae

$$\left.\begin{array}{cc} w^{(r)} = r^{(r)} + \beta_{r-1}w^{(r-1)}, & \beta_{r-1} = -r^{(r)\prime}Aw^{(r-1)}/w^{(r-1)\prime}Aw^{(r-1)} \\ x^{(r+1)} = x^{(r)} + \alpha_r w^{(r)}, & \alpha_r = r^{(r)\prime}w^{(r)}/w^{(r)\prime}Aw^{(r)} \\ r^{(r+1)} = b - Ax^{(r+1)} = r^{(r)} - \alpha_r Aw^{(r)} \end{array}\right\}, \quad (90)$$

with $x^{(0)}$ arbitrary and $w^{(-1)} = o$ so that $w^{(0)} = r^{(0)}$.

29. Several important consequences follow from the choices of α_r and β_{r-1}. First we see that

$$r^{(r+1)'}w^{(r)} = w^{(r)'}r^{(r+1)} = w^{(r)'}(r^{(r)} - \alpha_r A w^{(r)}), \tag{91}$$

which vanishes with our choice of α_r so that $r^{(r+1)}$ and $w^{(r)}$ are orthogonal. Second, we have

$$r^{(r+1)'}r^{(r)} = (r^{(r)'} - \alpha_r w^{(r)'}A)(w^{(r)} - \beta_{r-1}w^{(r-1)})$$
$$= r^{(r)'}w^{(r)} - \alpha_r w^{(r)'}Aw^{(r)}, \tag{92}$$

the other terms vanishing in virtue of (91) and (88), and again our choice of α_r makes $r^{(r+1)}$ and $r^{(r)}$ orthogonal.

More important are the facts that

$$w^{(r+1)'}Aw^{(q)} = r^{(r+1)'}w^{(q)} = r^{(r+1)'}r^{(q)} = 0, \tag{93}$$

not only for $q = r$ but for all previous $q = 0, 1, \ldots, r-1$. They are clearly zero for $r = 0$, and induction will supply the general result. Suppose (93) to be true for $r = r$ and consider the quantities

$$w^{(r+2)'}Aw^{(q)}, \qquad r^{(r+2)'}w^{(q)} \quad \text{and} \quad r^{(r+2)'}r^{(q)}.$$

We have
$$r^{(r+2)'}w^{(q)} = (r^{(r+1)'} - \alpha_{r+1}w^{(r+1)'}A)w^{(q)} = 0, \tag{94}$$

by (91) and (88) for $q = r$, by our inductive assumption for $q = 0$, $1, \ldots, r-1$, and by the formula for α_{r+1} for $q = r+1$. Also

$$r^{(r+2)'}r^{(q)} = (r^{(r+1)'} - \alpha_{r+1}w^{(r+1)'}A)(w^{(q)} - \beta_{q-1}w^{(q-1)}) = 0, \tag{95}$$

by (91), (88) and our inductive assumption for $q = r$, by our inductive assumption for $q = 0, 1, \ldots, r-1$, and by (91), (88) and the formula for α_{r+1} for $q = r+1$. We note here also that

$$r^{(r+2)'}r^{(q)} = r^{(r+2)'}(r^{(q+1)} + \alpha_q Aw^{(q)}), \tag{96}$$

and the vanishing of this for $q = r+1$ gives an alternative definition

$$\alpha_{r-1} = -r^{(r)'}r^{(r)}/r^{(r)'}Aw^{(r-1)}, \tag{97}$$

while its vanishing for smaller values of q gives the result

$$r^{(r+2)'}Aw^{(q)} = 0, \qquad q = 0, 1, \ldots, r. \tag{98}$$

Finally we have

$$w^{(r+2)'}Aw^{(q)} = (r^{(r+2)'} + \beta_{r+1}w^{(r+1)'})Aw^{(q)} = 0, \tag{99}$$

by (98), (88) and our inductive assumption for $q = 0, 1, \ldots, r$, and by our formula for β_{r+1} for $q = r+1$. This completes the inductive proof but we also notice the simplification

$$\beta_{r-1} = \frac{-r^{(r)'}(r^{(r-1)}-r^{(r)})/\alpha_{r-1}}{\{r^{(r-1)'}+\beta_{r-2}w^{(r-2)'}\}\{r^{(r-1)}-r^{(r)}\}/\alpha_{r-1}} = \frac{r^{(r)'}r^{(r)}}{r^{(r-1)'}r^{(r-1)}}. \tag{100}$$

Now the n vectors $w^{(0)}$, $w^{(1)}$,..., $w^{(n-1)}$ are independent, that is there is no relation of the form

$$p_0 w^{(0)} + p_1 w^{(1)} + \ldots + p_n w^{(n-1)} = o, \tag{101}$$

unless every $p_r = 0$ or unless some $w^{(r)}$ is null. In fact premultiplication with $w^{(r)'}A$ would give $p_r = 0$ directly from the conjugacy relations. Since the vectors are independent, and since the residual vector $r^{(n)}$ is orthogonal to all of them, it follows that $r^{(n)} \equiv o$ and the process must terminate after n steps.

30. As example we take the equations defined by the array

$$
\begin{matrix}
A & & & & b \\
\begin{bmatrix} 4 & -1 & 1 \\ -1 & 2 & 1 \\ 1 & 1 & 3 \end{bmatrix} & , & & & \begin{bmatrix} 4 \\ 2 \\ 5 \end{bmatrix} ,
\end{matrix}
\tag{102}
$$

with solution $x' = (1, 1, 1)$. Computation gives the results of Table 2.

The process terminates at the correct answer with the determination of $x^{(3)}$, and we can verify that all the various orthogonality and conjugate properties are satisfied. In practice, of course, we work throughout with rounded decimal or binary numbers, so that the process does not give residuals which are exactly zero and does not terminate. In fact the process is iterative in practice and we can repeat the whole computation starting with the calculated $x^{(n)}$ as the first approximation. Those experienced with the method report that $x^{(n+1)}$ is usually a much better result, the rounding errors in well-conditioned problems being largely eliminated by this single further step.

31. It may happen that the process terminates before we reach $x^{(n)}$. For example with equations (31) and the first Gauss-Seidel approximation our computation proceeds as shown in Table 3.

The vanishing of $r^{(2)}$ gives $\beta_1 = 0$ and $w^{(2)} = r^{(2)} = o$.

†32. We consider now a connexion between this conjugate gradient method and the method of conjugate vectors, illustrated in §§ 3–5 of

TABLE 2

r	$x^{(r)\prime}$			$r^{(r)\prime}$			$w^{(r)\prime}$			$(Aw^{(r)})^{\prime}$			β_r	α_r
0	0	0	1	3	1	2	3	1	2	13	1	10	$\frac{1}{20}$	$\frac{7}{30}$
1	$\frac{21}{30}$	$\frac{7}{30}$	$\frac{44}{30}$	$-\frac{1}{30}$	$\frac{23}{30}$	$-\frac{10}{30}$	$\frac{7}{60}$	$\frac{49}{60}$	$-\frac{14}{60}$	$-\frac{35}{60}$	$\frac{77}{60}$	$\frac{14}{60}$	$\frac{1352}{2023}$	$\frac{90}{119}$
2	$\frac{201}{255}$	$\frac{217}{255}$	$\frac{329}{255}$	$\frac{104}{255}$	$-\frac{52}{255}$	$-\frac{130}{255}$	$\frac{702}{1445}$	$\frac{494}{1445}$	$-\frac{962}{1445}$	$\frac{1352}{1445}$	$-\frac{676}{1445}$	$-\frac{1690}{1445}$		$\frac{17}{39}$
3	1	1	1	0	0	0								

TABLE 3

r	$x^{(r)\prime}$			$r^{(r)\prime}$			$w^{(r)\prime}$			$(Aw^{(r)})^{\prime}$			β_r	α_r
0	2	1	$\frac{1}{2}$	$-\frac{3}{4}$	$-\frac{1}{4}$	0	$-\frac{3}{4}$	$-\frac{1}{4}$	0	$-\frac{7}{8}$	$-\frac{5}{8}$	$-\frac{4}{8}$	$\frac{56}{169}$	$\frac{10}{13}$
1	$\frac{37}{26}$	$\frac{21}{26}$	$\frac{13}{26}$	$-\frac{1}{13}$	$\frac{3}{13}$	$\frac{5}{13}$	$-\frac{55}{169}$	$\frac{25}{169}$	$\frac{65}{169}$	$-\frac{10}{169}$	$\frac{30}{169}$	$\frac{50}{169}$		$\frac{13}{10}$
2	1	1	1	0	0	0								

Chapter 5 for a symmetric matrix. With a change in notation the conjugate vector method forms vectors in the sequence

$$\left.\begin{array}{l} z^{(0)} = y^{(0)}, \qquad z^{(1)} = y^{(1)} + \alpha_{10} z^{(0)} \\ z^{(2)} = y^{(2)} + \alpha_{21} z^{(1)} + \alpha_{20} z^{(0)}, \dots \end{array}\right\}, \tag{103}$$

where the $y^{(r)}$ are independent vectors, and chooses the coefficients α_{rs} so that the vectors z are mutually conjugate, so that

$$\alpha_{rs} = -y^{(r)\prime} A z^{(s)} / z^{(s)\prime} A z^{(s)}. \tag{104}$$

The solution of $Ax = b$ is assumed to be expressible in the form

$$x = \sum_{r=0}^{n-1} \lambda_r z^{(r)}, \tag{105}$$

and the conjugacy conditions give

$$\lambda_r = z^{(r)\prime} b / z^{(r)\prime} A z^{(r)}. \tag{106}$$

In Chapter 5 we took the vector $y^{(p)}$ to be the corresponding column of the unit matrix and observed some resulting simplifications in the computation. We show now a choice of vectors which makes the two methods identical and in which the conjugate gradient vector $x^{(p+1)}$ is the partial sum $\sum_{r=0}^{p} \lambda_r z^{(r)}$ of the solution (105) of the conjugate vector method.

33. Starting in (90) with $x^{(0)} = 0$ we have $r^{(0)} = b = w^{(0)}$. Take $z^{(0)} = w^{(0)}$, $y^{(0)} = r^{(0)}$ in (103). Then, from (90),

$$\alpha_0 = r^{(0)\prime} w^{(0)} / w^{(0)\prime} A w^{(0)} = \lambda_0, \qquad x^{(1)} = \alpha_0 w^{(0)} = \lambda_0 z^{(0)}. \tag{107}$$

Next we have $r^{(1)} = b - A x^{(1)} = r^{(0)} - \alpha_0 A w^{(0)}$, and take $y^{(1)} = r^{(1)}$. Then $z^{(1)} = y^{(1)} + \alpha_{10} z^{(0)}$, and (104) gives

$$\alpha_{10} = -y^{(1)\prime} A z^{(0)} / z^{(0)\prime} A z^{(0)} = -r^{(1)\prime} A w^{(0)} / w^{(0)\prime} A w^{(0)} = \beta_0. \tag{108}$$

Hence $z^{(1)} = r^{(1)} + \beta_0 w^{(0)} = w^{(1)}$, and

$$\alpha_1 = r^{(1)\prime} w^{(1)} / w^{(1)\prime} A w^{(1)} = (b' - \alpha_0 w^{(0)\prime} A) w^{(1)} / w^{(1)\prime} A w^{(1)}$$
$$= b' z^{(1)} / z^{(1)\prime} A z^{(1)} = \lambda_1, \tag{109}$$

the term $\alpha_0 w^{(0)\prime} A w^{(1)}$ vanishing in virtue of the conjugacy relation. Then

$$x^{(2)} = x^{(1)} + \alpha_1 w^{(1)} = \lambda_0 z^{(0)} + \lambda_1 z^{(1)}. \tag{110}$$

The next step is the crucial point of the demonstration. We form

$$r^{(2)} = b - Ax^{(2)} = r^{(1)} - \alpha_1 Aw^{(1)}, \qquad (111)$$

and take $y^{(2)} = r^{(2)}$. Then $z^{(2)} = y^{(2)} + \alpha_{21}z^{(1)} + \alpha_{20}z^{(0)}$, and (104) gives

$$\left. \begin{aligned} \alpha_{21} &= -y^{(2)\prime}Az^{(1)}/z^{(1)\prime}Az^{(1)} = -r^{(2)\prime}Aw^{(1)}/w^{(1)\prime}Aw^{(1)} = \beta_1 \\ \alpha_{20} &= -y^{(2)\prime}Az^{(0)}/z^{(0)\prime}Az^{(0)} = -r^{(2)\prime}Aw^{(0)}/w^{(0)\prime}Aw^{(0)} = 0 \end{aligned} \right\}, \qquad (112)$$

the second result following from (98). This gives

$$z^{(2)} = y^{(2)} + \alpha_{21}z^{(1)} = r^{(2)} + \beta_1 w^{(1)} = w^{(2)}, \qquad (113)$$

and there follows

$$\begin{aligned} \alpha_2 = r^{(2)\prime}w^{(2)}/w^{(2)\prime}Aw^{(2)} &= \{b' - (\lambda_0 z^{(0)\prime} + \lambda_1 z^{(1)\prime})A\}w^{(2)}/w^{(2)\prime}Aw^{(2)} \\ &= \{b' - (\lambda_0 w^{(0)\prime} + \lambda_1 w^{(1)\prime})A\}w^{(2)}/w^{(2)\prime}Aw^{(2)} \\ &= b'w^{(2)}/w^{(2)\prime}Aw^{(2)} \\ &= b'z^{(2)}/z^{(2)\prime}Az^{(2)} \\ &= \lambda_2. \end{aligned} \qquad (114)$$

The continuation is now clear, with $z^{(r)} = w^{(r)}$, $y^{(r)} = r^{(r)}$, and the final solution (105) is the $x^{(n)}$ of the conjugate gradient method.

34. There is perhaps little to choose between the two methods for general symmetric matrices of small order, but the conjugate gradient method has the advantage, for sparse matrices of large order, of conserving storage with the vanishing of all terms α_{rs}, in (103), for $s \neq r - 1$. It also has the advantage, already mentioned, that we can carry the process further and eliminate some of the rounding error. The use of conjugacy relations, moreover, is basic in certain methods, discussed in Chapter 10, for the determination of latent roots and vectors.

Additional notes and bibliography

§ 2. The iterative methods of §§ 2–22, with extensions and full theoretical and practical details, are treated, with particular reference to the solution of elliptic partial differential equations, in

VARGA, R. S., 1962, *Matrix iterative analysis*. Prentice-Hall, New York.

This book has a comprehensive bibliography, including recent original papers in this field.

§ 19. This point about the accuracy of iterative methods was noted by

WILKINSON (1961) *loc. cit.*, p. 170.

§ 23. The gradient methods also have a considerable literature. We refer here to

FORSYTHE, G. E., 1953, Solving linear algebraic equations can be interesting. *Bull. Amer. math. Soc.* **59**, 299–329.

This gives a very large bibliography covering all methods, iterative and other, discovered up to 1953. Later work is mainly concerned with error analysis and details of electronic computation, but extensions of gradient methods and numerical experiments, particularly for solving partial differential equations, are given by

ENGELI, M., GINSBURG, TH., RUTISHAUSER, H. and STIEFEL, E., 1959, *Refined iterative methods for computation of the solution and the eigenvalues of self-adjoint boundary-value problems.* Birkhauser, Basle.

This book also has a useful bibliography of recent works in the field of iterative and gradient solution of linear systems.

§ 32. The connexion between the conjugate gradient method and the method of conjugate vectors was noted by FORSYTHE (1953) *loc. cit.*, this page.

9

Iterative Methods for Latent Roots
and Vectors

Introduction

1. IN THE next two chapters we shall discuss methods for determining the latent roots and vectors of general real matrices. As with linear equations, so here we have two types of method, the direct process and the indirect, but the decision which to use may depend not so much on the form of the matrix as on what solutions are required. For example there are n latent roots and generally n latent vectors for a matrix of order n, and the practical problem may require us to find all the solutions, or only a small number involving a few of the roots of largest or smallest modulus, and so on. If only one or a few solutions are needed we might well apply an iterative method of the type discussed in this chapter, whereas if all or most of the roots and vectors are required we should favour one of the direct methods, mainly involving similarity transformations, which we shall treat in Chapter 10.

†**2.** At the outset we make one important remark. The latent roots λ_r and latent vectors $x^{(r)}$ of the matrix A satisfy the equations

$$(A - I\lambda_r)x^{(r)} = o, \qquad r = 1, 2, \ldots, n, \tag{1}$$

and $x^{(r)}$ is non-trivial if λ_r is a root of the determinantal equation

$$|A - I\lambda| = 0. \tag{2}$$

It is tempting to expand the determinant and to produce the characteristic polynomial, a polynomial of order n, whose roots are the required values λ_r. Except for matrices of small order, however, this is usually unsatisfactory. The coefficients of the polynomial will have errors due to rounding of products, and it is known that many polynomials are ill-conditioned with regard to their zeros, that is small changes in the coefficients can cause large changes in the zeros.

We can, however, use (2) to find real zeros by evaluating the determinant for various values of λ, say by a Gauss elimination process, and looking for changes of sign in its value. This method is quite practicable, but without a large amount of computation gives little evidence of the position in the spectrum $\lambda_1 > \lambda_2 > ... > \lambda_n$ of any root so obtained. The iterative methods provide this evidence, at least for some of the roots.

Direct iteration

3. We assume throughout that the roots are distinct, so that there are n independent vectors $x^{(r)}$, $r = 1, 2, ..., n$, and any arbitrary vector y can be expressed in the form

$$y = \sum_{r=1}^{n} \alpha_r x^{(r)}. \tag{3}$$

We consider then the iterative scheme defined by

$$y^{(p)} = Ay^{(p-1)}, \qquad y^{(0)} \text{ arbitrary.} \tag{4}$$

If $y^{(0)}$ is expressible in the form (3), we have

$$y^{(p)} = Ay^{(p-1)} = A^2 y^{(p-2)} = ... = A^p y^{(0)} = \sum_{r=1}^{n} \alpha_r \lambda_r^p x^{(r)}, \tag{5}$$

since $Ax^{(r)} = \lambda_r x^{(r)}$, $A^2 x^{(r)} = \lambda_r^2 x^{(r)}$, and so on.

Suppose first that the root λ_1 of largest modulus is real and that $|\lambda_r| \neq |\lambda_1|$, $r = 2, 3, ..., n$. Then

$$y^{(p)} = \lambda_1^p \left\{ \alpha_1 x^{(1)} + \sum_{r=2}^{n} \alpha_r \left(\frac{\lambda_r}{\lambda_1} \right)^p x^{(r)} \right\}, \tag{6}$$

and if α_1 is not zero and p is sufficiently large this becomes

$$y^{(p)} = \lambda_1^p (\alpha_1 x^{(1)} + \epsilon^{(p)}), \tag{7}$$

where $\epsilon^{(p)}$ is a vector with very small components.

When p is so large that ϵ is negligible to required precision it follows that $y^{(p)}$ is the unnormalized latent vector $x^{(1)}$ corresponding to the root λ_1. Moreover the ratios of corresponding components of successive vectors $y^{(p)}$ and $y^{(p-1)}$ will have the constant value λ_1, and the required latent root is determined from this fact. The rate of convergence depends partly on the constants α_r, but more effectively on the ratios $|\lambda_2/\lambda_1|$, $|\lambda_3/\lambda_1|, ...$. The smaller these ratios the faster will be our rate of convergence.

4. If α_1 is zero in the selected $y^{(0)}$ we apparently shall not converge to the required root and vector, but in practice we work with a fixed number of figures, and successive vectors are rounded, so that

instead of producing exactly $y^{(p)}$ from $y^{(p-1)}$ we produce $y^{(p)}+\eta$, where η is a vector formed from the rounding errors. If this vector contains a non-zero multiple of $x^{(1)}$ its effect will grow in later iterations and will ultimately be the dominant part of the vector. For this reason the iterative process is less valuable for the computation of any but the most dominant latent root and corresponding vector.

Even if the matrix has exact integer coefficients, and the latent vector corresponding to λ_1 is completely lacking in our $y^{(0)}$, we shall almost certainly introduce $x^{(1)}$ by our computational method if we normalize each $y^{(p)}$ to a vector $z^{(p)}$ with its largest component unity, a device which is particularly valuable with a fixed-point machine. Moreover the factor by which we divide $y^{(p)}$ to produce $z^{(p)}$ is our current estimate of λ_1, at least by the time the position of the largest component of $y^{(p)}$ has become fixed.

5. As an example we seek the root of largest modulus of the symmetric matrix

$$A = \begin{bmatrix} 1 & 1 & 3 \\ 1 & -2 & 1 \\ 3 & 1 & 3 \end{bmatrix}, \tag{8}$$

the roots being real though not necessarily positive. The computing scheme is represented by the equation

$$z^{(p)} = k_p^{-1}Az^{(p-1)}, \tag{9}$$

where k_p is the largest element of $y^{(p)} = Az^{(p-1)}$. Successive results, rounded to two decimals, are shown in Table 1. The number k_p is the current estimate of the latent root of largest modulus. The results obtained after five steps are clearly close to the required values.

TABLE 1

p	$y^{(p)'}$			$z^{(p)}$			k_p
0	1	1	1	1	1	1	
1	5	0	7	0·71	0·00	1·00	7·00
2	3·71	1·71	5·13	0·72	0·33	1·00	5·13
3	4·05	1·06	5·49	0·74	0·19	1·00	5·49
4	3·93	1·36	5·41	0·73	0·25	1·00	5·41
5	3·98	1·23	5·44	0·73	0·23	1·00	5·44

6. For the matrix

$$A = \begin{bmatrix} -2 & -1 & 4 \\ 2 & 1 & -2 \\ -1 & -1 & 3 \end{bmatrix} \tag{10}$$

the root of largest modulus is 2, with vector (1, 0, 1). If we start the
iteration with $y^{(0)'} = (3, 1, 2)$, and use $y^{(p)} = Ay^{(p-1)}$ without normalization, we get

$$y^{(1)'} = (1, 3, 2), \qquad y^{(2)'} = (3, 1, 2), \tag{11}$$

and we are back where we started. Our guess $y^{(0)}$ has no component
of the vector corresponding to the largest root, and since the other
roots are $+1$ and -1 we show no convergence to any latent vector,
but merely to a combination of the two sub-dominant latent vectors.

On the other hand if we normalize as before we find the results of
Table 2, the figures being rounded to one decimal. We are clearly
converging, after an uncertain start, to the vector corresponding to
the largest root.

TABLE 2

p	$y^{(p)'}$			$z^{(p)'}$		
0	3	1	2	1·0	0·3	0·7
1	0·5	0·9	0·8	0·6	1·0	0·9
2	1·4	0·4	1·1	1·0	0·3	0·8
3	0·9	0·7	1·1	0·8	0·6	1·0
4	1·8	0·2	1·6	1·0	0·1	0·9
5	1·5	0·3	1·6	0·9	0·2	1·0
6	2·0	0·0	1·9	1·0	0·0	0·9

We note in passing that it is not essential to take ratios of components of *successive* vectors, and that the formula

$$y_i^{(p+q)}/y_i^{(p)} \rightarrow \lambda_1^q \tag{12}$$

will often provide a better result for λ_1, the taking of the qth root
removing some local irregularities, particular for $|\lambda_1| > 1$.

7. If the root of largest modulus is complex there is a corresponding
complex conjugate solution $\bar{\lambda}_1$ and $x^{(1)}$. In this case we cannot express
$y^{(p)}$ in the ultimate form (7), since there are two roots λ_1 and $\bar{\lambda}_1$ of
equal modulus, and their vectors $x^{(1)}$ and $\bar{x}^{(1)}$ persist in growing at the
same absolute rate in the iteration. But

$$\alpha_1 \lambda_1^p x^{(1)} + \bar{\alpha}_1 \bar{\lambda}_1^p \bar{x}^{(1)}$$

will not show any consistent trend, since each element is of the form
$R^p \cos p\theta$, which oscillates as p increases.

Instead of (7) we can write

$$y^{(p)} = \alpha_1 \lambda_1^p x^{(1)} + \bar{\alpha}_1 \bar{\lambda}_1^p \bar{x}^{(1)} + \epsilon^{(p)}, \tag{13}$$

where $\epsilon^{(p)}$ is a small vector, and if this is negligible we can find λ_1

and $x^{(1)}$ by considering three successive vectors $y^{(p)}$, $y^{(p+1)}$ and $y^{(p+2)}$. If λ_1 and $\bar{\lambda}_1$ are the roots of the quadratic equation $\lambda^2 + b\lambda + c = 0$, and ϵ is negligible, it follows that

$$y^{(p+2)} + by^{(p+1)} + cy^{(p)} = 0, \tag{14}$$

and the constants b and c can be found from any two equations of the set (14), that is using any two corresponding components of three successive vectors. Or one might use all the components and find b and c so that the sum of squares of the residuals of the equations (14) is a minimum.

Once λ_1 and $\bar{\lambda}_1$ have been found we can compute the corresponding latent vectors from any two successive vectors $y^{(p)}$ and $y^{(p+1)}$. For if

$$\alpha_1\lambda_1^p x^{(1)} = \xi + i\eta, \quad \bar{\alpha}_1\bar{\lambda}_1^p \bar{x}^{(1)} = \xi - i\eta, \quad \lambda_1 = \mu + i\nu, \quad \bar{\lambda}_1 = \mu - i\nu, \tag{15}$$

the equations (13) for $y^{(p)}$ and $y^{(p+1)}$, with ϵ assumed negligible, give

$$y^{(p)} = 2\xi, \quad y^{(p+1)} = 2(\mu\xi - \nu\eta), \quad \eta = \{\mu y^{(p)} - y^{(p+1)}\}/2\nu, \tag{16}$$

so that ξ and η, the real and imaginary parts of the unnormalized latent vectors, are easily calculable.

8. Consider for example the matrix

$$A = \begin{bmatrix} 8 & -1 & -5 \\ -4 & 4 & -2 \\ 18 & -5 & -7 \end{bmatrix}, \tag{17}$$

and perform an iteration with a fairly good starting vector

Successive vectors are $\quad y^{(0)\prime} = (1, 0\cdot8, 1)$.

$$(2\cdot2, -2\cdot8, 7\cdot0), \quad (-14\cdot6, -34\cdot0, 4\cdot6),$$

$$(-105\cdot8, -86\cdot8, -125\cdot0), \tag{18}$$

and there is no sign of a constant ratio between successive vectors. We suspect a complex pair of roots, and try the equations

$$\left.\begin{array}{r} 2\cdot2c - 14\cdot6b - 105\cdot8 = 0 \\ -2\cdot8c - 34\cdot0b - 86\cdot8 = 0 \\ 7\cdot0c + 4\cdot6b - 125\cdot0 = 0 \end{array}\right\} \tag{19}$$

for the computation of the coefficients in (14).

The first two of (19) give $b = -4 \cdot 21...$, $c = 20 \cdot 14...$, with corresponding roots

$$\lambda_1, \bar{\lambda}_1 = 2 \cdot 10... \pm i 3 \cdot 96..., \tag{20}$$

and the least squares solution gives $b = -4 \cdot 22...$, $c = 20 \cdot 55...$,

$$\lambda_1, \bar{\lambda}_1 = 2 \cdot 11... \pm i 4 \cdot 01... . \tag{21}$$

We have clearly not carried the iteration far enough for the effective elimination of sub-dominant roots and vectors, but the results bear some resemblance to the true solution

$$\lambda_1, \bar{\lambda}_1 = 2 \pm 4i. \tag{22}$$

Taking (21) as our approximation the corresponding vectors, computed from (16) and the last two of (18) and normalized to have their largest components unity in modulus, are found to be

$$x', \bar{x}' = 0 \cdot 48 \pm 0 \cdot 50i, \qquad 1 \cdot 0, \qquad -0 \cdot 03 \pm 0 \cdot 99i, \tag{23}$$

compared with the true solution $0 \cdot 5 \pm 0 \cdot 5i$, $1 \cdot 0$, $0 \pm 1i$.

Acceleration of convergence

†9. If all the roots are real we can make the direct iteration process converge either to the largest or to the smallest root, in the algebraic sense, by using the modified matrix $(A - qI)$. In place of (5) we then have

$$y^{(p)} = (A - qI)y^{(p-1)} = ... = \sum_{r=1}^{n} \alpha_r (\lambda_r - q)^p x^{(r)}, \tag{24}$$

and we have effectively shifted the origin of the latent root spectrum a distance q to the right. We shall converge to the root λ_k and corresponding vector for which $\lambda_k - q$ has largest absolute value. If the roots are in *algebraic* order $\lambda_1 > \lambda_2 > ... > \lambda_n$, it is clear that either $\lambda_1 - q$ or $\lambda_n - q$ has the largest modulus and no choice of q can cause convergence to any other root.

With the first possibility equation (6) is replaced by

$$y^{(p)} = (\lambda_1 - q)^p \left\{ \alpha_1 x^{(1)} + \alpha_2 \left(\frac{\lambda_2 - q}{\lambda_1 - q} \right)^p x^{(2)} + ... + \alpha_n \left(\frac{\lambda_n - q}{\lambda_1 - q} \right)^p x^{(n)} \right\}, \tag{25}$$

the rate of convergence depends mainly on the ratios $(\lambda_r - q)/(\lambda_1 - q)$, and the largest of these will be as small as possible if we choose q so that

$$\lambda_2 - q = -(\lambda_n - q), \qquad q = \tfrac{1}{2}(\lambda_2 + \lambda_n). \tag{26}$$

Similarly we can write

$$y^{(p)} = (\lambda_n - q)^p \left\{ \alpha_n x^{(n)} + \alpha_1 \left(\frac{\lambda_1 - q}{\lambda_n - q} \right)^p x^{(1)} + \ldots + \alpha_{n-1} \left(\frac{\lambda_{n-1} - q}{\lambda_n - q} \right)^p x^{(n-1)} \right\},$$

(27)

and we shall converge most rapidly to the root corresponding to λ_n if we choose q so that

$$(\lambda_1 - q) = -(\lambda_{n-1} - q), \qquad q = \tfrac{1}{2}(\lambda_1 + \lambda_{n-1}).$$

(28)

For (25) we find that $y^{(p)} \to \alpha_1(\lambda_1 - q)^p x^{(1)}$, and the ratios

$$y_i^{(p+1)}/y_i^{(p)} \to \lambda_1 - q.$$

For (27) we have $y^{(p)} \to \alpha_n(\lambda_n - q)^p x^{(n)}$ and $y_i^{(p+1)}/y_i^{(p)} \to \lambda_n - q$, and again rounding errors will secure convergence even if α_1 or α_n is initially zero.

10. There are two special cases of this acceleration method. If the two real roots of largest modulus are equal but of opposite sign the direct iteration with A will not give convergence to either root, successive vectors $y^{(p)}$ being ultimately linear combinations of the two latent vectors corresponding to these roots. We can of course recognize this by noting that, eventually, $y^{(p+2)} = \lambda^2 y^{(p)}$, so that ratios $y_i^{(p+2)}/y_i^{(p)}$ of components of successive vectors will settle to the constant λ^2. A suitable choice of q, however, will cause convergence to the required root. In a similar way we can 'separate' the complex roots of largest modulus by iterating with the matrix $(A - iqI)$ with a suitably chosen value of q, though here we might converge to a solution other than one of the largest complex pair.

Whether these special cases necessitate the separation device depends on our requirements and to some extent on our computing machine. In particular the separation of the complex pair requires a considerable amount of complex arithmetic, and complex multiplication must be at least four times as lengthy as real multiplication.

In fact it is not certain that the acceleration in general will have any significant advantage, since for best application we need some knowledge of the distribution of the roots, and this is often lacking in practical problems. One can perhaps watch the convergence of the iteration, either by eye or by machine programme, and vary q to obtain something approaching the optimum, and this method has had success with machines with this facility.

11. A second acceleration device is that of Aitken, which we have used already in connexion with iterative methods for solving linear equations. To investigate this we write the iteration equation (5) in the form

$$y^{(p)} = \sum_{r=1}^{n} \lambda_r^p x^{(r)},$$ (29)

incorporating the coefficients α_r in the $x^{(r)}$ and thereby performing a rather arbitrary and non-uniform type of normalization. Comparing corresponding components of successive vectors we find

$$\frac{y_i^{(p+1)}}{y_i^{(p)}} = \lambda_1 \left\{ \frac{x_i^{(1)} + \left(\frac{\lambda_2}{\lambda_1}\right)^{p+1} x_i^{(2)} + \ldots}{x_i^{(1)} + \left(\frac{\lambda_2}{\lambda_1}\right)^{p} x_i^{(2)} + \ldots} \right\}.$$ (30)

If the root λ_1 of largest modulus is real this ratio will converge, as we have seen, to the value of this root.

Suppose now that λ_2, the root of next largest modulus, is also real, and that p is so large that $(\lambda_r/\lambda_1)^p$ is negligible for $r = 3, 4, \ldots, n$. From (30) we then find

$$\frac{y_i^{(p+1)}}{y_i^{(p)}} = \lambda_1 + \left(\frac{\lambda_2}{\lambda_1}\right)^p (\lambda_2 - \lambda_1) \frac{x_i^{(2)}}{x_i^{(1)}} + \epsilon,$$ (31)

where ϵ includes terms of the form $(\lambda_2/\lambda_1)^{2p}, \ldots$, and will ultimately be negligible. Now writing $y_i^{(p+1)}/y_i^{(p)} = \mu_p$, our current estimate of the required latent root, we see from (31) that

$$\mu_p - \lambda_1 = \left(\frac{\lambda_2}{\lambda_1}\right)^p (\lambda_2 - \lambda_1) \frac{x_i^{(2)}}{x_i^{(1)}}, \quad \frac{\mu_{p+1} - \lambda_1}{\mu_p - \lambda_1} = \left(\frac{\lambda_2}{\lambda_1}\right) = \frac{\mu_p - \lambda_1}{\mu_{p-1} - \lambda_1},$$ (32)

and the usual Aitken formula follows for a better estimate of λ_1.

12. A similar process can improve the approximate latent vector whose current estimate is $y^{(p)}$. For at the stage when $(\lambda_r/\lambda_1)^p$ is negligible for $r > 2$ we have

$$y^{(p)} = x^{(1)} + \epsilon x^{(2)},$$ (33)

where the $x^{(r)}$ are suitably normalized, and we normalize $y^{(p)}$ to form $z^{(p)}$ whose largest component, say the kth, is unity. We assume also that this position is maintained in subsequent iterations, and consider the resulting vectors

$$z^{(p)} = \frac{x^{(1)} + \epsilon x^{(2)}}{x_k^{(1)} + \epsilon x_k^{(2)}}, \quad z^{(p+1)} = \frac{\lambda_1 x^{(1)} + \epsilon \lambda_2 x^{(2)}}{\lambda_1 x_k^{(1)} + \epsilon \lambda_2 x_k^{(2)}}, \quad z^{(p+2)} = \frac{\lambda_1^2 x^{(1)} + \epsilon \lambda_2^2 x^{(2)}}{\lambda_1^2 x_k^{(1)} + \epsilon \lambda_2^2 x_k^{(2)}}.$$ (34)

For any component of these vectors, say the ith, the Aitken formula gives, after a little manipulation, the result

$$\frac{z_i^{(p)}z_i^{(p+2)}-z_i^{(p+1)^2}}{z_i^{(p)}-2z_i^{(p+1)}+z_i^{(p+2)}} = \frac{\lambda_1^2 x_i^{(1)}x_k^{(1)}-\epsilon^2\lambda_2^2 x_i^{(2)}x_k^{(2)}}{\lambda_1^2 x_k^{(1)^2}-\epsilon^2\lambda_2^2 x_k^{(2)^2}}. \tag{35}$$

This is $x_i^{(1)}/x_k^{(1)}+0(\epsilon^2)$, so that the Aitken formula has reduced the error in the vector from $0(\epsilon)$ to $0(\epsilon^2)$.

As in previous applications of the Aitken formula we have a natural uncertainty about the stage at which it is best applied. Again, therefore, we regard the device as a method for accelerating convergence, for example by producing a better approximate vector (35) for subsequent use with the standard iterative method, rather than as the final 'clearing-up' stage of our process.

Other roots and vectors. Inverse iteration

13. By the methods outlined we can find, for real roots, the two extremes of the spectrum, no other roots being calculable easily by direct iteration. There are various possibilities for modifying the basic method, but they may need frequent interruptions in the process of computation.

For example if we have one vector $x^{(1)}$ and corresponding root λ_1 of largest modulus, we can repeat the iteration with a starting approximation $y^{(0)}$ which contains no component of $x^{(1)}$. In the symmetric case we can remove the component $x^{(1)}$ in the formula

$$y^{(0)} = z^{(0)}+\alpha_1 x^{(1)}, \tag{36}$$

where $z^{(0)}$ is arbitrary, by making $y^{(0)}$ orthogonal to $x^{(1)}$, so that

$$\alpha_1 = -x^{(1)'}z^{(0)}/x^{(1)'}x^{(1)}, \tag{37}$$

and in the unsymmetric case the use of orthogonality properties for this purpose needs a knowledge of the corresponding vectors of the transposed matrix.

Or we could annihilate $x^{(1)}$ with the formula

$$y^{(0)} = (A-I\lambda_1)z^{(0)}, \tag{38}$$

since $Ax^{(1)} = \lambda_1 x^{(1)}$ whether A is symmetric or not. Unfortunately, with either method, both the error in λ_1 and subsequent rounding errors will reintroduce a multiple of $x^{(1)}$, and from time to time we shall have to repeat the processes of orthogonalisation or annihilation. In fact (38) does not even eliminate $x^{(1)}$ temporarily unless λ_1 and $x^{(1)}$ are known exactly.

A second possibility, for computing the root of second largest modulus and corresponding vector, is to carry out the basic iteration to the point at which the effect of λ_2 is still present whereas the contribution from other roots and vectors has become negligible. For then, having ultimately computed λ_1 and $x^{(1)}$, we can extract their contributions from successive iterates and determine at least an approximation for λ_2 and $x^{(2)}$ from equations (32). The success of this method depends on the dominance of λ_2 over $\lambda_3, \lambda_4, \ldots$, and is arithmetically tedious unless the roots are real. Moreover there is a natural loss of significant figures and consequent poor precision.

†14. It is generally preferable to compute sub-dominant roots either from a different matrix, obtained by some form of 'deflation' which is discussed below in § 19 *et seq.*, or to use a process of *inverse iteration* with which we can converge to any desired root. This method is best reserved for real roots, since we cannot avoid much complex arithmetic if the roots are complex.

We consider first the iteration defined by

$$Ay^{(p)} = y^{(p-1)}, \qquad y^{(0)} \text{ arbitrary.} \qquad (39)$$

Since the roots of the matrix A^{-1} are the reciprocals of those of A, and its vectors are the same as those of A, it follows that

$$y^{(p)} = A^{-1}y^{(p-1)} = \ldots = (A^{-1})^p y^{(0)} = \sum_{r=1}^{n} \alpha_r \lambda_r^{-p} x^{(r)}, \qquad (40)$$

and we shall converge to the root of smallest modulus of A and the corresponding vector.

But we can converge to any root, for example the root nearest to the number k, by iterating inversely with $(A-Ik)$. For then

$$y^{(p)} = (A-Ik)^{-1}y^{(p-1)} = \ldots = \{(A-Ik)^{-1}\}^p y^{(0)} = \sum_{r=1}^{n} \alpha_r (\lambda_r-k)^{-p} x^{(r)}, \qquad (41)$$

and the dominant term on the right corresponds ultimately to the smallest value λ_r-k. This quantity is ultimately the ratio of components of the vector $y^{(p)}$ to those of $y^{(p+1)}$.

15. In practice it is faster not to compute the inverse of $(A-Ik)$, but to solve successively systems of linear equations, with the same matrix on the left but with different right-hand sides. For example we seek the root nearest to -2 of the matrix in (8), and the corresponding vector. We start with $y^{(0)\prime} = (1, 1, 1)$ and proceed to solve

by Gauss elimination the equations $(A+2I)y^{(p)} = y^{(p-1)}$. The elimination is here particularly simple, a single step producing the upper triangular form. In Table 3 we show the elimination with the first few successive right-hand sides.

TABLE 3

m	$A+2I$			$y^{(0)}$	$y^{(1)}$	$y^{(2)}$	$y^{(3)}$	$y^{(4)}$	$y^{(5)}$
	3	1	3	1	1	$-\frac{3}{2}$	$\frac{13}{2}$	$-\frac{39}{2}$	$\frac{263}{4}$
$-\frac{1}{3}$	1	0	1	1	-2	7	$-\frac{45}{2}$	74	$-\frac{483}{2}$
-1	3	1	5	1	0	$-\frac{1}{2}$	$\frac{1}{2}$	-3	$\frac{33}{4}$
	U								
	3	1	3	1	1	$-\frac{3}{2}$	$\frac{13}{2}$	$-\frac{39}{2}$	
		$-\frac{1}{3}$	0	$\frac{2}{3}$	$-\frac{7}{3}$	$\frac{15}{2}$	$-\frac{74}{3}$	$\frac{161}{2}$	
			2	0	-1	1	-6	$\frac{33}{2}$	

The ratios of $y^{(3)}$ to $y^{(4)}$, and $y^{(4)}$ to $y^{(5)}$, are respectively -0.33, -0.30, -0.17 and -0.30, -0.31, -0.36, so that we are beginning to converge well, and the required root is given by $\lambda+2 = -0.3$ approximately, so that $\lambda \doteq -2.3$. The corresponding vectors $y^{(4)}$ and $y^{(5)}$, normalized to have their largest components unity, are respectively $(-0.26, 1.00, -0.04)$ and $(-0.27, 1.00, -0.03)$, and we have the corresponding vector correct to about two decimals.

In practice, of course, we work with rounded decimal or binary numbers, and it is also desirable, at least in fixed-point arithmetic, to normalize each vector so that its largest component is unity before proceeding with the iteration. The reciprocal of the largest component is then the current estimate of $\lambda-k$.

16. Convergence is here fairly fast, the nearest interfering root being $\lambda = -1.12$ so that $\lambda-k$ is about 0.88, and convergence depends on the rate at which $|0.3/0.88|^r$ tends to zero. With a value of k nearer to -2.3 we should have a quite rapid rate of convergence.

If the chosen k is exactly equal to a latent root the matrix $A-kI$ is singular, and there is an apparent difficulty in solving the linear equations. If k is very close to a root the matrix is nearly singular, its inverse has large elements and the problem is apparently very ill-conditioned.

In fact, however, these situations cause no trouble. Consider the matrix

$$A = \begin{bmatrix} 5 & -1 & -2 \\ -1 & 3 & -2 \\ -2 & -2 & 5 \end{bmatrix}, \tag{42}$$

one of whose roots is exactly $\lambda = 5$. If we use the matrix $A - 5I$ in inverse iteration, starting with $y^{(0)'} = (1, 1, 1)$, we carry out the elimination as shown in Table 4.

TABLE 4

m		$A-5I$		$y^{(0)}$
0	0	−1	−2	1
−½	−1	−2	−2	1
	−2	−2	0	1
	0	−1	−2	1
−1	0	−1	−2	½
	−2	−2	0	1
	0	−1	−2	1
	0	0	0	−½
	−2	−2	0	1

The permuted triangular set of equations is given by the array

$$-2 \quad -2 \quad 0 \quad 1$$
$$-1 \quad -2 \quad 1 , \qquad (43)$$
$$0 \quad -\tfrac{1}{2}$$

and the matrix is singular. We cannot perform the back-substitution exactly, but if we replace the zero pivot by the element ϵ we can find

$$x_3 = -\frac{1}{2\epsilon}, \quad x_2 = -\left(1 - \frac{1}{\epsilon}\right), \quad x_1 = -\tfrac{1}{2}\left\{1 - 2\left(1 - \frac{1}{\epsilon}\right)\right\}, \quad (44)$$

and as $\epsilon \to 0$ we have the normalized vector $x_3 = -\tfrac{1}{2}$, $x_2 = +1$, $x_1 = -1$, obtained by multiplying by ϵ and letting $\epsilon \to 0$. This is the exact latent vector corresponding to the latent root $\lambda = 5$, and we have achieved the result in one step of the iteration.

17. In the computing machine we cannot easily perform this limiting process, but we get effectively the same result by taking ϵ to be any very small number, such as 10^{-t} in a t-digit decimal machine. In practice, moreover, this pivot is very unlikely to be zero since both our estimate of k and the solving procedure involve small errors. The important point is that all the components of the solution, though possibly very inaccurate as solutions of the linear equations, have errors in the same ratio, and the normalized vector is very accurate. This we can show rigorously by considering the error analysis of Chapter 6 for the solution of linear equations by Gauss elimination.

To keep numbers within range we normalize each successive vector in the iteration so that its largest component is unity. We then examine the solution of the iteration equation

$$(A - kI)z^{(r+1)} = y^{(r)},\tag{45}$$

where $y^{(r)}$ is of order unity, and we measure the accuracy of the resulting $z^{(r+1)}$ by the smallness of the residual vector

$$r = p^{-1}\{(A - kI)z^{(r+1)} - y^{(r)}\},\tag{46}$$

where p is the largest component of $z^{(r+1)}$. In §§ 32–35 of Chapter 6 we concluded that the Gauss elimination method produces the exact solution of the equations

$$(A - kI + E)z^{(r+1)} = y^{(r)} + \delta b,\tag{47}$$

where $|E_{rs}| < \frac{1}{2}(n-1)10^{-t}, \qquad |\delta b_r| < \frac{1}{2}n\,10^{s-t} + \frac{1}{2}(n-1)10^{-t},\tag{48}$

and t is the working accuracy, s the exponent of the largest element of the solution so that $p \doteqdot 10^s$.

Now if k is close to a latent root s will be large, but for the residual we have

$$\begin{rcases}|r| < 10^{-s}\{|\delta b| + |E|\,|z^{(r+1)}|\} \\ |r|_{\max} < \frac{1}{2}n10^{-t} + \frac{1}{2}(n-1)10^{-s-t} + \frac{1}{2}n(n-1)10^{-t}\end{rcases},\tag{49}$$

and this is clearly small if s is large. With a small residual the solution is good unless the latent vector problem is inherently ill-conditioned.

18. We also note that if $y^{(r)}$ is a linear combination $\sum\limits_{r=1}^{n} \alpha_r x^{(r)}$ of latent vectors of A, then the solution of (45) is expressible in the form

$$z^{(r+1)} = \sum_{r=1}^{n} \frac{\alpha_r x^{(r)}}{\lambda_r - k},\tag{50}$$

a statement easily verified by premultiplying (50) with $(A - Ik)$ and equating $Ax^{(r)}$ with $\lambda_r x^{(r)}$. If then k is near to a particular root λ_s, so that $\lambda_s - k = \epsilon$, (50) gives

$$z^{(r+1)} = \frac{\alpha_s x^{(s)}}{\epsilon} + \sum_{r \neq s} \frac{\alpha_r x^{(r)}}{\lambda_r - \lambda_s + \epsilon}.\tag{51}$$

If ϵ is small $z^{(r+1)}$ is dominated by the term $\alpha_s x^{(s)}$ and we have the required latent vector. Exceptions to this statement occur if α_s is also very small, that is if $y^{(r)}$ is deficient in the required latent vector, and if any $\lambda_r - \lambda_s$ is also very small, so that at least one other denominator in (51) is of order ϵ.

The first exception will lose its force in successive iterations, as rounding errors introduce multiples of $x^{(s)}$, but the nearness of λ_r to λ_s is significant, and is the major factor governing the ill-conditioning of the latent vector problem. We return to the first point, which is significant in certain other processes for finding roots and vectors, in §§ 24–5 of Chapter 10, and to the second in Chapter 11.

Matrix deflation

†19. Although we can converge to any root with the inverse iteration we may 'miss' a root by a poor choice of the number k. A common problem, for example, is the determination of a few of the roots of largest modulus, and for the unsymmetric case in particular we generally prefer direct iteration to produce the root or roots of largest modulus, followed by similar treatment of a modified matrix which automatically converges to the next smaller root or roots.

We consider two main methods. The first, for symmetric matrices, modifies the matrix after the determination of each root and vector but leaves it symmetric and of the same order. The second method, for any matrix, reduces the order of the matrix but destroys any symmetry which the original may have enjoyed.

20. Suppose that we have computed any root λ_1 and vector $x^{(1)}$ for the symmetric matrix A. If we had used direct iteration λ_1 would be the root of largest modulus. We also normalize the vector so that $x^{(1)'}x^{(1)} = 1$, and consider the matrix

$$B = (A - \lambda_1 x^{(1)} x^{(1)'}), \tag{52}$$

asserting that B has roots $0, \lambda_2, \lambda_3, \ldots, \lambda_n$ and corresponding vectors $x^{(1)}, x^{(2)}, \ldots, x^{(n)}$. This is easily verified by substitution in the equation

$$(A - \lambda_1 x^{(1)} x^{(1)'} - I\mu)y = o. \tag{53}$$

For if y is $x^{(1)}$ the left of (53) becomes

$$Ax^{(1)} - \lambda_1 x^{(1)}(x^{(1)'}x^{(1)}) - I\mu x^{(1)}, \tag{54}$$

and the first two terms cancel giving $\mu = 0$, $y = x^{(1)}$ as one solution of (53). If y is any other latent vector $x^{(r)}$ of A, then the left of (53) becomes

$$Ax^{(r)} - \lambda_1 x^{(1)}(x^{(1)'}x^{(r)}) - I\mu x^{(r)}, \tag{55}$$

the middle term vanishes in view of the orthogonal properties of latent vectors, and (55) then vanishes only if $\mu = \lambda_r$.

The matrix B is clearly symmetric, and we now work with B to produce directly another latent solution. Consider, for example, the matrix of (42), and assume that we have already found the latent root $\lambda = 5$ and the corresponding normalized vector $x' = (-\frac{2}{3}, \frac{2}{3}, -\frac{1}{3})$. (This is the 'middle' solution, not to be found by direct iteration, but its use simplifies the arithmetic.) We find

$$B = \begin{bmatrix} 5 & -1 & -2 \\ -1 & 3 & -2 \\ -2 & -2 & 5 \end{bmatrix} - 5 \begin{bmatrix} -\frac{2}{3} \\ \frac{2}{3} \\ -\frac{1}{3} \end{bmatrix} [-\frac{2}{3} \quad \frac{2}{3} \quad -\frac{1}{3}] = \frac{1}{9} \begin{bmatrix} 25 & 11 & -28 \\ 11 & 7 & -8 \\ -28 & -8 & 40 \end{bmatrix}, \quad (56)$$

and we can confirm our predictions about its roots and vectors.

21. In the second method, applicable to both symmetric and unsymmetric matrices, we suppose that we have computed λ_1 and $x^{(1)}$, and normalized the vector so that its largest component is unity. The matrix analysis is simplified if we assume that this is the first component, the general case becoming obvious after this analysis. The vector $x^{(1)}$ then has its first component unity, and we suppose the remaining components to form the $(n-1)$ vector $\xi^{(1)}$. All the other vectors are also normalized in this way, $x^{(r)}$ having unity in its first component and the remainder forming the vector $\xi^{(r)}$.

We then consider the matrix A to be partitioned in a conformable way, so that with obvious notation our problem is expressible by the equation

$$\begin{bmatrix} a_{11} & R_1' \\ \hline C_1 & B \end{bmatrix} \begin{bmatrix} 1 \\ \hline \xi \end{bmatrix} = \lambda \begin{bmatrix} 1 \\ \hline \xi \end{bmatrix}, \quad (57)$$

of which we have one solution $\lambda = \lambda_1, \xi = \xi^{(1)}$. Carrying out the matrix operations in (57) we find the two equations

$$a_{11} + R_1'\xi = \lambda, \qquad C_1 + B\xi = \lambda\xi. \quad (58)$$

Elimination of a_{11} and C_1, from (58) and the corresponding equations for λ_1 and $\xi^{(1)}$, gives

$$R_1'(\xi - \xi^{(1)}) = \lambda - \lambda_1, \qquad B(\xi - \xi^{(1)}) = \lambda\xi - \lambda_1\xi^{(1)}. \quad (59)$$

Premultiplication of the first of (59) by $\xi^{(1)}$, and subtraction from the second of (59), then produces the desired result

$$(B - \xi^{(1)}R_1')(\xi - \xi^{(1)}) = \lambda(\xi - \xi^{(1)}). \quad (60)$$

This equation says that the latent roots of the matrix

$$A_1 = (B - \xi^{(1)}R_1'),$$

of order one less than $A = A_0$, are λ_r, and its unnormalized latent vectors $y^{(r)}$ are proportional to $\xi^{(r)} - \xi^{(1)}$, for $r = 2, 3, ..., n$. Then $ky^{(r)} = \xi^{(r)} - \xi^{(1)}$, and we must choose k so that the vector $x^{(r)'} = [1 \mid \xi^{(r)'}]$ really is a latent vector of A. The constant comes from the first of (59), for if $\xi^{(r)} - \xi^{(1)} = ky^{(r)}$, then

$$k = (\lambda_r - \lambda_1)/R_1'y^{(r)}, \tag{61}$$

and the required latent vector of A is

$$x^{(r)} = \begin{bmatrix} 1 \\ \xi^{(r)} \end{bmatrix}, \quad \xi^{(r)} = \xi^{(1)} + \left(\frac{\lambda_r - \lambda_1}{R_1'y^{(r)}} \right) y^{(r)}. \tag{62}$$

22. The computation proceeds as follows. First we produce, presumably by direct iteration, a root and vector of A, then compute a new matrix $A_1 = B - \xi^{(1)}R_1'$ and find one of its roots and vectors. Continuing in this way we produce finally the matrix A_{n-1}, consisting of a single element, so that we know its root and vector. We then have all the roots of $A = A_0$, and one vector of each of the matrices $A_0, A_1, ..., A_{n-1}$. The vector of A_1 is built up into a vector of A_0 by the method of (62), and for the other matrices A_r the building-up process has r stages, through $A_{r-1}, A_{r-2}, ...,$ up to A_0.

This is rather similar to a process of back-substitution in the solution of linear equations, and the production of the reduced matrices $A_1, A_2, ...,$ involving the subtraction of multiples of the first row of the matrix from other rows, is reminiscent of the elimination part of the solution of linear equations. In the analysis we chose the first row for this purpose, but in practice we normalize the vector so that its largest component, say the ith, is unity, and the 'pivotal' row and column are respectively the ith row and column of the matrix, the single element in (57) being a_{ii}.

23. Consider for example the matrix of (42), and assume that we have computed the root $\lambda = 5$, with vector $x' = (-1, 1, -0.5)$. (Again this is not the largest root but this fact does not affect the rest of the computation.) We take the middle element to be the pivotal element, so that in the notation corresponding to (57) we have

$$B = \begin{bmatrix} 5 & -2 \\ -2 & 5 \end{bmatrix}, \quad R_1' = (-1, \quad -2), \quad C_1 = \begin{bmatrix} -1 \\ -2 \end{bmatrix}, \quad a_{11} = 3. \tag{63}$$

Also $\xi^{(1)'} = (-1, -0.5)$, and is lacking its *middle* component in relation to the corresponding vector of A. The reduced matrix

$A_1 = B - \xi^{(1)} R_1'$ is

$$A_1 = \begin{bmatrix} 4 & -4 \\ -2 \cdot 5 & 4 \end{bmatrix}, \tag{64}$$

and direct iteration gives a root of about $7 \cdot 16$, with vector $[1, -0 \cdot 79]$. Then

$$A_2 = [4] - [-0 \cdot 79][-4] = [0 \cdot 84], \tag{65}$$

so that the single root of A_2 is $0 \cdot 84$, and its vector has the single component unity.

We now have all the roots of A, and one vector, and we proceed to build up the vectors of A_1 and A_2 into vectors of A. From that of A_1 we see that the corresponding vector of A has unity for its second element, and the other elements are those of

$$\xi^{(1)} + (\lambda_r - \lambda_1) y^{(r)} / R_1' y^{(r)},$$

where

$\lambda_r = 7 \cdot 16, \lambda_1 = 5, y^{(r)'} = (1, -0 \cdot 79),$

$$R_1' = (-1, -2), \xi^{(1)'} = (-1, -0 \cdot 5). \tag{66}$$

Computation gives the approximate resulting vector with components $(2 \cdot 72, 1 \cdot 00, -3 \cdot 44)$, corresponding to the root $7 \cdot 16$.

For the remaining vector, corresponding to the root $0 \cdot 84$, we first build up the vector of A_2 into the other vector of A_1. This will have unity in its first component, and the other component is obtained as before with $\xi^{(1)} = -0 \cdot 79, \lambda_r = 0 \cdot 84, \lambda_1 = 7 \cdot 16, y^{(r)'} = 1, R_1' = -4$, giving the vector $(1, 0 \cdot 79)$. Finally the build-up of this vector into the remaining vector of A uses (62) with $\lambda_r = 0 \cdot 84, \lambda_1 = 5, y^{(r)'} = (1, 0 \cdot 79)$, $R_1' = (-1, -2), \xi^{(1)'} = (-1, -0 \cdot 5)$, and we have to insert a unit element in the middle position. We find the vector $(0 \cdot 61, 1 \cdot 00, 0 \cdot 77)$, and this is in fact correct to two decimal places.

24. It is perhaps a little difficult to keep track of the position of the unit component in the various vectors, and in building up from a vector y of A_k to the corresponding vector of A_{k-1} it probably pays to extend y, of length $n-k$, to 'full length' $n-k+1$ by adding a zero in that component which will ultimately be unity. The row R_1' is then regarded as the complete row of A_{k-1} and the scalar product $R_1' y$ is of course unaffected.

This suggests another partitioning of A, in the form

$$A = A_0 = \left[\frac{R_1'}{C} \right], \tag{67}$$

where R_1' is now a complete row of A, here assumed to be the first for simplicity of analysis, and C the remaining matrix of shape $(n-1, n)$. Again we assume that the vectors $x^{(r)}$ are normalized to have one of their elements unity, and here this element is the first, corresponding to our choice of R_1'. We then consider the matrix

$$A_1 = A_0 - x^{(1)} R_1', \tag{68}$$

corresponding to the previous calculation of one root λ_1 and vector $x^{(1)}$.

It is easy to verify that A_1 has roots $0, \lambda_2, \ldots, \lambda_n$, and corresponding vectors $x^{(1)}, x^{(1)} - x^{(2)}, \ldots, x^{(1)} - x^{(n)}$. For

$$
\begin{aligned}
A_1(x^{(1)} - x^{(r)}) &= A_0(x^{(1)} - x^{(r)}) - x^{(1)} R_1'(x^{(1)} - x^{(r)}) \\
&= \lambda_1 x^{(1)} - \lambda_r x^{(r)} - x^{(1)}(\lambda_1 - \lambda_r) \\
&= \lambda_r(x^{(1)} - x^{(r)}),
\end{aligned} \tag{69}
$$

the substitutions $R_1' x^{(1)} = \lambda_1$, $R_1' x^{(r)} = \lambda_r$ following from the first algebraic equation of the set $Ax = \lambda x$ and the fact that the first component of x is unity. Moreover

$$A_1 x^{(1)} = A_0 x^{(1)} - x^{(1)} R_1' x^{(1)} = \lambda_1 x^{(1)} - \lambda_1 x^{(1)} = 0, \tag{70}$$

verifying that $x^{(1)}$ is a vector of A_1, with latent root zero.

The first row of the matrix A_1 consists of zeros, each latent vector of A_1 has a zero in its first component, and the matrix obtained from A_1 by omitting its first row and column is the matrix of that name in the previous analysis.

Connexion with similarity transformation

†25. It is also interesting to see that we can express the theory in the form of a similarity transformation. Returning to the partitioning in (57), with the same notation, we consider the transformation $S^{-1}AS$, where

$$
S = \left[\begin{array}{c|c} 1 & o' \\ \hline \xi^{(1)} & I \end{array} \right], \quad
S^{-1} = \left[\begin{array}{c|c} 1 & o' \\ \hline -\xi^{(1)} & I \end{array} \right]. \tag{71}
$$

This matrix S is of exactly the same form as the matrix J_1 of § 12, Chapter 3, which was used in the Gauss elimination process and for which we noted the simple form of its inverse. We find, with use of equations (58), that

$$
S^{-1}AS = \left[\begin{array}{c|c} \lambda_1 & R_1' \\ \hline o & B - \xi^{(1)} R_1' \end{array} \right] = \bar{A}, \text{ say}, \tag{72}
$$

and this matrix has the same latent roots as A. Moreover if z is a vector of $\bar{A}$, then Sz is a corresponding vector of A.

Now z' has the form $[1 \mid ky']$, where k is a constant and y is a vector of $B-\xi^{(1)}R_1'$, the matrix A_1 of § 21. Then

$$\begin{bmatrix} \lambda_1 & R_1' \\ \hline o & A_1 \end{bmatrix} \begin{bmatrix} 1 \\ \hline ky \end{bmatrix} = \lambda \begin{bmatrix} 1 \\ \hline ky \end{bmatrix}, \tag{73}$$

so that
$$\lambda_1 + kR_1'y = \lambda, \qquad k = (\lambda-\lambda_1)/R_1'y. \tag{74}$$

Finally the vector of A is

$$\begin{bmatrix} 1 & o' \\ \hline \xi^{(1)} & I \end{bmatrix} \begin{bmatrix} 1 \\ \hline ky \end{bmatrix} = \begin{bmatrix} 1 \\ \hline \xi^{(1)}+ky \end{bmatrix} = \begin{bmatrix} 1 \\ \hline \xi^{(1)}+\left(\dfrac{\lambda-\lambda_1}{R_1'y}\right)y \end{bmatrix}, \tag{75}$$

agreeing with the previous result (62).

Additional notes and bibliography

§ 2. The ill-conditioned nature of the determination of the roots of polynomials is discussed in

WILKINSON, J. H., 1959, The evaluation of the zeros of ill-conditioned polynomials, Parts I and II. *Numerische Mathematik* **1**, 150–180.

He shows, for example, that some of the roots of the polynomial

$$(x+1)(x+2)...(x+20)+2^{-23}x^{19}$$

differ considerably from those of the unperturbed polynomial. For example the root of largest modulus is changed from -20 to $-20\cdot8469...$, and the root -17 is changed to one of the complex pair $-16\cdot73... \pm i2\cdot81...$.

The *distribution* of the roots is apparently an important factor, and for some distributions the perturbations have much smaller effect.

§ 9. Another process, which in its simplest form can be made to converge to the largest or smallest root of a symmetric matrix, is that of

HESTENES, M. R. and KARUSH, W., 1951, A method of gradients for the calculation of latent roots and vectors of a real symmetric matrix. *J. Res. Nat. Bur. Stand.* **47**, 45–61.

The method is based on the fact that the Rayleigh quotient

$$\mu(y) = \frac{y'Ay}{y'y}$$

has maximum and minimum values when y is a latent vector of A. The gradient of $\mu(y)$ is $Ay-\mu y$, so that it is reasonable to consider the iteration

$$y^{(r+1)} = y^{(r)} \pm \alpha_r(Ay^{(r)}-\mu_r y^{(r)}),$$

similar to that of our gradient methods for linear equations. If α_r is taken to be constant, and less than $2/(\lambda_1 - \lambda_n)$, where $\lambda_1 > \lambda_2 > \ldots > \lambda_n$, then with the positive sign we converge to the largest root and corresponding vector and with the negative sign we converge to the smallest root and vector.

Wilkinson (in the next reference) has shown that the best α gives a rate of convergence which is ultimately the same as that for the optimum q in equations (26) and (28).

§ 14. In the more general case $(A - \lambda B)x = o$ we can apply either direct or inverse iteration without explicit reduction to any simple form, possibilities for which were mentioned in the notes on Chapter 2. Direct iteration involves the recurrence

$$By^{(p)} = Ay^{(p-1)},$$

and inverse iteration the recurrence

$$Ay^{(p)} = By^{(p-1)}.$$

Both processes involve the solution of linear equations at each step, but this is preferably performed without inverting the relevant matrix. For convergence to any particular root the inverse iteration becomes

$$(A - kB)y^{(p)} = By^{(p-1)},$$

and again we need not invert a matrix.

§§ 19–24. These and other methods of direct iteration and matrix deflation have been discussed in some detail in

WILKINSON, J. H., 1954, The calculation of the latent roots and vectors of matrices on the pilot model of the ACE. *Proc. Camb. Phil. Soc.* **50**, 536–566.

In particular he analyses the deflation process of § 21 in the case when λ_1 and $\xi^{(1)}$, the first root and vector obtained by iteration, are complex, so that we also know another solution $\bar{\lambda}_1$ and $\bar{\xi}^{(1)}$ and can therefore carry out a double deflation. If $\xi^{(1)}$ is $\alpha^{(1)} + i\beta^{(1)}$, $\bar{\xi}^{(1)} = \alpha^{(1)} - i\beta^{(1)}$, where $\alpha^{(1)}$ and $\beta^{(1)}$ are real vectors, we normalize so that the component of maximum modulus, say the rth, is unity. We find the first reduced matrix A_1 in the usual way from the formula

$$A_1 = B - (\alpha^{(1)} + i\beta^{(1)})R_r',$$

and one vector of A_1 is proportional to $(\alpha^{(1)} + i\beta^{(1)}) - (\alpha^{(1)} - i\beta^{(1)})$, which is the real vector $\beta^{(1)}$ with its rth component equal to zero. This can be removed immediately to give A_2.

Now if $a_{ij}^{(p)}$ denotes the (i, j) element of A_p, where a_{ij} is the corresponding element of $A_0 = A$, we find

$$a_{ij}^{(1)} = a_{ij} - (\alpha_i^{(1)} + i\beta_i^{(1)})a_{rj},$$

and

$$a_{ij}^{(2)} = a_{ij}^{(1)} - \frac{\beta_i^{(1)}}{\beta_s^{(1)}} a_{sj}^{(1)},$$

where s is the position of the largest element of $\beta^{(1)}$. From these equations we deduce

$$a_{ij}^{(2)} = (a_{ij} - \alpha_i^{(1)}a_{rj}) - \frac{\beta_i^{(1)}}{\beta_s^{(1)}} (a_{sj} - \alpha_s^{(1)}a_{rj}),$$

so that the matrix A_2 can be obtained by removing from A_0 the *real* vector $\alpha^{(1)}$, ignoring the imaginary part $i\beta^{(1)}$, and then removing from the matrix A_1^* so obtained the *real* vector $\beta^{(1)}$.

In the 'back-substitution' process, however, for building-up from a vector of A_2 to a vector of A, we must remember that the corresponding 'pivotal row' of the true intermediate matrix A_1 is not that of our 'bogus' intermediate A_1^*, and we must either use a further analysis of Wilkinson or perform some complex arithmetic with the pivotal row of the true intermediate A_1.

For example the matrix
$$A = \begin{bmatrix} 8 & -1 & -5 \\ -4 & 4 & -2 \\ 18 & -5 & -7 \end{bmatrix}$$

has complex roots $2 \pm 4i$ and corresponding vectors $(0\cdot5 \pm 0\cdot5i, \pm i, 1)$. Removing the vector with the positive sign we find the true first reduced matrix

$$A_1 = \begin{bmatrix} 8 & -1 \\ -4 & 4 \end{bmatrix} - \begin{bmatrix} 0\cdot5+0\cdot5i \\ i \end{bmatrix} [18 \quad -5] = \begin{bmatrix} -1-9i & 1\cdot5+2\cdot5i \\ -4-18i & 4+5i \end{bmatrix},$$

and this we know has a root $2-4i$ with vector $[0\cdot5, 1]$, lacking its last component in relation to the corresponding vector of A. Removal of this vector gives

$$A_2 = [-1-9i] - [0\cdot5][-4-18i] = [1],$$

which has the root 1 and vector $[1]$, lacking its last two components in relation to a vector of A.

The corresponding vector of A_1 has its second component unity, and its first component is

$$0\cdot5 + \frac{1-(2-4i)}{[-4-18i][1]} = \frac{3+5i}{4+18i}.$$

The corresponding vector of A has its third component unity, and its first two are

$$[0\cdot5+0\cdot5i, \, i] + \frac{1-(2+4i)}{[18, \, -5] \begin{bmatrix} \dfrac{3+5i}{4+18i} \\ 1 \end{bmatrix}} \begin{bmatrix} \dfrac{3+5i}{4+18i}, \, 1 \end{bmatrix} = [1, \, 2],$$

and the vector $[1, 2, 1]$ is the third vector of A with root $\lambda_3 = 1$.

The 'bogus' reduced matrix A_1^* is

$$A_1^* = \begin{bmatrix} 8 & -1 \\ -4 & 4 \end{bmatrix} - \begin{bmatrix} 0\cdot5 \\ 0 \end{bmatrix} [18 \quad -5] = \begin{bmatrix} -1 & 1\cdot5 \\ -4 & 4 \end{bmatrix},$$

and the consequent removal of the vector $[0\cdot5, 1]$ gives A_2 as before.

The 'further analysis' of Wilkinson avoids the calculation of the intermediate vector of the true A_1. He shows that if $\alpha \pm i\beta$ are vectors of A, with roots $\mu \pm i\nu$, and if the rth element is the one of maximum modulus of $\alpha + i\beta$ and the sth element that of β, then the vector $x^{(3)}$ of A, corresponding to the vector $z^{(3)}$ of A_2 with root λ_3, is given by
$$x^{(3)} = \alpha + p\beta + qz^{(3)},$$

where p and q are constants. He shows that their values are obtained from the two simultaneous equations
$$\left. \begin{aligned} \lambda_3 &= \mu + p\nu + qR'z^{(3)} \\ \nu\beta_s &= p(\mu-\lambda_3)\beta_s + q(R^*)' \, z^{(3)} \end{aligned} \right\},$$

where R' is the pivotal row of A and $(R^*)'$ is the 'bogus' pivotal row of A_1^*.

Here we have

$$1 = 2 + 4p + q[18, -5]\begin{bmatrix} 1 \\ 0 \end{bmatrix}\Bigg\},$$

$$4(1) = p(2-1)(1) + q[-4][1]$$

giving $p = 2$, $q = -\frac{1}{2}$ and

$$x^{(3)\prime} = [0\cdot5, 0, 1] + 2[0\cdot5, 1, 0] - \tfrac{1}{2}[1, 0, 0] = [1, 2, 1].$$

§ 25. Various other methods of deflation, based on similarity transformations, are given in

FELLER, W. and FORSYTHE, G. E., 1951, New matrix transformations for obtaining characteristic vectors. *Quart. Appl. Math.* **VIII**, 325–331.

One of their transformations for a symmetric matrix both preserves symmetry and reduces the order by unity. If one root $x^{(1)}$ and vector λ_1 are known, we consider the vector $x^{(1)}$ to be normalized so that $x^{(1)\prime} x^{(1)} = 1$, and that it is partitioned into the form $\begin{bmatrix} \xi \\ x_n \end{bmatrix}$, where x_n is its last component and ξ the $n-1$ vector formed from the remaining components. In the appropriate similarity transformation $A = S^{-1}AS$ we take

$$S = \left[\begin{array}{c|c} I_{n-1} - \dfrac{1}{1+x_n}\,\xi\xi' & \xi \\ \hline -\xi' & x_n \end{array}\right],$$

and S^{-1} is the same with $-\xi$ replacing ξ and ξ' replacing $-\xi'$. This is easily proved with the remark that $\xi'\xi = 1 - x_n^2$. If we partition A in a conformable way,

$$A = \left[\begin{array}{c|c} B & c_n \\ \hline c'_n & a_{nn} \end{array}\right],$$

and perform the multiplications in $\bar{A} = S^{-1}AS$, we find, with judicious use of the formulae

$$B\xi + c_n x_n = \lambda_1 \xi, \qquad c'_n \xi + a_{nn} x_n = \lambda_1 x_n,$$

that $\bar{A}$ is given by

$$\bar{A} = \left[\begin{array}{c|c} \bar{B} & o \\ \hline o' & \lambda_1 \end{array}\right],$$

and the matrix $\bar{B}$, of order $n-1$, is symmetric with latent roots $\lambda_2, \dots, \lambda_n$. After some manipulation we find

$$\bar{B} = B - \frac{1}{1+x_n}\,(c_n\xi' + \xi c'_n) + \frac{a_{nn}-\lambda_1}{(1+x_n)^2}\,\xi\xi'.$$

For the example of (42), with

$$A = \begin{bmatrix} 5 & -1 & -2 \\ -1 & 3 & -2 \\ -2 & -2 & 5 \end{bmatrix},$$

we know a root $\lambda_1 = 5$ with normalized vector $x' = (\frac{2}{3}, -\frac{2}{3}, \frac{1}{3})$. Then

$$B = \begin{bmatrix} 5 & -1 \\ -1 & 3 \end{bmatrix}, \quad c_n = \begin{bmatrix} -2 \\ -2 \end{bmatrix}, \quad \xi = \frac{1}{3}\begin{bmatrix} 2 \\ -2 \end{bmatrix}, \quad x_n = \frac{1}{3}, \quad a_{nn} - \lambda_1 = 0,$$

and we find

$$B = \begin{bmatrix} 7 & -1 \\ -1 & 1 \end{bmatrix},$$

which may be compared with equation (56) for the previous transformation of a symmetric matrix.

10

Transformation Methods for Latent Roots and Vectors

Introduction

†1. WHEN ALL or many of the roots and vectors are needed it is generally preferable to use methods of a direct rather than of an iterative nature, which effectively reduce the matrix to a simple form by a similarity transformation which leaves the roots unchanged. In this chapter we consider various transformations, for both symmetric and unsymmetric matrices, based on the fact that if we can find a matrix Y, such that

$$Y^{-1}AY = B, \tag{1}$$

then A and B have the same latent roots, and if the vectors of A form the matrix X, and the roots form the diagonal matrix D, then $AX = XD$ and

$$B(Y^{-1}X) = (Y^{-1}X)D, \tag{2}$$

so that the modal matrix of B is $Y^{-1}X$.

In the symmetric case the matrix Y will generally be orthogonal, so that $Y^{-1} = Y'$, and is often obtained as the product of simple orthogonal matrices. The similarity transformation is then an orthogonal transformation. In the unsymmetric case Y may also be the final result of a sequence of elementary similarity transformations.

Method of Jacobi, symmetric matrices

2. We consider first the classical method of Jacobi, in which A is transformed into D by a convergent infinite sequence of successive transformations of the form

$$A_r = Y'_r A_{r-1} Y_r, \qquad A_0 = A, \tag{3}$$

in which each Y_r is an orthogonal matrix. The product of orthogonal matrices is itself an orthogonal matrix, for if $Y_r^{-1} = Y'_r$, then

$$(Y_1 Y_2 \dots Y_{r-1} Y_r)' = (Y'_r Y'_{r-1} \dots Y'_2 Y'_1)$$
$$= (Y_r^{-1} Y_{r-1}^{-1} \dots Y_2^{-1} Y_1^{-1}) = (Y_1 Y_2 \dots Y_{r-1} Y_r)^{-1}. \tag{4}$$

We have already met one simple orthogonal matrix, the 'row-permuting' matrix previously called I_r. Here we use an extension of the simple two-dimensional 'axis-rotating' matrix

$$P = \begin{bmatrix} \cos\theta & \sin\theta \\ -\sin\theta & \cos\theta \end{bmatrix}, \tag{5}$$

which clearly satisfies $PP' = I$. The corresponding $(n \times n)$ orthogonal matrix, analogous to a rotation of axes in the (p, q) plane, has most of the elements of the unit matrix, except that

$$P_{pp} = \cos\theta, \qquad P_{qq} = \cos\theta, \qquad P_{pq} = \sin\theta, \qquad P_{qp} = -\sin\theta, \tag{6}$$

and it has the appearance typified by

$$P = \begin{bmatrix} 1 & 0 & 0 & 0 & 0 \\ 0 & 1 & 0 & 0 & 0 \\ 0 & 0 & c & 0 & s \\ 0 & 0 & 0 & 1 & 0 \\ 0 & 0 & -s & 0 & c \end{bmatrix}, \tag{7}$$

where $c = \cos\theta$, $s = \sin\theta$, and in (7) we are rotating in the $(3, 5)$ plane of five-dimensional space.

3. If a typical matrix Y_r' in (3) has this form we can compute the matrix $Y_r'A_{r-1}Y_r$ and choose θ so that an off-diagonal element is zero, our ultimate aim being the annihilation of all off-diagonal elements. We note that $S_r = Y_r'A_{r-1}$ is the same as A_{r-1} in all rows except the pth and the qth, and here

$$\left. \begin{aligned} R_p(S_r) &= cR_p(A_{r-1}) + sR_q(A_{r-1}) \\ R_q(S_r) &= -sR_p(A_{r-1}) + cR_q(A_{r-1}) \end{aligned} \right\}. \tag{8}$$

The matrix A_r is then identical with S_r in all except the pth and qth columns, and here

$$\left. \begin{aligned} C_p(A_r) &= cC_p(S_r) + sC_q(S_r) \\ C_q(A_r) &= -sC_p(S_r) + cC_q(S_r) \end{aligned} \right\}. \tag{9}$$

We deduce some important properties of A_r. First its 'trace', the sum of its diagonal terms, is the same as that of A_{r-1}. This must be true since the trace is the sum of the latent roots, but we can also verify this algebraically. For if $a_{ij}^{(r-1)}$ is the (i, j) element of A_{r-1} we find

$$(S_r)_{pp} = ca_{pp}^{(r-1)} + sa_{qp}^{(r-1)}, \qquad (S_r)_{pq} = ca_{pq}^{(r-1)} + sa_{qq}^{(r-1)} \\ (S_r)_{qp} = -sa_{pp}^{(r-1)} + ca_{qp}^{(r-1)}, \qquad (S_r)_{qq} = -sa_{pq}^{(r-1)} + ca_{qq}^{(r-1)}$$

(10)

and then

$$a_{pp}^{(r)} = c(S_r)_{pp} + s(S_r)_{pq} = c^2 a_{pp}^{(r-1)} + 2cs a_{pq}^{(r-1)} + s^2 a_{qq}^{(r-1)} \\ a_{qq}^{(r)} = -s(S_r)_{qp} + c(S_r)_{qq} = s^2 a_{pp}^{(r-1)} - 2cs a_{pq}^{(r-1)} + c^2 a_{qq}^{(r-1)}$$

(11)

since $a_{pq}^{(r-1)} = a_{qp}^{(r-1)}$. The sum of these diagonal terms, the only ones differing from those of A_{r-1}, is $a_{pp}^{(r-1)} + a_{qq}^{(r-1)}$. Also

$$a_{pq}^{(r)} = -s(S_r)_{pp} + c(S_r)_{pq} = cs(a_{qq}^{(r-1)} - a_{pp}^{(r-1)}) + (c^2 - s^2) a_{pq}^{(r-1)} \\ a_{qp}^{(r)} = c(S_r)_{qp} + s(S_r)_{qq} = cs(a_{qq}^{(r-1)} - a_{pp}^{(r-1)}) + (c^2 - s^2) a_{pq}^{(r-1)}$$

(12)

and symmetry is maintained.

Now we can make the symmetric off-diagonal terms $a_{pq}^{(r)}$ and $a_{qp}^{(r)}$ vanish if we choose θ so that

$$\tan 2\theta = \frac{2cs}{c^2 - s^2} = \frac{2a_{pq}^{(r-1)}}{a_{pp}^{(r-1)} - a_{qq}^{(r-1)}} = \alpha,$$

(13)

and c and s can be evaluated separately from the formulae

$$c^2 = \tfrac{1}{2} + \frac{1}{2\sqrt{\alpha^2 + 1}}, \quad s^2 = \tfrac{1}{2} - \frac{1}{2\sqrt{\alpha^2 + 1}}.$$

(14)

†4. With this transformation the sum of squares of the elements of S_r is the same as that of A_{r-1}, for the only changes between A_{r-1} and S_r are in the rows p and q, and from (8) we see that the sum of squares of these elements of S_r is

$$(cR_p' + sR_q')(cR_p + sR_q) + (-sR_p' + cR_q')(-sR_p + cR_q),$$

(15)

where R refers to a row of A_{r-1}, and this sum is simply $R_p' R_p + R_q' R_q$. A similar argument shows that A_r has no further change in the sum of squares of its elements.

But the distribution of this quantity is different, the sum of squares of the diagonal elements of A_r being greater than that of A_{r-1} by the amount $2\{a_{pq}^{(r-1)}\}^2$. The only diagonal terms which are different are $a_{pp}^{(r)}$ and $a_{qq}^{(r)}$, defined in (11), and a little trivial manipulation gives the required result.

At each step we therefore 'transfer' to the diagonal some part of the contribution from the off-diagonal terms, and the process converges to a diagonal matrix. It would seem that the best rate of convergence is secured by choosing successive transformations so that the largest off-diagonal term a_{pq} is temporarily reduced to zero, though in practice the 'scanning' required in the computing machine occupies non-negligible time, and a cyclic order of rotation is generally preferred, in successive planes $(1, 2), (1, 3),..., (1, n); (2, 3),..., (2, n)$; etc, starting again at the beginning after each complete cycle. We note also that every (p, q) rotation changes all the elements in rows and columns p and q, so that an a_{pq} previously reduced to zero will become non-zero again as a result of any subsequent rotation in planes (p, s) or (s, q).

The sequence of rotations is therefore infinite, though convergent, and we would like to know when we can terminate the sequence for a given precision in the roots and vectors. This we shall discuss in Chapter 11. We note here that when this decision is made the required latent roots are the diagonal elements of the accepted A_s, and that the determination of the vectors of A is particularly simple, involving only successive accumulation of products $Y_1 Y_2 Y_3...$, in which each matrix multiplication changes only two of the columns.

Method of Givens, symmetric matrices

†5. In the method of Givens we carry out a finite sequence of orthogonal transformations with matrices of type (7), but reduce the matrix to triple-diagonal form by suitable choices of the angles θ. We therefore try to liquidate in succession, and in cyclic order, the elements of successive matrices A_r which lie outside the diagonal and the two sloping lines on each side of the diagonal, and in such a way that the zeros so obtained are unchanged in subsequent transformations.

We consider typically a matrix of order four, and rotate first in the $(2, 3)$ plane, for which

$$Y_1' = \begin{bmatrix} 1 & 0 & 0 & 0 \\ 0 & c & s & 0 \\ 0 & -s & c & 0 \\ 0 & 0 & 0 & 1 \end{bmatrix}. \tag{16}$$

In the matrix $A_1 = Y_1' A Y_1$ the $(1, 3)$ and $(3, 1)$ elements have the same value

$$-sa_{12} + ca_{13}, \tag{17}$$

and we make this vanish by choosing $\tan \theta = a_{13}/a_{12}$, that is

$$s = a_{13}(a_{12}^2 + a_{13}^2)^{-\frac{1}{2}}, \qquad c = a_{12}(a_{12}^2 + a_{13}^2)^{-\frac{1}{2}}, \tag{18}$$

and we note in passing that this is an easier computation than that of the Jacobi method.

†6. We then make successive rotations in the $(2, 4)$, $(2, 5)$,..., $(2, n)$ planes, choosing θ at each stage so that the new $(1, 4)$, $(1, 5)$,..., $(1, n)$ elements are reduced to zero. Since these rotations do not affect other elements in the first row and column in successive positions $3, 4,...,$ $n-1$ we retain the previous zeros, and after $(n-2)$ rotations of this kind we have the matrix A_{n-2}, typified for $n = 4$ by the form

$$A_2 = \begin{bmatrix} \times & \times & 0 & 0 \\ \times & \times & \times & \times \\ 0 & \times & \times & \times \\ 0 & \times & \times & \times \end{bmatrix}. \tag{19}$$

The second row is treated in exactly the same way, with rotations in planes $(3, 4)$, $(3, 5)$,..., $(3, n)$. The important fact is that the leading (2×2) matrix of A_{n-2} is unaffected by these transformations, and so are the zeros in the first row and column since the new elements in these positions are just linear combinations of the zeros which already exist there.

Proceeding in this way we produce, after

$$(n-2)+(n-3)+...+1 = \tfrac{1}{2}(n-1)(n-2)$$

transformations, a triple-diagonal matrix typified for $n = 4$ by the form

$$A_3 = \begin{bmatrix} \times & \times & 0 & 0 \\ \times & \times & \times & 0 \\ 0 & \times & \times & \times \\ 0 & 0 & \times & \times \end{bmatrix}. \tag{20}$$

The roots of this matrix are the same as those of A, and its vectors form the matrix $Y^{-1}X$, where X is the modal matrix of A, so that X is computed easily, as before, by successive multiplication of these vectors by the rotation matrices Y_r.

7. It remains to compute the roots and vectors of the triple-diagonal matrix typified by (20). In the general case the roots are those of the determinantal equation

$$\begin{vmatrix} a_1-\lambda & b_2 & & & & \\ b_2 & a_2-\lambda & b_3 & & & \\ \cdot & \cdot & \cdot & \cdot & \cdot & \cdot & \cdot & \cdot & \cdot & \cdot & \cdot \\ & & & b_{n-1} & a_{n-1}-\lambda & b_n \\ & & & & b_n & a_n-\lambda \end{vmatrix} = 0, \qquad (21)$$

and we proceed to discuss an appropriate method for their determination. We can find the characteristic equation, obtained by expanding the determinant, by a process of recurrence. Suppose $f_r(\lambda)$ is the determinant of the matrix formed by the first r rows and columns, so that in particular $f_1(\lambda) = a_1-\lambda$, $f_2(\lambda) = (a_1-\lambda)(a_2-\lambda)-b_2^2$. Then we prove that

$$f_{r+1}(\lambda) = (a_{r+1}-\lambda)f_r(\lambda)-b_{r+1}^2 f_{r-1}(\lambda), \qquad (22)$$

which is certainly true for $r = 1$ if we take $f_0(\lambda) = 1$. Now

$$f_{r+1}(\lambda) = \begin{vmatrix} a_1-\lambda & b_2 & & & \\ b_2 & a_2-\lambda & b_3 & & \\ \cdot & \cdot & \cdot & \cdot & \cdot & \cdot & \cdot & \cdot & \cdot & \cdot \\ & & b_r & a_r-\lambda & b_{r+1} \\ & & & b_{r+1} & a_{r+1}-\lambda \end{vmatrix}, \qquad (23)$$

and if we expand the determinant along the last row we find

$$f_{r+1}(\lambda) = (a_{r+1}-\lambda)f_r(\lambda)-b_{r+1}\begin{vmatrix} a_1-\lambda & b_2 & & & \\ b_2 & a_2-\lambda & b_3 & & \\ \cdot & \cdot & \cdot & \cdot & \cdot & \cdot & \cdot & \cdot & \cdot \\ & & b_{r-1} & a_{r-1}-\lambda & 0 \\ & & & b_r & b_{r+1} \end{vmatrix}, \qquad (24)$$

and the value of the determinant in (24) is just $b_{r+1}f_{r-1}(\lambda)$, so that (22) is verified. The characteristic polynomial is finally $f_n(\lambda)$.

We do not in fact compute explicitly the coefficients of $f_r(\lambda)$, but substitute particular values of λ into (22) and inspect the sequence

$f_0(\lambda), f_1(\lambda),\ldots,f_n(\lambda)$, which has certain important properties which facilitate the calculation of the roots.

†8. We digress a little to inspect a general property of the latent roots of successive principal minors of any symmetric matrix. Suppose we have the latent roots and vectors of the principal sub-matrix of order r, which we denote by C_r, and consider the 'bordering' of C_r by the next row and column of the given matrix, producing the next submatrix C_{r+1}. Partitioning of C_{r+1} gives

$$\left[\begin{array}{c|c} C_r & p \\ \hline p' & a_{r+1} \end{array}\right]\left[\begin{array}{c} \xi \\ \hline \eta \end{array}\right] = \mu\left[\begin{array}{c} \xi \\ \hline \eta \end{array}\right], \tag{25}$$

where μ is a latent root of C_{r+1} and its latent vector consists of the vector ξ and a last single element η. We know λ_s and $x^{(s)}$ in the equation $C_r x^{(s)} = \lambda_s x^{(s)}$, for $s = 1, 2,\ldots, r$. Expanding (25) we obtain

$$\left.\begin{array}{c} C_r\xi + p\eta = \mu\xi \\ p'\xi + a_{r+1}\eta = \mu\eta \end{array}\right\}, \tag{26}$$

the first of (26) gives $\xi = -(C_r - I\mu)^{-1}p\eta$, and the second of (26) produces

$$\{-p'(C_r - I\mu)^{-1}p + a_{r+1} - \mu\}\eta = 0. \tag{27}$$

The expression in braces is clearly proportional to the determinant $|C_{r+1} - I\mu|$, and its zeros are the latent roots of C_{r+1}. Now

$$(C_r - I\mu)^{-1} = \sum_{s=1}^{r} \frac{x^{(s)}x^{(s)'}}{\lambda_s - \mu}, \tag{28}$$

for premultiplication with $(C_r - I\mu)$ gives on the right the sum

$$\sum_{s=1}^{r} x^{(s)}x^{(s)'},$$

which is equal to XX', where X is the modal matrix, and this is equal to the unit matrix if the latent vectors are suitably normalized.

The characteristic equation (27) then becomes

$$f(\mu) = \sum_{s=1}^{r} \frac{(p'x^{(s)})^2}{\lambda_s - \mu} - (a_{r+1} - \mu) = 0, \tag{29}$$

and the important fact is that the coefficient of every $(\lambda_s - \mu)^{-1}$ is positive. Using this fact, and inspecting the signs of $f(\mu)$ near the points $\mu = \lambda_s$, we find that there is a root μ between any two successive roots λ_s, λ_{s+1}, and one smaller than the least and one greater than the largest of the λ_s. In other words the latent roots of the submatrix C_{r+1} separate those of the submatrix C_r.

This gives a fairly easy process, by the Newton method of successive approximation, for computing the zeros of $f(\mu)$, and the vectors of C_{r+1} follow simply from the formula

$$\xi = -(C_r - I\mu)^{-1}p\eta = -\sum_{s=1}^{r} \frac{x^{(s)}x^{(s)\prime}}{\lambda_s - \mu}\, p\eta. \tag{30}$$

The method of computing by these means all roots and vectors of all the leading submatrices is called the 'one-step escalator method', and was used quite extensively in desk-machine days.

†9. Returning to the triple-diagonal form, the results of § 8 enable us to assert that for a given value, $\lambda = p$, the sequence $f_r(p)$ defined by (22) is a Sturm sequence, that is the number of changes of sign in the sequence is equal to the number of roots of $f_n(\lambda)$ smaller than p in algebraic value.

We can use this result to compute all the roots, or any particular root or roots which we may require. Suppose for example that we seek λ_k in the sequence $\lambda_1 > \lambda_2 > \ldots > \lambda_k > \ldots > \lambda_n$. The process starts with two values a and b of p, such that $a > \lambda_1$ and $b < \lambda_n$. For a the number of sign changes in the sequence is n, and for b this number is zero. We then try $p = \frac{1}{2}(a+b)$, and bisect the appropriate interval containing λ_k, and so on.

By this process we shall obtain at each step one extra correct binary digit in our estimate of λ_k, so that for a given precision the amount of work is strictly calculable. The choice of a and b is easy, for we have the theorem that the root of largest modulus is smaller than any norm of the matrix, so that we can take

$$-b = a = \operatorname*{Max}_{r}(|a_r| + |b_r| + |b_{r+1}|). \tag{31}$$

If the matrix is known to be positive definite, so that the roots are positive, we can take $b = 0$.

10. Consider for example the matrix

$$\begin{bmatrix} 2 & -1 & 0 \\ -1 & 2 & -1 \\ 0 & -1 & 1 \end{bmatrix}, \tag{32}$$

for which the row norm is 4 so that the roots lie between $+4$ and -4.

We will compute the middle root, and Table 1 shows some steps of the computation. (When any f is zero its sign is taken to be that of the previous member of the sequence.)

TABLE 1

p	f_0	f_1	f_2	f_3	Comment
-4	$+1$	$+6$	$+35$	$+169$	No change of sign, no root <-4.
$+4$	$+1$	-2	$+3$	-7	Three changes of sign, three roots $<+4$.
0	$+1$	$+2$	$+3$	$+1$	No change of sign, no root <0.
$+2$	$+1$	$+0$	-1	$+1$	Two changes of sign, two roots $<+2$.
$+1$	$+1$	$+1$	$+0$	-1	One change of sign, one root $<+1$, middle root between $+1$, $+2$.
$+1\cdot5$	$+1$	$+0\cdot5$	$-0\cdot75$	$-0\cdot125$	One change of sign, middle root between $+1\cdot5$ and $+2$.

At this stage the required root is well isolated and we can complete its calculation. It is worth noting, however, that we can now obtain quadratic convergence with Newton's process, for which the computation is here quite simple. We seek to make in the current p a change δp given by

$$\delta p = -f_n(p)/f'_n(p), \qquad (33)$$

and we can compute the derivative $f'_n(p)$ from a recurrence relation derived by differentiation from that of $f_r(p)$. From (22) we find the required formula

$$f'_{r+1}(p) = (a_{r+1}-p)f'_r(p) - b^2_{r+1}f'_{r-1}(p) - f_r(p), \qquad (34)$$

and its starting values are $f'_0(p) = 0$, $f'_1(p) = -1$.

In our example, for $p = 1\cdot5$, we find

$$f'_0(p) = 0, \qquad f'_1(p) = -1, \qquad f'_2(p) = -1\cdot0,$$
$$f_3(p) = +2\cdot25, \qquad \delta p = -(-0\cdot125)/2\cdot25, \qquad (35)$$

so that $p = 1\cdot556$ is our next estimate. It implies that $p = 1\cdot5$ was correct to two figures, so that we would expect $1\cdot556$ to be already good in most of its figures. In fact the third decimal is incorrect by about 1 unit.

11. When a root has been computed it remains to determine the corresponding vector. This we can obtain directly from the original matrix by solving the resulting set of homogeneous algebraic equations, or preferably, involving less work, by going via the corresponding vector of the triple-diagonal matrix. There are, however, certain problems of stability and error analysis which we need to consider in this connexion, and this we defer until §§ 24–25.

Method of Householder, symmetric matrices

†12. We have seen that the method of Givens, though producing a triple-diagonal matrix which is slighly less convenient than a diagonal matrix, performs this transformation with considerably less work. In particular it produces one zero after another in the final matrix and no successive transformation affects the previous zeros.

The method of Householder also produces a triple-diagonal form, and with a further increase in efficiency and economy. The basic feature is that each transformation produces a complete row of zeros in the appropriate positions, again without affecting previous rows, so that only $n-2$ transformations are required though each involves rather more calculation than that of the Givens method.

Consider the transformation sequence

$$A_r = P_r A_{r-1} P_r, \qquad A_0 = A, \tag{36}$$

where

$$P_r = I - 2w^{(r)}w^{(r)\prime}, \qquad w^{(r)\prime}w^{(r)} = 1, \tag{37}$$

and apart from this normalization the elements of the vector $w^{(r)}$ are chosen so that A_r will have zeros, except in the 'triple-diagonal positions', in the whole of a particular row.

The matrix P_r is clearly symmetric, and it is also orthogonal, since

$$P_r P_r' = (I - 2w^{(r)}w^{(r)\prime})(I - 2w^{(r)}w^{(r)\prime})$$
$$= I - 4w^{(r)}w^{(r)\prime} + 4w^{(r)}(w^{(r)\prime}w^r)w^{(r)\prime} = I, \tag{38}$$

by virtue of the normalization of $w^{(r)}$. We therefore have the desirable orthogonal transformation, and we show, typically for a matrix of order four, the choice of suitable $w^{(r)}$ to produce the required zeros.

In the first transformation, $A_1 = P_1 A_0 P_1$, we try to produce zeros in positions (1, 3), (1, 4) of the first row of A_1. For this purpose we take $w^{(1)\prime} = (0, w_2, w_3, w_4)$, such that $w_2^2 + w_3^2 + w_4^2 = 1$. The corresponding matrix P_1 is

$$P_1 = \begin{bmatrix} 1 & 0 & 0 & 0 \\ 0 & 1-2w_2^2 & -2w_2w_3 & -2w_2w_4 \\ 0 & -2w_2w_3 & 1-2w_3^2 & -2w_3w_4 \\ 0 & -2w_2w_4 & -2w_3w_4 & 1-2w_4^2 \end{bmatrix}. \tag{39}$$

13. Now $P_1 A_0$ leaves unchanged the first row of A_0, and the $(1, 3)$ and $(1, 4)$ elements of $A_1 = P_1 A_0 P_1$ are respectively

$$\left. \begin{aligned} a_{13}^{(1)} &= -2w_2 w_3 a_{12} + (1-2w_3^2)a_{13} - 2w_3 w_4 a_{14} \\ &= a_{13} - 2w_3(w_2 a_{12} + w_3 a_{13} + w_4 a_{14}) \\ a_{14}^{(1)} &= -2w_2 w_4\, a_{12} - 2w_3 w_4 a_{13} + (1-2w_4^2)a_{14} \\ &= a_{14} - 2w_4(w_2 a_{12} + w_3 a_{13} + w_4 a_{14}) \end{aligned} \right\}. \tag{40}$$

The $(1,2)$ element is given by

$$\begin{aligned} a_{12}^{(1)} &= (1-2w_2^2)a_{12} - 2w_2 w_3 a_{13} - 2w_2 w_4 a_{14} \\ &= a_{12} - 2w_2(w_2 a_{12} + w_3 a_{13} + w_4 a_{14}). \end{aligned} \tag{41}$$

It is easy to verify that the sum of squares of these elements is $a_{12}^2 + a_{13}^2 + a_{14}^2$, so that if $a_{13}^{(1)}$ and $a_{14}^{(1)}$ are reduced to zero we must have

$$\left. \begin{aligned} a_{13} - 2pw_3 &= 0 \\ a_{14} - 2pw_4 &= 0 \end{aligned} \right\}, \tag{42}$$

where $p = w_2 a_{12} + w_3 a_{13} + w_4 a_{14}$, and then

$$a_{12} - 2pw_2 = \pm(a_{12}^2 + a_{13}^2 + a_{14}^2)^{\frac{1}{2}}. \tag{43}$$

Multiplying equation (43) and the two equations (42) respectively by w_2, w_3, w_4, and adding, we obtain

$$p = \mp w_2(a_{12}^2 + a_{13}^2 + a_{14}^2)^{\frac{1}{2}}, \tag{44}$$

so that from (43) and (42) we find

$$\left. \begin{aligned} w_2^2 &= \tfrac{1}{2}\left(1 \mp \frac{a_{12}}{(a_{12}^2 + a_{13}^2 + a_{14}^2)^{\frac{1}{2}}}\right), \quad w_3 = \mp\frac{a_{13}}{2w_2(a_{12}^2 + a_{13}^2 + a_{14}^2)^{\frac{1}{2}}} \\ w_4 &= \mp\frac{a_{14}}{2w_2(a_{12}^2 + a_{13}^2 + a_{14}^2)^{\frac{1}{2}}} \end{aligned} \right\}. \tag{45}$$

In the interests of accuracy we choose the sign in the first of (45) to be that of a_{12}, so that w_2 is as large as possible and we are saved embarrassment in the division by w_2 in the last two of (45).

14. The matrix A_1 now has the form

$$\begin{bmatrix} \times & \times & 0 & 0 \\ \times & \times & \times & \times \\ 0 & \times & \times & \times \\ 0 & \times & \times & \times \end{bmatrix}, \tag{46}$$

and we proceed to eliminate in exactly the same way the required elements in the second row. The vector $w^{(2)'}$ is $(0, 0, w_3, w_4)$, with a notation whose ambiguity should not cause trouble, the matrix P_2 is given by

$$P_2 = \begin{bmatrix} 1 & 0 & 0 & 0 \\ 0 & 1 & 0 & 0 \\ 0 & 0 & 1-2w_3^2 & -2w_3w_4 \\ 0 & 0 & -2w_3w_4 & 1-2w_4^2 \end{bmatrix}, \tag{47}$$

and $P_2A_1P_2$ leaves unchanged both the leading sub-matrix of order two and the zeros in the first row and column. The continuation in the general case is obvious.

When we have the final triple-diagonal matrix A_{n-2} we compute its roots by the technique of the Givens method, and the vectors of A_{n-2} are obtained, with devices discussed in §§ 24–25 below, in exactly the same way. We recover the vectors of A by premultiplying with the successive P_r in order $P_{n-2}, P_{n-3}, \ldots, P_1$, which is faster than forming explicitly the product $P_1P_2 \ldots P_{n-2}$ and operating with this on the vector of A_{n-2}.

Example of Givens and Householder

15. To illustrate the Givens and Householder transformations we take the matrix

$$A = \begin{bmatrix} 1 & \sqrt{2} & \sqrt{2} & 2 \\ \sqrt{2} & -\sqrt{2} & -1 & \sqrt{2} \\ \sqrt{2} & -1 & \sqrt{2} & \sqrt{2} \\ 2 & \sqrt{2} & \sqrt{2} & -3 \end{bmatrix}, \tag{48}$$

and perform the computation with exact arithmetic. The first Y_1' in the Givens process has the form (16), with s and c defined by (18), so that $s = 1/\sqrt{2}$, $c = 1/\sqrt{2}$, and multiplication gives the first transformed matrix

$$A_1 = \begin{bmatrix} 1 & 2 & 0 & 2 \\ 2 & -1 & \sqrt{2} & 2 \\ 0 & \sqrt{2} & 1 & 0 \\ 2 & 2 & 0 & -3 \end{bmatrix}. \tag{49}$$

Then Y_2' has the form

$$Y_2' = \begin{bmatrix} 1 & 0 & 0 & 0 \\ 0 & c & 0 & s \\ 0 & 0 & 1 & 0 \\ 0 & -s & 0 & c \end{bmatrix}, \tag{50}$$

where $s = a_{14}^{(1)}/(a_{12}^{(1)2}+a_{14}^{(1)2})^{\frac{1}{2}}$, $c = a_{12}^{(1)}/(a_{12}^{(1)2}+a_{14}^{(1)2})^{\frac{1}{2}}$, so that again

$$s = 1/\sqrt{2}, \qquad c = 1/\sqrt{2},$$

and we find

$$A_2 = \begin{bmatrix} 1 & 2\sqrt{2} & 0 & 0 \\ 2\sqrt{2} & 0 & 1 & -1 \\ 0 & 1 & 1 & -1 \\ 0 & -1 & -1 & -4 \end{bmatrix}. \tag{51}$$

For the last matrix Y_3 we have

$$Y_3' = \begin{bmatrix} 1 & 0 & 0 & 0 \\ 0 & 1 & 0 & 0 \\ 0 & 0 & c & s \\ 0 & 0 & -s & c \end{bmatrix}, \tag{52}$$

where $s = a_{24}^{(2)}/(a_{23}^{(2)2}+a_{24}^{(2)2})^{\frac{1}{2}}$, $c = a_{23}^{(2)}/(a_{23}^{(2)2}+a_{24}^{(2)2})^{\frac{1}{2}}$, so that $s = -\dfrac{1}{\sqrt{2}}$, $c = \dfrac{1}{\sqrt{2}}$, and we find the final triple-diagonal form

$$A_3 = \begin{bmatrix} 1 & 2\sqrt{2} & 0 & 0 \\ 2\sqrt{2} & 0 & \sqrt{2} & 0 \\ 0 & \sqrt{2} & -\frac{1}{2} & \frac{5}{2} \\ 0 & 0 & \frac{5}{2} & -\frac{5}{2} \end{bmatrix}. \tag{53}$$

16. For the corresponding Householder transformation our first matrix P_1 has the form (39), with w_2, w_3 and w_4 defined by (45), so that $w_2 = \frac{1}{2}\sqrt{3}$, $w_3 = 1/2\sqrt{3}$, $w_4 = 1/\sqrt{6}$, and

$$P_1 = \begin{bmatrix} 1 & 0 & 0 & 0 \\ 0 & -\frac{1}{2} & -\frac{1}{2} & -\frac{1}{\sqrt{2}} \\ 0 & -\frac{1}{2} & \frac{5}{6} & -\frac{1}{3\sqrt{2}} \\ 0 & -\frac{1}{\sqrt{2}} & -\frac{1}{3\sqrt{2}} & \frac{2}{3} \end{bmatrix}. \tag{54}$$

The matrix multiplication then gives

$$
A_1 = \begin{bmatrix}
1 & -2\sqrt{2} & 0 & 0 \\[2mm]
-2\sqrt{2} & 0 & -\dfrac{1}{3}-\dfrac{4}{3\sqrt{2}} & -\dfrac{1}{3}+\dfrac{4}{3\sqrt{2}} \\[2mm]
0 & -\dfrac{1}{3}-\dfrac{4}{3\sqrt{2}} & \dfrac{4}{9}+\dfrac{8}{9\sqrt{2}} & -\dfrac{7}{9}+\dfrac{20}{9\sqrt{2}} \\[2mm]
0 & -\dfrac{1}{3}+\dfrac{4}{3\sqrt{2}} & -\dfrac{7}{9}+\dfrac{20}{9\sqrt{2}} & \dfrac{31}{9}-\dfrac{8}{9\sqrt{2}}
\end{bmatrix}, \tag{55}
$$

and we have completed the production of zeros in the first row.

A further single step, with P_2 having the form (47), will produce the final triple-diagonal form. We have $(a_{23}^{(1)2}+a_{24}^{(1)2}) = 2$, and we take the negative sign in (47) so that

$$
w_3^2 = \tfrac{1}{2}\left\{1+\frac{1}{\sqrt{2}}\left(\frac{1}{3}+\frac{4}{3\sqrt{2}}\right)\right\} = \frac{5}{6}+\frac{1}{6\sqrt{2}}, \quad w_4 = \left(\frac{1}{3}-\frac{4}{3\sqrt{2}}\right)\Big/2\sqrt{2}\,w_3, \tag{56}
$$

and the matrix P_2 simplifies remarkably to

$$
P_2 = \begin{bmatrix}
1 & 0 & 0 & 0 \\[2mm]
0 & 1 & 0 & 0 \\[2mm]
0 & 0 & -\dfrac{2}{3}-\dfrac{1}{3\sqrt{2}} & \dfrac{2}{3}-\dfrac{1}{3\sqrt{2}} \\[2mm]
0 & 0 & \dfrac{2}{3}-\dfrac{1}{3\sqrt{2}} & \dfrac{2}{3}+\dfrac{1}{3\sqrt{2}}
\end{bmatrix}. \tag{57}
$$

The multiplication $P_2A_1P_2$ gives the final result

$$
A_2 = \begin{bmatrix}
1 & -2\sqrt{2} & 0 & 0 \\[1mm]
-2\sqrt{2} & 0 & \sqrt{2} & 0 \\[1mm]
0 & \sqrt{2} & -\tfrac{1}{2} & -\tfrac{5}{2} \\[1mm]
0 & 0 & -\tfrac{5}{2} & -\tfrac{5}{2}
\end{bmatrix}. \tag{58}
$$

Uniqueness of triple-diagonal form

†17. It is interesting to observe that the final triple-diagonal forms in the two methods are identical except for the signs of the off-diagonal elements, though the intermediate matrices, with zeros in the appropriate rows, may be quite different. It can be shown in general, for any orthogonal transformation of the form

$$
Y'AY = B, \tag{59}
$$

in which B is triple-diagonal and Y is an orthogonal matrix whose first column is $e^{(1)}$, the first column of the unit matrix, that both Y and B are unique except for the signs of the off-diagonal elements. Consider for example a matrix of order four, and write (59) in the form $AY = YB$, so that

$$\begin{bmatrix} a_{11} & a_{12} & a_{13} & a_{14} \\ a_{12} & a_{22} & a_{23} & a_{24} \\ a_{13} & a_{23} & a_{33} & a_{34} \\ a_{14} & a_{24} & a_{34} & a_{44} \end{bmatrix} \begin{bmatrix} 1 & 0 & 0 & 0 \\ 0 & y_{22} & y_{23} & y_{24} \\ 0 & y_{32} & y_{33} & y_{34} \\ 0 & y_{42} & y_{43} & y_{44} \end{bmatrix} = \begin{bmatrix} 1 & 0 & 0 & 0 \\ 0 & y_{22} & y_{23} & y_{24} \\ 0 & y_{32} & y_{33} & y_{34} \\ 0 & y_{42} & y_{43} & y_{44} \end{bmatrix} \begin{bmatrix} a_1 & b_2 & 0 & 0 \\ b_2 & a_2 & b_3 & 0 \\ 0 & b_3 & a_3 & b_4 \\ 0 & 0 & b_4 & a_4 \end{bmatrix} . \quad (60)$$

The elements of the first column of the product matrices give

$$a_{11} = a_1, \qquad a_{12} = b_2 y_{22}, \qquad a_{13} = b_2 y_{32}, \qquad a_{14} = b_2 y_{42}, \quad (61)$$

and a_1 is uniquely determined from the first of (61). Since Y is an orthogonal matrix its columns are orthogonal, and the sum of the squares of the elements in each column is equal to unity, so that in particular the first row of Y has zeros in the off-diagonal positions and $y_{22}^2 + y_{32}^2 + y_{42}^2 = 1$. From the squares of the last three of (61) we then find $b_2^2 = a_{12}^2 + a_{13}^2 + a_{14}^2$, so that b_2 is unique apart from its sign, and the corresponding uniqueness of y_{22}, y_{32}, y_{42}, the elements of the second column of Y, then follows also from (61).

Looking at the elements other than the first of the second column we have the three equations

$$\left. \begin{aligned} a_{22}y_{22} + a_{23}y_{32} + a_{24}y_{42} &= a_2 y_{22} + b_3 y_{23} \\ a_{23}y_{22} + a_{33}y_{32} + a_{34}y_{42} &= a_2 y_{32} + b_3 y_{33} \\ a_{24}y_{22} + a_{34}y_{32} + a_{44}y_{42} &= a_2 y_{42} + b_3 y_{43} \end{aligned} \right\}, \quad (62)$$

and if we multiply respectively by y_{22}, y_{32} and y_{42} and add we isolate a_2 on the right and determine its uniqueness, even with respect to sign. The sum of squares of the equations (62) then gives b_3^2 uniquely, and hence b_3 and y_{23}, y_{33}, y_{43} are unique except for sign. The continuation is obvious and proves the theorem.

Method of Lanczos, symmetric matrices

†18. We consider now the method of Lanczos, which also produces a triple-diagonal form by a finite sequence of orthogonal transformations. In this case, however, we obtain the transformation by building-up explicitly the sequence of vectors which form the orthogonal

matrix Y, the corresponding elimination of off-diagonal elements of A arising implicitly through a particular choice of these vectors.

The vectors are built up in the sequence $y^{(0)} = o$, $y^{(1)}$ is arbitrary, and

$$y^{(r+1)} = Ay^{(r)} - \alpha_r y^{(r)} - \beta_r y^{(r-1)}, \qquad r = 1, 2, ..., n, \qquad (63)$$

where we choose α_r and β_r so that $y^{(r+1)}$ is orthogonal to the two previous vectors $y^{(r)}$ and $y^{(r-1)}$. This gives immediately

$$\alpha_r = y^{(r)\prime} A y^{(r)} / y^{(r)\prime} y^{(r)}, \qquad \beta_r = y^{(r-1)\prime} A y^{(r)} / y^{(r-1)\prime} y^{(r-1)}, \qquad (64)$$

by respective premultiplication of (63) by $y^{(r)\prime}$ and $y^{(r-1)\prime}$ and the observation that $y^{(r)\prime} y^{(r-1)}$ has been made to vanish at the previous stage.

The important consequence of the definition (63) of the 'next' vector and the choices (64) of α and β is that $y^{(r+1)}$ is orthogonal to all previous vectors $y^{(s)}$, $s = 0, 1, ..., r$.

This is clearly true for $r = 1$, for $y^{(2)}$ is orthogonal to $y^{(0)} = o$. Suppose it is true for $r = p$. Then

$$y^{(p+2)} = Ay^{(p+1)} - \alpha_{p+1} y^{(p+1)} - \beta_{p+1} y^{(p)}, \qquad (65)$$

and α_{p+1} and β_{p+1} are chosen so that $y^{(p+2)}$ is orthogonal to $y^{(p+1)}$ and $y^{(p)}$. Premultiplication with $y^{(s)\prime}$, for $s = 0, 1, ..., p-1$, gives

$$y^{(s)\prime} y^{(p+2)} = y^{(s)\prime} A y^{(p+1)} - \alpha_{p+1} y^{(s)\prime} y^{(p+1)} - \beta_{p+1} y^{(s)\prime} y^{(p)}, \qquad (66)$$

of which the last two terms vanish in virtue of the inductive hypothesis. Also

$$y^{(s)\prime} A y^{(p+1)} = y^{(p+1)\prime} A y^{(s)} = y^{(p+1)\prime} (y^{(s+1)} + \alpha_s y^{(s)} + \beta_s y^{(s-1)}), \qquad (67)$$

and again our hypothesis makes each term vanish and verifies the theorem.

Except in special cases the vectors $y^{(1)}, y^{(2)}, ..., y^{(n)}$ will all be non-null, and therefore $y^{(n+1)}$ will necessarily be null, being orthogonal to all the $y^{(r)}$, $r = 1, 2, ..., n$.

19. Consider now the general relations (63). If the vectors $y^{(r)}$, $r = 1, 2, ..., n$, form the matrix Y, we find

$$AY = YC, \qquad (68)$$

where

$$C = \begin{bmatrix} \alpha_1 & \beta_2 & & & & \\ 1 & \alpha_2 & \beta_3 & & & \\ & 1 & \alpha_3 & \beta_4 & & \\ & \cdot & \cdot & \cdot & \cdot & \cdot & \cdot & \cdot \\ & & & 1 & \alpha_{n-1} & \beta_n \\ & & & & 1 & \alpha_n \end{bmatrix}, \tag{69}$$

so that $Y^{-1}AY = C$, and we have a similarity transformation.

The lack of symmetry in C is of no great consequence. Indeed if we take the vector $y^{(1)}$ to be the first column of the unit matrix we shall find, in analogy with the discussion of § 17, that the α_r in (69) is the same as the a_r of the Givens and Householder transformations, and the β_r in (69) is the square of the corresponding element b_r in the previous methods. Since b_r^2 is the important quantity in the Sturm sequence (22) it follows that the Sturm theory still applies, and the process of computing the roots of (69) is exactly that of the previous methods.

We note a simplification in the second of (64) for β_r. Since

$$y^{(r-1)'}Ay^{(r)} = y^{(r)'}Ay^{(r-1)} = y^{(r)'}(y^{(r)} + \alpha_{r-1}y^{(r-1)} + \beta_{r-1}y^{(r-2)}) = y^{(r)'}y^{(r)}, \tag{70}$$

in virtue of the orthogonality relations, we obtain

$$\beta_r = y^{(r)'}y^{(r)}/y^{(r-1)'}y^{(r-1)}. \tag{71}$$

20. For the matrix (48), starting with $y^{(1)'} = (1, 0, 0, 0)$, we find the results of Table 2 and note the relations with the other methods.

TABLE 2

$y^{(1)'}$	1	0	0	0	
$\{Ay^{(1)}\}'$	1	$\sqrt{2}$	$\sqrt{2}$	2	$\alpha_1 = 1$
$y^{(2)'}$	0	$\sqrt{2}$	$\sqrt{2}$	2	
$\{Ay^{(2)}\}'$	8	$\sqrt{2}-2$	$\sqrt{2}+2$	-2	$\alpha_2 = 0, \beta_2 = 8$
$y^{(3)'}$	0	$\sqrt{2}-2$	$\sqrt{2}+2$	-2	
$\{Ay^{(3)}\}'$	0	$-\sqrt{2}-4$	$-\sqrt{2}+4$	10	$\alpha_3 = -\frac{1}{2}, \beta_3 = 2$
$y^{(4)'}$	0	$-\frac{5}{2}\sqrt{2}-5$	$-\frac{5}{2}\sqrt{2}+5$	5	
$\{Ay^{(4)}\}'$	0	$\frac{25}{2}\sqrt{2}$	$\frac{25}{2}\sqrt{2}$	-25	$\alpha_4 = -\frac{5}{2}, \beta_4 = \frac{25}{4}$
$y^{(5)'}$	0	0	0	0	

The relations become still more obvious if we replace the general equation (63) by

$$y^{(r+1)} = k_{r+1}(Ay^{(r)} - \alpha_r y^{(r)} - \beta_r y^{(r-1)}), \tag{72}$$

the constant being chosen so that $y^{(r+1)'}y^{(r+1)} = 1$. In this case we have α_r as before but now

$$\beta_r = y^{(r-1)'}Ay^{(r)} = y^{(r)'}Ay^{(r-1)} = k_r^{-1}. \tag{73}$$

The matrix Y is here truly orthogonal and

$$Y'AY = \begin{bmatrix} \alpha_1 & k_2^{-1} & \\ k_2^{-1} & \alpha_2 & k_3^{-1} \\ & \cdot & \cdot & \cdot \end{bmatrix}. \tag{74}$$

Method of Lanczos, unsymmetric matrices

†21. If the matrix is unsymmetric we can still produce a triple-diagonal form by similarity transformations by an extension of the method of Lanczos, in which we produce two sets of bi-orthogonal vectors. With $y^{(0)}$ and $z^{(0)}$ both null, and with $y^{(1)}$ and $z^{(1)}$ arbitrary, we compute successive vectors in the sequence

$$y^{(r+1)} = Ay^{(r)} - \alpha_r y^{(r)} - \beta_r y^{(r-1)}, \qquad z^{(r+1)} = A'z^{(r)} - \alpha_r' z^{(r)} - \beta_r' z^{(r-1)}, \tag{75}$$

choosing the constants so that $y^{(r+1)}$ is orthogonal to the two previous vectors $z^{(r)}$ and $z^{(r-1)}$, and $z^{(r+1)}$ is orthogonal to the two previous vectors $y^{(r)}$ and $y^{(r-1)}$. Then

$$\left. \begin{aligned} \alpha_r &= z^{(r)'}Ay^{(r)}/z^{(r)'}y^{(r)}, & \beta_r &= z^{(r-1)'}Ay^{(r)}/z^{(r-1)'}y^{(r-1)} \\ \alpha_r' &= y^{(r)'}A'z^{(r)}/y^{(r)'}z^{(r)}, & \beta_r' &= y^{(r-1)'}A'z^{(r)}/y^{(r-1)'}z^{(r-1)} \end{aligned} \right\}. \tag{76}$$

We can show as before that $y^{(r+1)}$, $z^{(r+1)}$ are respectively orthogonal to all previous vectors $z^{(p)}$, $y^{(p)}$, $p = 1, 2, ..., r$. Moreover it is clear that $\alpha_r = \alpha_r'$, the expression for the latter being the transpose of that for the former, and

$$\beta_r = \frac{y^{(r)'}A'z^{(r-1)}}{y^{(r-1)'}z^{(r-1)}} = \frac{y^{(r)'}(z^{(r)} + \alpha_{r-1}'z^{(r-1)} + \beta_{r-1}'z^{(r-2)})}{y^{(r-1)'}z^{(r-1)}} = \frac{y^{(r)'}z^{(r)}}{y^{(r-1)'}z^{(r-1)}} = \beta_r' \tag{77}$$

by a similar argument. The final vectors $y^{(n+1)}$ and $z^{(n+1)}$ are both null if the vectors $z^{(1)}, ..., z^{(n)}$ and $y^{(1)}, ..., y^{(n)}$ are linearly independent.

The relations (75) can then be written in the form

$$AY = YC, \qquad A'Z = ZC, \tag{78}$$

where C has the form of (69) but with α and β defined by (76). Moreover with suitable normalization we can take $Y_1 = YD^{-1}$, where D is diagonal with elements $y^{(r)\prime}z^{(r)}$, so that $Z'Y_1 = I$, and $Z'AY_1 = DCD^{-1} = C'$.

†22. The matrix C, of course, may have complex roots and there are no Sturm sequence properties to assist in their calculation. One possibility is to use the equation corresponding to (22) to calculate explicitly the polynomial $f_n(\lambda)$, though the ill-conditioned nature of the problem of evaluating roots of polynomials usually necessitates double- or multiple-length arithmetic at this stage.

Or we can use the method of Muller, which here effectively computes the value of $f(\lambda) = |C - I\lambda|$ from the recurrence relation of type (22), and from three values $f(p_1)$, $f(p_2)$ and $f(p_3)$ finds a zero of the quadratic passing through the points $\{p_r, f(p_r)\}$, $r = 1, 2, 3$. The chosen zero is taken as p_4, and the process is repeated with p_4 replacing one of the previous p_r. Some complex arithmetic is necessary, but this is of course inevitable when some roots are complex, and the Muller method does not involve a prohibitive amount of complex arithmetic. Convergence to a latent root is apparently guaranteed when any p_r is close enough to that particular root.

23. Consider for example the matrix

$$A = \begin{bmatrix} 8 & -1 & -5 \\ -4 & 4 & -2 \\ 18 & -5 & -7 \end{bmatrix}, \tag{79}$$

treated in § 8 of Chapter 9 by iterative methods. The computation of the vectors $y^{(r)}$ and $z^{(r)}$ is shown in Table 3.

TABLE 3

$y^{(1)\prime}$	1	0	0	$z^{(1)\prime}$	1	0	0	
$\{Ay^{(1)}\}'$	8	−4	18	$\{A'z^{(1)}\}'$	8	−1	−5	$\alpha_1 = 8$
$y^{(2)\prime}$	0	−4	18	$z^{(2)\prime}$	0	−1	−5	
$\{Ay^{(2)}\}'$	−86	−52	−106	$\{A'z^{(2)}\}'$	−86	21	37	$\alpha_2 = -\frac{291}{43}$
$y^{(3)\prime}$	0	$-\frac{3400}{43}$	$\frac{680}{43}$	$z^{(3)\prime}$	0	$\frac{612}{43}$	$\frac{136}{43}$	$\beta_2 = -86$
$\{Ay^{(3)}\}'$	0	$-\frac{14960}{43}$	$\frac{12240}{43}$	$\{A'z^{(3)}\}'$	0	$\frac{1768}{43}$	$-\frac{2176}{43}$	$\alpha_3 = \frac{168}{43}$
$y^{(4)\prime}$	0	0	0	$z^{(4)\prime}$	0	0	0	$\beta_3 = \frac{23120}{1849}$

The triple-diagonal matrix, with coefficients rounded to one decimal, is then

$$\begin{bmatrix} 8 & -86 & \\ 1 & -6\cdot8 & 12\cdot5 \\ & 1 & 3\cdot8 \end{bmatrix}. \tag{80}$$

The sequence for the successive minors of $|C-I\lambda|$ is

$$f_0(\lambda) = 1, \qquad f_1(\lambda) = 8-\lambda, \qquad f_2(\lambda) = (-6\cdot8-\lambda)f_1(\lambda)+86,$$

$$f_3(\lambda) = (3\cdot8-\lambda)f_2(\lambda)-12\cdot5f_1(\lambda), \tag{81}$$

and with $\lambda = 0, 2, 4$, we find

$$f_3(0) = 20\cdot08, \qquad f_3(2) = -15\cdot24, \qquad f_3(4) = -58\cdot56.$$

The quadratic through these points is

$$-f(\lambda) = \lambda^2+15\cdot66\lambda-20\cdot08,$$

with an approximate zero at $\lambda = 1\cdot2$. Choosing $1\cdot2$ instead of 4 in the previous triad we find $f_3(1\cdot2) = -2\cdot84$, and the quadratic is

$$f(\lambda) = 1\cdot8\lambda^2-21\cdot26\lambda+20\cdot08,$$

with a zero near $\lambda = 1\cdot03$, which corresponds to the true $\lambda = 1$ of the original matrix.

The remaining roots form a complex pair and the same method will work. We must, however, avoid the possibility of converging again to the root λ_1 already obtained by 'dividing it out', that is we now want a root or roots of $g_3(\lambda) = f_3(\lambda)/(\lambda-\lambda_1)$. It is clearly unnecessary to expand this into a polynomial and all we have to do is to evaluate this expression in place of $f_3(\lambda)$. For example with $\lambda = 0, 2, 4$ as before, we find, corresponding to $\lambda_1 = 1$, the respective values

$$g_3(0) = -20\cdot08, \qquad g_3(2) = -15\cdot24, \qquad g_3(4) = -19\cdot52.$$

The quadratic through these points is

$$-f(\lambda) = 1\cdot14\lambda^2-4\cdot7\lambda+20\cdot08,$$

with approximate roots $2\cdot1\pm3\cdot7i$. These correspond to the exact roots $2\pm4i$ of the original matrix.

In practice and in general we should take one of these roots as a new value of λ in the iterative sequence, and some complex arithmetic, including the determination of zeros of a complex quadratic, is now necessary.

Vectors of triple-diagonal matrices

†24. The methods of Givens, Householder and Lanczos all reduce the problem to the determination of latent roots and vectors of a triple-diagonal matrix, and we have already discussed various possibilities for the calculation of the latent roots. At first sight the calculation of the vectors is then particularly simple, since we merely have to substitute a particular λ in the homogeneous equations

$$(C - I\lambda)x = o, \tag{82}$$

and solve them by any respectable method such as Gauss elimination. Indeed if our computed λ were exact, and no rounding errors were committed in the solving process, the matrix $C - I\lambda$ would be singular. We could then omit an equation from the set (82), solve the rest as a set of non-homogeneous equations in the $(n-1)$ ratios of the vector components (if the resulting matrix is non-singular) and the result would automatically satisfy the omitted equation.

The quoted deficiencies in the computed λ, and perhaps in our solving process, however, mean that even if we can solve exactly the selected $n-1$ equations the omitted equation will not be satisfied exactly, and we may normalize our computed vector so that we have effectively solved the full set of equations

$$(C - I\lambda)x = e^{(i)}, \tag{83}$$

where $e^{(i)}$ is the ith column of the unit matrix and the ith equation was omitted in the solving process.

If $e^{(i)}$ is expressed as a linear combination of latent vectors,

$$e^{(i)} = \sum_{r=1}^{n} \alpha_r x^{(r)}, \tag{84}$$

the solution of (83) is given by

$$x = \sum_{r=1}^{n} \frac{\alpha_r x^{(r)}}{\lambda_r - \lambda}, \tag{85}$$

and if λ is a close approximation to a root λ_k, so that $\lambda_k - \lambda = \epsilon$, (85) becomes

$$x = \frac{\alpha_k x^{(k)}}{\epsilon} + \sum_{r \neq k} \frac{\alpha_r x^{(r)}}{\lambda_r - \lambda_k + \epsilon}. \tag{86}$$

Now if no other root is close to λ_k, and if α_k is not small, (86) shows that we obtain correctly the required latent vector $x^{(k)}$, since this term dominates the right-hand side of (86). If two roots are close together the problem is ill-conditioned with respect to the determination of the latent vectors, and this we here exclude. But if α_k is small, of order ϵ, the right of (86) is not dominated by a single latent vector, and our result may be very poor. If $\alpha_k = 0$, that is $e^{(i)}$ is 'defective' in the latent vector $x^{(k)}$, then $x^{(k)\prime}e^{(i)} = 0$ and $x^{(k)}$ has its ith component zero.

25. This, unfortunately, turns out to be quite possible in the particular triple-diagonal matrices obtained by the methods of this chapter. It happens that some of the off-diagonal terms may be very small, and if any is identically zero the matrix 'splits up' into two parts. For example if b_2 in (21) is zero, one root is a_1, and the corresponding vector is $(1, 0, 0, \ldots, 0)$. In general, if b_{r+1} is zero, the first r roots come from the leading sub-matrix of order r, and the last $(n-r)$ components of the corresponding vectors are zero; the last $(n-r)$ roots come from the matrix of order $(n-r)$ in the bottom right-hand corner, and the first r components of the corresponding vectors are zero.

Now we have already seen, in §§ 16–18 of Chapter 9, the possibility of performing inverse iteration when the matrix $(C-I\lambda)$ is singular or nearly singular, and we propose to use this method here, solving in succession the equations

$$(C-I\lambda)y^{(r+1)} = y^{(r)}, \tag{87}$$

the vector $y^{(r)}$ being suitably normalized at each stage. In theory, with the correct λ, we need only find $y^{(1)}$ from an arbitrary $y^{(0)}$, but we have seen that $y^{(0)}$ must not be defective in the vector we seek. To ensure this we could take some random digits for the components of $y^{(0)}$, but it is easier and equally satisfactory not to choose $y^{(0)}$ explicitly, but to reduce $(C-I\lambda)$ to upper triangular form by Gauss elimination, and then to back-substitute taking the vector $(1, 1, \ldots, 1)$ on the right-hand side.

This of course is not a complete guarantee that the corresponding $y^{(0)}$ is not defective in our required vector, but the new $y^{(1)}$ should certainly now have some component of $x^{(k)}$, and the production of $y^{(2)}$ from (87) should then produce a latent vector with an accuracy as great as we could expect in view of the possible error in the selected λ.

Other similarity transformations. The L-R method

†**26.** Various other similarity transformations have been discovered in recent years, and the most useful are discussed in detail in J. H.

Wilkinson's forthcoming book. Here we mention briefly two or three methods and in the bibliography we list other developments.

The L-R (left-right) transformation of Rutishauser starts with the matrix $A = A_1$, and performs the triangular decomposition $A_1 = L_1 R_1$, where L_1 is a unit lower triangle and R_1 an upper triangle. These triangles are then multiplied in reverse order, giving $A_2 = R_1 L_1$, and the process is repeated with the equations

$$A_2 = L_2 R_2,\; R_2 L_2 = A_3,\ldots,\quad A_p = L_p R_p,\; R_p L_p = A_{p+1},\ldots. \quad (88)$$

This is obviously a similarity transformation because

$$A_2 = R_1 L_1 = R_1 A_1 R_1^{-1},\ldots, A_{p+1} = (R_p R_{p-1}\ldots R_1) A_1 (R_p R_{p-1}\ldots R_1)^{-1}, \quad (89)$$

and all the matrices A_p have the same latent roots. We also have

$$A_{p+1} = (L_1 L_2 \ldots L_p)^{-1} A_1 (L_1 L_2 \ldots L_p), \quad (90)$$

and of course $B_p = L_1 L_2 \ldots L_p$ and $C_p = R_p R_{p-1} \ldots R_1$ are respectively unit lower triangles and upper triangles for all p. Moreover, since

$$L_p R_p = R_{p-1} L_{p-1}, \quad (91)$$

it follows that

$$B_p C_p = A^p, \quad (92)$$

and the final triangular matrices represent the decomposition of a power of A.

27. Rutishauser proves theorems of which the following is a selection.

(i) If B_p converges as $p \to \infty$ then A_p converges to an upper triangular matrix.

(For if B_p has a limit, then $L_p = B_{p-1}^{-1} B_p$ also has a limit, and this is the unit matrix. Then $R_p = B_p^{-1} A_1 B_{p-1}$, by judicious use of (91), so that R_p also has a limit, and then $A_p = L_p R_p$ also has a limit, the limiting upper triangle R_p, since $L_p \to I$, the unit matrix.)

(ii) If A has real distinct latent roots and no leading sub-matrix of X or of X^{-1} is singular in the transformation $A = X D X^{-1}$, where D is a diagonal matrix whose elements are the latent roots, then $\lim_{p \to \infty} B_p$ is the lower triangle in the triangular decomposition of the matrix X, and $\lim_{p \to \infty} A_p$ is upper triangular with the latent roots on the diagonal in descending order of modulus.

(iii) Convergence is guaranteed if A is symmetric and positive definite, though the diagonal elements of lim A_p are not necessarily in order of modulus unless the second condition in (ii) is satisfied.

(iv) If we carry out the Cholesky-type decomposition $A = LL'$ for the symmetric positive-definite case, and replace R by L' in the L-R formulae (88), then symmetry is preserved in successive A_p, and A_p converges to a diagonal matrix.

(v) If some of the roots are complex then, under certain conditions similar to those of (ii), some of the diagonal elements of A_p converge to the values of the real roots. The lower part of $A_p \to 0$ except for some elements on the sloping line below the diagonal. Any such value, say $a_{q,q-1}^{(p)}$, may not converge to zero, but the latent roots of the (2×2) matrix

$$\begin{bmatrix} a_{q-1,q-1}^{(p)} & a_{q-1,q}^{(p)} \\ a_{q,q-1}^{(p)} & a_{q,q}^{(p)} \end{bmatrix}$$

converge to a complex pair of roots of A.

One very important use for the L-R method is that for matrices of 'band-type', with typical form

$$A = \begin{bmatrix} \times & \times & \times & & & & \\ \times & \times & \times & \times & & & \\ \times & \times & \times & \times & \times & & \\ & \times & \times & \times & \times & \times & \\ & & \times & \times & \times & \times & \times \\ & & & \times & \times & \times & \times \\ & & & & \times & \times & \times \end{bmatrix}, \qquad (93)$$

many of which occur in the solution of ordinary differential equations of eigenvalue type. The 'band form' is preserved in all the transformations and this is economic in terms of time and machine storage. Moreover, for the symmetric positive-definite case, solved by the method of (iv), it turns out that convergence is fastest to the smallest latent roots, and in practical problems these are the roots most often required.

28. One disadvantage of the L-R method is the inconvenience or inaccuracy of performing the triangular decomposition for a general matrix, in which a leading sub-matrix may be singular or nearly singular. We would in fact like to perform triangular decomposition with interchanges, and we noted in § 30 of Chapter 4 the possibility of splitting a row permutation of A into triangles in such a way that no element of L exceeds unity, giving the effect of 'taking largest element as pivot'.

The corresponding similarity transform, corresponding to the L-R method, is given by the equations

$$A_1 = I_1^{-1}L_1R_1, \; A_2 = R_1I_1^{-1}L_1 = I_2^{-1}L_2R_2, \; A_3 = R_2I_2^{-1}L_2, \text{ etc,} \quad (94)$$

where I_r is a row-permuting matrix, and we easily find

$$A_2 = R_1A_1R_1^{-1}, \qquad A_3 = (R_2R_1)A_1(R_2R_1)^{-1}, \dots . \quad (95)$$

29. As we begin to converge to the solution, that is as A_p approaches its triangular form, it becomes increasingly likely that the row-permuting matrix is just the unit matrix, and the largest pivot is on the diagonal.

For example with the matrix

$$A_1 = \begin{bmatrix} 1 & 1 & 3 \\ 1 & -2 & 1 \\ 3 & 1 & 3 \end{bmatrix}$$

we find, rounded after the second stage, the results

$$A_1 = \overset{I_1^{-1}L_1}{\begin{bmatrix} \frac{1}{3} & -\frac{2}{7} & 1 \\ \frac{1}{3} & 1 & 0 \\ 1 & 0 & 0 \end{bmatrix}} \overset{R_1}{\begin{bmatrix} 3 & 1 & 3 \\ & -\frac{7}{3} & 0 \\ & & 2 \end{bmatrix}},$$

$$A_2 = R_1I_1^{-1}L_1 = \begin{bmatrix} \frac{13}{3} & \frac{1}{7} & 3 \\ -\frac{7}{9} & -\frac{7}{3} & 0 \\ 2 & 0 & 0 \end{bmatrix} =$$

$$= \overset{I_2^{-1}L_2}{\begin{bmatrix} 1 & & \\ -\frac{7}{39} & 1 & \\ \frac{6}{13} & \frac{1}{35} & 1 \end{bmatrix}} \overset{R_2}{\begin{bmatrix} \frac{13}{3} & \frac{1}{7} & 3 \\ & -\frac{30}{13} & \frac{7}{13} \\ & & -\frac{7}{5} \end{bmatrix}},$$

$$A_3 = R_2 I_2^{-1} L_2 = \begin{bmatrix} 5 \cdot 69 & 0 \cdot 23 & 3 \cdot 00 \\ 0 \cdot 66 & -2 \cdot 29 & 0 \cdot 54 \\ -0 \cdot 65 & -0 \cdot 04 & -1 \cdot 40 \end{bmatrix} =$$

$$= \overset{\textstyle I_3^{-1} L_3}{\begin{bmatrix} 1 & & \\ 0 \cdot 12 & 1 & \\ -0 \cdot 11 & 0 \cdot 01 & 1 \end{bmatrix}} \overset{\textstyle R_3}{\begin{bmatrix} 5 \cdot 69 & 0 \cdot 23 & 3 \cdot 00 \\ & -2 \cdot 32 & 0 \cdot 18 \\ & & -1 \cdot 07 \end{bmatrix}},$$

$$A_4 = R_3 I_3^{-1} L_3 = \begin{bmatrix} 5 \cdot 39 & 0 \cdot 26 & 3 \cdot 00 \\ -0 \cdot 30 & -2 \cdot 32 & 0 \cdot 18 \\ 0 \cdot 12 & -0 \cdot 01 & -1 \cdot 07 \end{bmatrix},$$

and already the diagonal terms give the roots with an error less than five units in the second decimal place. Here $I_2 = I_3 = I$.

The Q-R method

†30. The Q-R method of Francis also avoids the problem of large numbers by decomposing the matrix $A = A_1$ into the product $Q_1 R_1$, where R_1 is an upper triangle and Q_1 is orthogonal. The successive transformations, similar to those of the L-R method, are then given by

$$A_1 = Q_1 R_1, \quad A_2 = R_1 Q_1 = Q_2 R_2, \quad A_3 = R_2 Q_2 = Q_3 R_3, \ldots, \quad (96)$$

and we have the similarity transformations

$$A_2 = R_1 A_1 R_1^{-1}, \qquad A_3 = R_2 A_2 R_2^{-1} = (R_2 R_1) A_1 (R_2 R_1)^{-1}, \ldots . \quad (97)$$

At each stage the orthogonal matrix Q_k is built up from the product of simple orthogonal matrices of the type used in the methods of Jacobi and Givens. With each simple matrix we 'eliminate' a single element, and proceed column by column to eliminate all the sub-diagonal elements. For a (3×3) matrix, for example, we have

$$\overset{\textstyle T_1'}{\begin{bmatrix} c & s & 0 \\ -s & c & 0 \\ 0 & 0 & 1 \end{bmatrix}} \begin{bmatrix} a_{11} & a_{12} & a_{13} \\ a_{21} & a_{22} & a_{23} \\ a_{31} & a_{32} & a_{33} \end{bmatrix} = \begin{bmatrix} a_{11}' & a_{12}' & a_{13}' \\ 0 & a_{22}' & a_{23}' \\ a_{31} & a_{32} & a_{33} \end{bmatrix}, \quad (98)$$

provided that $-sa_{11}+ca_{21} = 0$, so that

$$s = \frac{a_{21}}{\sqrt{(a_{11}^2+a_{21}^2)}}, \quad c = \frac{a_{11}}{\sqrt{(a_{11}^2+a_{21}^2)}}. \tag{99}$$

Then

$$\begin{matrix} & & T_2' \\ \begin{bmatrix} c & 0 & s \\ 0 & 1 & 0 \\ -s & 0 & c \end{bmatrix} & \begin{bmatrix} a_{11}' & a_{12}' & a_{13}' \\ 0 & a_{22}' & a_{23}' \\ a_{31} & a_{32} & a_{33} \end{bmatrix} & = \begin{bmatrix} a_{11}'' & a_{12}'' & a_{13}'' \\ 0 & a_{22}' & a_{23}' \\ 0 & a_{32}'' & a_{33}'' \end{bmatrix}, \end{matrix} \tag{100}$$

if

$$s = \frac{a_{31}}{\sqrt{(a_{11}'^2+a_{31}^2)}}, \quad c = \frac{a_{11}'}{\sqrt{(a_{11}'^2+a_{31}^2)}}, \tag{101}$$

and further elimination is performed on the second column with a matrix T_3 which has the elements of the unit matrix in its first row and column.

The continuation in the general case is obvious and we finally produce an upper triangle R. Then

$$T_3'T_2'T_1'A = R, \quad A = T_1T_2T_3R = QR. \tag{102}$$

The reverse product $RQ = RT_1T_2T_3$ is obtained by successive post-multiplications of R with the T matrices, each of which is a simple operation.

Reduction to Hessenberg form

†31. The similarity transformation to upper triangular form of the L-R and Q-R methods is analogous to the transformation to diagonal form by the Jacobi method, involving a truncated infinite sequence of operations. We mention finally a method analogous to that of Givens and Householder in which the matrix is transformed in a finite sequence of operations to an almost upper (lower) triangular form, that is a matrix with one extra sloping line below (above) the diagonal, a so-called upper (lower) *Hessenberg matrix*.

Obvious possibilities for a general unsymmetric matrix are the methods of Givens and Householder, which would in fact produce a Hessenberg matrix, and this would also be triple-diagonal in the symmetric case. But the number of arithmetic operations can be reduced somewhat with the use of *elementary* similarity transformations, using simple matrices M and the formulae

$$A_2 = M_2^{-1}A_1M_2, \quad A_3 = M_3^{-1}A_2M_3,..., \tag{103}$$

the matrices M_r being chosen to 'eliminate' the unwanted elements in particular rows or columns. They are in fact very similar to the matrices of multipliers in the Gauss elimination.

Consider, for example, for a matrix of order four, the operation

$$
A_2 = \overset{M_2^{-1}}{\begin{bmatrix} 1 & 0 & 0 & 0 \\ 0 & 1 & 0 & 0 \\ 0 & -m_{32} & 1 & 0 \\ 0 & -m_{42} & 0 & 1 \end{bmatrix}} \overset{A_1}{\begin{bmatrix} a_{11} & a_{12} & a_{13} & a_{14} \\ a_{21} & a_{22} & a_{23} & a_{24} \\ a_{31} & a_{32} & a_{33} & a_{34} \\ a_{41} & a_{42} & a_{43} & a_{44} \end{bmatrix}} \overset{M_2}{\begin{bmatrix} 1 & 0 & 0 & 0 \\ 0 & 1 & 0 & 0 \\ 0 & m_{32} & 1 & 0 \\ 0 & m_{42} & 0 & 1 \end{bmatrix}}, \quad (104)
$$

where $m_{32} = a_{31}/a_{21}$, $m_{42} = a_{41}/a_{21}$, and in which we have already previously verified the relative forms of M_2^{-1} and M_2. The matrix A_2 has the appearance

$$
A_2 = \begin{bmatrix} a_{11} & a_{12}^{(2)} & a_{13} & a_{14} \\ a_{21} & a_{22}^{(2)} & a_{23} & a_{24} \\ 0 & a_{32}^{(2)} & a_{33}^{(2)} & a_{34}^{(2)} \\ 0 & a_{42}^{(2)} & a_{43}^{(2)} & a_{44}^{(2)} \end{bmatrix}, \quad (105)
$$

where the notation indicates the distribution of old and new elements. Then we take

$$
A_3 = \overset{M_3^{-1}}{\begin{bmatrix} 1 & 0 & 0 & 0 \\ 0 & 1 & 0 & 0 \\ 0 & 0 & 1 & 0 \\ 0 & 0 & -m_{43} & 1 \end{bmatrix}} \overset{A_2}{\begin{bmatrix} a_{11} & a_{12}^{(2)} & a_{13} & a_{14} \\ a_{21} & a_{22}^{(2)} & a_{23} & a_{24} \\ 0 & a_{32}^{(2)} & a_{33}^{(2)} & a_{34}^{(2)} \\ 0 & a_{42}^{(2)} & a_{43}^{(2)} & a_{44}^{(2)} \end{bmatrix}} \overset{M_3}{\begin{bmatrix} 1 & 0 & 0 & 0 \\ 0 & 1 & 0 & 0 \\ 0 & 0 & 1 & 0 \\ 0 & 0 & m_{43} & 1 \end{bmatrix}}, \quad (106)
$$

where $m_{43} = a_{42}^{(2)}/a_{32}^{(2)}$, and the resulting appearance is the upper Hessenberg form

$$
A_3 = \begin{bmatrix} a_{11} & a_{12}^{(2)} & a_{13}^{(3)} & a_{14} \\ a_{21} & a_{22}^{(2)} & a_{23}^{(3)} & a_{24} \\ 0 & a_{32}^{(2)} & a_{33}^{(3)} & a_{34}^{(2)} \\ 0 & 0 & a_{43}^{(3)} & a_{44}^{(3)} \end{bmatrix}, \quad (107)
$$

the zeros of A_2 being unaffected by the later transformation. The continuation in the general case is obvious.

32. Like Gauss elimination, however, the process breaks down or may be very inaccurate if any of the 'pivots' is zero or very small, that is if any multiplier is infinite or very large. Wilkinson has shown that this can be avoided by 'row interchanges' and the introduction of the row-permuting matrices.

In this process we select the largest of the elements $a_{21}^{(1)}$, $a_{31}^{(1)}$,..., $a_{n1}^{(1)}$ in the original matrix, and if this is in row r we interchange the rows 2 and r and also the columns 2 and r, so that we have performed the transformation

$$B_1 = I_1 A_1 I_1^{-1}, \tag{108}$$

where I_1 is the relevant row-permuting matrix, here not only orthogonal so that (108) is an orthogonal similarity transformation, but also symmetric.

We can now choose the M_2 as in (104), where A_1 is replaced by B_1, and its elements cannot exceed unity. The matrix A_2 is then given by

$$A_2 = M_2^{-1} B_1 M_2 = M_2^{-1} I_1 A_1 I_1^{-1} M_2, \tag{109}$$

and the transformation is still of similarity type.

At the next stage we choose the largest of the elements $a_{32}^{(2)}$, $a_{42}^{(2)}$,..., and if this is in row s we interchange rows and columns 3 and s, forming

$$B_2 = I_2 A_2 I_2^{-1}. \tag{110}$$

The M_3 can now be chosen as in (106), with B_2 replacing A_2, and is such that its elements cannot exceed unity. Then

$$B_3 = M_3^{-1} B_2 M_3 = M_3^{-1} I_2 A_2 I_2^{-1} M_3, \tag{111}$$

and the important fact is that this latest transformation preserves the zeros in the first column of A_2. The obvious continuation also preserves all previous zeros, and after $n-2$ transformations we have B_{n-1} as an upper Hessenberg matrix and a similarity transform of $A = A_1$.

Roots and vectors of Hessenberg matrix

33. To complete the solution of our problem we must next find roots and vectors of the Hessenberg matrix, and finally operate on the vectors with the matrix of the transformation to produce vectors of A. The latter is relatively trivial, and we now consider briefly two methods for computing the latent solutions of, say, an upper Hessenberg matrix.

The first possibility is to reduce it by further similarity transformations to a matrix of triple-diagonal form. Consider for example the Hessenberg matrix of type (107), in the sequel denoted by $A = A_1$ with obvious notation for its elements. It can easily be verified that the similarity transformation

$$A_2 = M_2 A_1 M_2^{-1}, \tag{112}$$

where

$$M_2^{-1} = \begin{bmatrix} 1 & 0 & 0 & 0 \\ 0 & 1 & -m_{23} & -m_{24} \\ 0 & 0 & 1 & 0 \\ 0 & 0 & 0 & 1 \end{bmatrix}, \tag{113}$$

M_2 is the same with the signs changed in the m_{rs}, and

$$m_{23} = a_{13}/a_{12}, \qquad m_{24} = a_{14}/a_{12}, \tag{114}$$

produces zeros in the required places in the top row without destroying those in the first two columns.

Finally the transformation

$$A_3 = M_3 A_2 M_3^{-1}, \tag{115}$$

with

$$M_3^{-1} = \begin{bmatrix} 1 & 0 & 0 & 0 \\ 0 & 1 & 0 & 0 \\ 0 & 0 & 1 & -m_{34} \\ 0 & 0 & 0 & 1 \end{bmatrix}, \tag{116}$$

where m_{34} is chosen to produce the single required zero in the second row, completes the reduction to triple-diagonal form.

The solution can then be completed by any suitable methods, such as those of §§ 22–23 for an unsymmetric triple-diagonal matrix.

†34. Alternatively we can use the Hessenberg form without further transformation, obtaining simultaneously a latent root and the corresponding latent vector. For the upper Hessenberg matrix of order four we have to solve the equations

$$\left.\begin{aligned} (a_{11}-\lambda)x_1 + a_{12}x_2 + a_{13}x_3 + a_{14}x_4 &= 0 \\ a_{21}x_1 + (a_{22}-\lambda)x_2 + a_{23}x_3 + a_{24}x_4 &= 0 \\ a_{32}x_2 + (a_{33}-\lambda)x_3 + a_{34}x_4 &= 0 \\ a_{43}x_3 + (a_{44}-\lambda)x_4 &= 0 \end{aligned}\right\}. \tag{117}$$

Now for any general value of λ we can solve the last three equations in succession, taking $x_4 = 1$ and working backwards to produce $x_3, x_2, \ldots, x_1$. Substituting in the first equation we find a number which is proportional to $|A - I\lambda|$ for that particular value of λ. By methods such as that of Muller in § 22 we can find a zero and the corresponding vector.

Additional notes and bibliography

§ 1. Any of the transformation methods for finding latent roots and vectors of a matrix A can clearly be used for the determination of A^{-1}. For if $Y^{-1}AY = B$, then $A^{-1} = YB^{-1}Y^{-1}$, and the computation of B^{-1} is relatively easy. It does not appear, however, that these methods have any advantages, either in speed or accuracy, over the more obvious processes for finding A^{-1}.

§ 4. The convergence of the Jacobi process is treated in

FORSYTHE, G. E. and HENRICI, P., 1960, The cyclic Jacobi method for computing the principal values of a complex matrix. *Trans. Amer. math. Soc.* **94**, 1–23.

§ 5. The method of Givens was first described in

GIVENS, W., 1954, Numerical computation of the characteristic values of a real symmetric matrix. *Oak Ridge National Laboratory, ORNL*-1574.

For matrices of large order, treated by a machine with a small high-speed memory, the matrix will commonly be located in an auxiliary 'backing store'. For the process described here, which at a typical stage replaces rows and columns i and j by linear combinations of these rows and columns, the storage of the matrix by rows necessitates the transfer of all rows i to n to the high-speed store. It is possible to run some rotations 'in parallel', thereby decreasing the transfer time by a factor of order n, and this method is described in

ROLLETT, J. S. and WILKINSON, J. H., 1961, An efficient scheme for the codiagonalisation of a symmetric matrix by Givens' method in a computer with a two-level store. *Computer J.* **4**, 177–80.

The paper also compares the storage requirements and the amounts of arithmetic for the Givens and Householder methods respectively. Apart from the computation of square roots, Givens takes about $\frac{4}{3}n^3$ multiplications and Householder about half as many.

§ 6. The idea of the Givens process can be used for inverting a general unsymmetric matrix A. Successive premultiplication of A by matrices $Y'_{r,s}$, corresponding to rotations in the (r, s) plane, will produce an upper triangular matrix if we take (r, s) in order $(1, 2), (1, 3), \ldots, (1, n); (2, 3), (2, 4), \ldots, (2, n)$; etc. If the product of these orthogonal matrices is Y', we have $Y'A = U$, so that $A^{-1} = U^{-1}Y'$, and the inversion of U is a fairly simple operation. An error analysis of this process is given by Wilkinson (1961) *loc. cit.*, p. 170, and again he finds no particular advantage over the more obvious methods.

§ 8. The escalator methods are described in

MORRIS, J., 1947, *The escalator process*. Chapman and Hall, London,
FOX, L., 1952, Escalator methods for latent roots. *Quart. J. Mech.* **5,** 178–90, and
FORSYTHE, G. E., 1952, Alternative derivations of Fox's escalator formulae for latent roots. *Quart. J. Mech.* **5,** 191–5.

We can produce equation (28) rather more easily by observing that if the normalized vectors x of C_r form the orthogonal matrix X, and its latent roots are λ_s, $s = 1, 2,..., r$, then

$$(C - I\mu)^{-1} = X D^{-1} X',$$

where D^{-1} is a diagonal matrix with elements $(\lambda_s - \mu)^{-1}$. The result (28) follows immediately.

These are some special cases which are not here taken into account, for example the possibility of equal roots which we have ignored throughout in this book, and the possibility of a null vector p, which is hardly important since the matrix C_{r+1} then splits into two parts which can be treated separately.

§ 9. A full theory of the Sturm sequence in this connexion is given in

GIVENS, W., 1953, A method of computing eigenvalues and eigenvectors suggested by classical results on symmetric matrices. *U.S. Bur. Stand. Appl. Math. Ser.* **29,** 117–22,

and in

ORTEGA, J. M., 1960, On Sturm sequences for tridiagonal matrices. Applied Mathematics and Statistics Laboratories, Stanford University, Tech. Rep. No. 4.

§ 12. Householder's method was first described in

HOUSEHOLDER, A. S. and BAUER, F. L., 1959, On certain methods for expanding the characteristic polynomial. *Numerische Mathematik* **1,** 29–37.

The computational process illustrated here is given in

WILKINSON, J. H., 1960, Householder's method for the solution of the algebraic eigenproblem. *Computer J.* **3,** 23–7.

We can also use a modification of the Householder method to find the inverse of a general unsymmetric A. Successive premultiplication by matrices of the general form of P, first with $w^{(1)'} = (w_1, w_2, w_3, ..., w_n)$, then with a zero in the first element of w, and so on, will reduce A to upper triangular form. This process is also analysed for error by Wilkinson (1961) *loc. cit.*, p. 170, again without significant advantage over other methods.

§ 17. The result about the uniqueness of the triple-diagonal form was proved by Rollett and Wilkinson (1961) *loc. cit.*, p. 268. Their analysis uses matrix algebra, but is otherwise effectively the same. If the columns of Y in $AY = YB$ are denoted by $e^{(1)}$, $c^{(2)}$, $c^{(3)},..., c^{(n)}$ we have, for the first column,

$$Ae^{(1)} = a_1 e^{(1)} + b_2 c^{(2)},$$

and

$$e^{(1)'}Ae^{(1)} = a_1 e^{(1)'}e^{(1)} + b_2 e^{(1)'}c^{(2)} = a_1,$$

in virtue of the orthonormal properties of the columns of Y. This determines a_1 uniquely.

Then $b_2 c^{(2)} = Ae^{(1)} - a_1 e^{(1)}$ and is unique. Also $c^{(2)}$ is of unit length, so that b_2 and $c^{(2)}$ are unique apart from sign, the signs of b_2 and $c^{(2)}$ being the same.

For the next column,

$$Ac^{(2)} = b_2 e^{(1)} + a_2 c^{(2)} + b_3 c^{(3)},$$

and premultiplication with $c^{(2)\prime}$ gives $a_2 = c^{(2)\prime} Ac^{(2)}$ which is unique. Then $b_3 c^{(3)} = Ac^{(2)} - b_2 e^{(1)} - a_2 c^{(2)}$, and the sign of the right-hand side is everywhere that of b_2. Then b_3 and $c^{(3)}$ are unique, their product having the sign of b_2, and so on.

§ 18. The method of Lanczos was first given in

LANCZOS, C., 1950, An iteration method for the solution of the eigenvalue problem of linear differential and integral operators. *J. Res. Nat. Bur. Stand.* **45**, 255–82.

The computational process illustrated here is given in

WILKINSON, J. H., 1958, The calculation of eigenvectors by the method of Lanczos. *Computer J.* **1**, 148–52.

This paper makes various important points about the computation. First, it is possible that a vector $y^{(r)}$ could be null for $r < n+1$. In this case we can take instead of $y^{(r)}$ *any* vector $\eta^{(r)}$ which is orthogonal to all previous $y^{(k)}$, $k = 1, 2, \ldots, r-1$. The next vector $y^{(r+1)}$ is then given by

$$y^{(r+1)} = A\eta^{(r)} - \alpha_r \eta^{(r)} - \beta_r y^{(r-1)},$$

and the orthogonality of $y^{(r+1)}$ to $\eta^{(r)}$ and $y^{(r-1)}$ gives

$$\beta_r = y^{(r-1)\prime} A\eta^{(r)} / y^{(r-1)\prime} y^{(r-1)}.$$

The numerator of β_r can be expressed in the form

$$y^{(r-1)\prime} A\eta^{(r)} = \eta^{(r)\prime} Ay^{(r-1)} = \eta^{(r)\prime} \{y^{(r)} + \alpha_{r-1} y^{(r-1)} + \beta_{r-1} y^{(r-2)}\} = \eta^{(r)\prime} y^{(r)},$$

which vanishes since $y^{(r)}$ is null. Hence $\beta_r = 0$, and with $r = 3$, for example, the matrix C of $AY = YC$, given in the general case by (69), now becomes

$$C = \begin{bmatrix} \alpha_1 & \beta_2 & & \\ 1 & \alpha_2 & 0 & \\ \hline & 0 & \alpha_3 & \beta_4 \\ & & 1 & \alpha_4 & \beta_5 \end{bmatrix},$$

splitting into two separate sub-matrices.

Second, it is common for any $y^{(r)}$, while not exactly null, to lose significant figures by cancellation, so that $y^{(r)}$ will not be orthogonal to previous vectors to the full number of significant figures. This orthogonality is strictly necessary, and it is therefore important to add the appropriate number of extra digits in

$y^{(r)}$, and to choose them so that $y^{(r)}$ is truly orthogonal, to working accuracy, to all previous vectors. That is we replace such a y_r by

$$\eta_r = y^{(r)} - \alpha_1 y^{(1)} - \alpha_2 y^{(2)} - \ldots - \alpha_{r-1} y^{(r-1)}, \text{ where}$$
$$\alpha_k = y^{(k)\prime} y^{(r)} / y^{(k)\prime} y^{(k)},$$

performing this 'reorthogonalising' process whenever necessary.

§ 21. Reorthogonalization is also necessary in the unsymmetric Lanczos method, and Wilkinson (1958) *loc. cit.*, p. 270, also discusses the effect of the exact vanishing of either a $y^{(r)}$ or a $z^{(r)}$. If $y^{(r)}$ is null we continue with any vector which is orthogonal to all previous $z^{(k)}$, and if $z^{(r)}$ is null we continue with any vector which is orthogonal to all previous $y^{(k)}$, $k = 1, 2, \ldots, r-1$. But if $y^{(r)}$ or $z^{(r)}$ is null, but not both, the corresponding β_r and β'_r do not have the same values. If $y^{(r)}$ is null and we continue with $\eta^{(r)}$, we have

$$y^{(r+1)} = A\eta^{(r)} - \alpha_r \eta^{(r)} - \beta_r y^{(r-1)},$$
$$z^{(r+1)} = A'z^{(r)} - \alpha'_r z^{(r)} - \beta'_r z^{(r-1)},$$

and $\beta_r = z^{(r-1)\prime} A\eta^{(r)} / z^{(r-1)\prime} y^{(r-1)}$. The numerator simplifies to

$$z^{(r-1)\prime} A\eta^{(r)} = \eta^{(r)\prime} A'z^{(r-1)} = \eta^{(r)\prime} \{z^{(r)} + \alpha'_{r-1} z^{(r-1)} + \beta'_{r-1} z^{(r-2)}\} = \eta^{(r)\prime} z^{(r)}.$$

But

$$\beta'_r = y^{(r-1)\prime} A'z^{(r)} / y^{(r-1)\prime} z^{(r-1)},$$

and the numerator simplifies to

$$y^{(r-1)\prime} A'z^{(r)} = z^{(r)\prime} Ay^{(r-1)} = z^{(r)\prime} \{y^{(r)} + \alpha_{r-1} y^{(r-1)} + \beta_{r-1} y^{(r-2)}\} = z^{(r)\prime} y^{(r)} = 0.$$

Hence $\beta'_r = 0$ while $\beta_r \neq 0$. Similarly if $z^{(r)}$ is null but not $y^{(r)}$, we have $\beta_r = 0$ but $\beta'_r \neq 0$.

The matrices corresponding to (69) are respectively typified by

$$C = \begin{bmatrix} \alpha_1 & \beta_2 & & & \\ 1 & \alpha_2 & \beta_3 & & \\ \hline & 0 & \alpha_3 & \beta_4 & \\ & & 1 & \alpha_4 & \beta_5 \\ & & & \cdots \cdots & \end{bmatrix}, \quad C = \begin{bmatrix} \alpha_1 & \beta_2 & & & \\ 1 & \alpha_2 & 0 & & \\ \hline & 1 & \alpha_3 & \beta_4 & \\ & & 1 & \alpha_4 & \beta_5 \\ & & & \cdots \cdots & \end{bmatrix}.$$

In the latter case the vectors corresponding to the roots of the second sub-matrix have zeros in the first two components, but those corresponding to the roots of the first sub-matrix do not have zeros in components 3, 4,... . The full set of zeros occurs only when $y^{(r)}$ and $z^{(r)}$ are simultaneously null, so that both β_r and β'_r are zero.

§ 22. The most useful iterative methods for finding the roots of explicit polynomials are those of Newton and Bairstow, of which a sufficiently detailed description is given in the paper by Wilkinson (1959) *loc. cit.*, p. 233. Some precautions, however, are required when any root already obtained is 'divided' out, producing a polynomial of lower degree. Wilkinson shows in his paper that it is essential, for the accuracy of later roots, to compute and divide them out in ascending order of magnitude.

He considers the polynomial $f(x) = (x+2^{-1})(x+2^{-2})...(x+2^{-20})$, and supposes that $-(2^{-1}+\delta)$ is a computed approximation to the largest root. The division process gives $f(x) = (x+2^{-1}+\delta)g(x)+r$, and r is zero if δ is zero. The zeros of $g(x)$ are those of $f(x)-r$, and he shows that if $\delta = 10^{-18}$ then $|r| \sim 2^{-81}$. The constant term of $f(x)$ is 2^{-210}, so that of $f(x)-r$ is greater by the factor 2^{129}. Since the constant term is the product of the roots at least some roots of $f(x)-r$ must be substantially different from those of $f(x)$. If the smallest root is eliminated first the remainder is very small and has negligible effect.

The method of Muller was first given in

MULLER, D. E., 1956, A method for solving algebraic equations using an automatic computer. *Math. Tab. Wash.* **10**, 208–15.

It is particularly advantageous in this context since we do not form the explicit polynomial, and the subsequent use of $f(\lambda)/(\lambda-\lambda_1)$ avoids explicit division so that the order of calculation of the roots is relatively unimportant.

§ 24. This analysis was first given, with particular reference to the Givens method, in

WILKINSON, J. H., 1958, The calculation of the eigenvectors of codiagonal matrices. *Computer J.* **1**, 90–6.

§ 26. The L-R method was first given in

RUTISHAUSER, H., 1958, Solution of eigenvalue problems with the L-R transformation. *U.S. Bur. Stand. Appl. Math. Ser.* **49**, 47–81.

§ 30. The Q-R method is given in

FRANCIS, J. G. F., 1961, The Q-R transformation, Parts I and II. *Computer J.* **4**, 265–71, 332–45.

§ 31–33. This analysis for the transformation to Hessenberg form is given in

WILKINSON, J. H., 1959, Stability of the reduction of a matrix to almost triangular and triangular forms by elementary similarity transformations. *J. Ass. comp. Mach.* **6**, 336–59.

He shows that, whereas stability can be guaranteed by the interchange process of §32 for the production of the Hessenberg form, it is not possible to find a guaranteed stable process for the further reduction to triple-diagonal form.

Similar analysis, with comparison with the Lanczos process, is given in

STRACHEY, C. and FRANCIS, J. G. F., 1961, The reduction of a matrix to co-diagonal form by eliminations. *Computer J.* **4**, 168–76,

and a detailed error analysis appears in

WILKINSON, J. H., 1962, Instability of the elimination method of reducing a matrix to tri-diagonal form. *Computer J.* **5**, 61–70.

§ 34. This suggestion is given in

HYMAN, H., 1957, Eigenvalues and eigenvectors of general matrices. *Twelfth National meeting A.C.M., Houston, Texas,*

and its accuracy is analysed in

WILKINSON, J. H., 1960, Error analysis of floating-point computation. *Numerisch Mathematik* **2**, 319–40.

An application of Bairstow's method, for producing successive roots of the Hessenberg matrix without explicit evaluation of the characteristic polynomial, and therefore without the problem and difficulty of explicit 'dividing out', is given in

HANDSCOMB D. C., 1962, Computation of the latent roots of a Hessenberg matrix by Bairstow's method. *Computer J.* **5**, 139–41.

The solution of the Hessenberg form can also be effected by a process of deflation, at each stage of which a latent root is computed, perhaps by iteration, and the matrix deflated to a similar Hessenberg form of one order lower.

Consider for example the Hessenberg form

$$A - I\lambda = \begin{bmatrix} a_{11}-\lambda & a_{12} & a_{13} & a_{14} \\ a_{21} & a_{22}-\lambda & a_{23} & a_{24} \\ & a_{32} & a_{33}-\lambda & a_{34} \\ & & a_{43} & a_{44}-\lambda \end{bmatrix},$$

in which we assume that no element $a_{r+1,r}$ is zero, and suppose we know a latent root λ_1. We seek a similarity transformation $H^{-1}AH$ such that the last column of the new matrix has zeros in the first three rows and λ_1 in the fourth, that is we seek a transformation $H^{-1}(A-I\lambda_1)H$ such that the new matrix has its fourth column null.

This we can effect by taking H to be the product of orthogonal matrices of the type used in the Givens process. In fact if we rotate in the successive planes $(4, 3)$, $(4, 2)$ and $(4, 1)$, choosing the angles so that $a_{44}-\lambda_1$, a_{34} and a_{24} are reduced successively to zero, the multiplications on the right transform $(A-I\lambda_1)$ to the shape

$$B = \begin{bmatrix} \times & \times & \times & b_{14} \\ b_{21} & \times & \times & 0 \\ & b_{32} & \times & 0 \\ & & b_{43} & 0 \end{bmatrix}.$$

Now if λ_1 is a root we have $|A-I\lambda_1| = 0$, and the determinant of H is unity so that $|B| = 0$. But $|B| = -b_{14}\,b_{21}\,b_{32}\,b_{43}$, and it can be shown, with our assumption about the elements of A, that none of $b_{r+1,r}$ can vanish. It follows that $b_{14} = 0$ and the last column of B is null.

We complete the similarity transformation by premultiplication with H^{-1}, and the transformed matrices of $A-I\lambda_1$ and of A have the respective forms

$$\begin{bmatrix} \times & \times & \times & 0 \\ \times & \times & \times & 0 \\ 0 & \times & \times & 0 \\ \times & \times & \times & 0 \end{bmatrix}, \quad \begin{bmatrix} \times & \times & \times & 0 \\ \times & \times & \times & 0 \\ 0 & \times & \times & 0 \\ \times & \times & \times & \lambda_1 \end{bmatrix}.$$

The remaining roots of A are those of the leading matrix of order $n-1$, which we note is also of Hessenberg form. The continuation is therefore obvious and at the end we have reduced the matrix A, by orthogonal similarity transformations, to a (lower) triangular form.

Complex arithmetic is of course introduced by a possibly complex λ_1, and the orthogonal transformation then becomes a unitary transformation $\bar{H}'AH$, where the bar denotes the complex conjugate. We take, typically, the forms

$$H = \begin{bmatrix} 1 & 0 & 0 & 0 \\ 0 & \bar{c} & -s & 0 \\ 0 & \bar{s} & c & 0 \\ 0 & 0 & 0 & 1 \end{bmatrix}, \quad \bar{H}' = \begin{bmatrix} 1 & 0 & 0 & 0 \\ 0 & c & s & 0 \\ 0 & -\bar{s} & \bar{c} & 0 \\ 0 & 0 & 0 & 1 \end{bmatrix},$$

and the values of s and c, though no longer representing respectively the sine and cosine of an angle, are still given by equations like (18).

The vector of A corresponding to the root λ_1 is obtained as a by-product of this method. For if $A_1 = \bar{H}'AH$, the form of A_1 shows that its vector corresponding to λ_1 is $e^{(n)}$, the last column of the unit matrix. The vector of A is then $He^{(n)}$, which is the last column of H.

Full details and extensions of this process are given in

GIVENS, W., 1958, Computation of plane unitary rotations transforming a general matrix to triangular form. *J. Soc. industr. appl. Math.* **6,** 26–50.

An analogous elementary similarity transformation is given by

WILKINSON, J. H., 1959, (*loc. cit.*, p. 272).

11

Notes on Error Analysis for Latent Roots
and Vectors

Introduction

1. THE error analysis of the latent root and vector problem has the same general features as that for linear equations. We have first the inherent errors relevant, at least for physical problems, to the ill-conditioned nature of the problem. We are concerned here about the extent to which small changes in the coefficients of the matrix change the latent roots and vectors. Second we should analyse each individual technique, and current work attempts, as in the case of linear equations, to find what problem we have actually solved, that is what is the perturbed matrix $A + \delta A$ for which our solution is exact. The methods are classified as good or bad according as the bounds for the elements of δA are small or large, and in the literature the words 'stable' and 'unstable', applied to any process, have this general significance.

Finally we examine our computed results and look for methods of improving them, at least for mathematical problems, in a way analogous to that used for linear equations in the computation of the residuals of a first approximation as a preliminary to improvement. In this chapter we shall consider first the general question of ill-conditioning, then some methods of improving approximate solutions, and finally the methods of perturbation analysis. Since the field is large and is discussed in comprehensive detail in J. H. Wilkinson's forthcoming book we shall here give little more than general indications of the methods. Again we restrict ourselves to the case of matrices with distinct latent roots.

Ill-conditioning

2. The problem of ill-conditioning is concerned with the difference between the latent solutions of a matrix A and a small perturbation $A + \delta A$. We suppose that A has latent roots $\lambda_1, \lambda_2, ..., \lambda_n$ and vectors $x^{(1)}, x^{(2)}, ..., x^{(n)}$, and that the vectors of A' are $y^{(1)}, y^{(2)}, ..., y^{(n)}$.

Assuming that $A + \delta A$ has roots $\lambda + \delta \lambda$, and that the latent vectors are $x^{(r)} + \sum_{s \neq r} \epsilon_{rs} x^{(s)}$, where ϵ_{rs} is small, we have

$$(A + \delta A)(x^{(r)} + \sum_{s \neq r} \epsilon_{rs} x^{(s)}) = (\lambda_r + \delta \lambda_r)(x^{(r)} + \sum_{s \neq r} \epsilon_{rs} x^{(s)}). \quad (1)$$

With neglect of second-order quantities we find

$$\delta A x^{(r)} + \sum_{s \neq r} \epsilon_{rs} A x^{(s)} = \delta \lambda_r x^{(r)} + \sum_{s \neq r} \epsilon_{rs} \lambda_r x^{(s)}, \quad (2)$$

and premultiplication with $y^{(r)\prime}$ gives

$$\delta \lambda_r = y^{(r)\prime} \, \delta A x^{(r)} / y^{(r)\prime} x^{(r)}, \quad (3)$$

the other terms vanishing in virtue of the orthogonality properties of the vectors.

If now the vectors are normalized so that $y^{(r)\prime} y^{(r)} = x^{(r)\prime} x^{(r)} = 1$, and no element of δA has modulus greater than ϵ, equation (3) gives

$$|\delta \lambda_r| < n\epsilon / |y^{(r)\prime} x^{(r)}|. \quad (4)$$

For unsymmetric matrices $y^{(r)\prime} x^{(r)}$ can be arbitrarily small, and the change $\delta \lambda_r$ is then arbitrarily large, but for symmetric matrices $y^{(r)\prime} x^{(r)} = x^{(r)\prime} x^{(r)} = 1$, and

$$|\delta \lambda_r| < n\epsilon. \quad (5)$$

It follows that the latent roots of a symmetric matrix are not greatly affected by small changes in the elements, whereas in the unsymmetric case the situation depends on the product $y^{(r)\prime} x^{(r)}$ of vectors of A' and A.

†3. We have seen previously that the condition of the vectors depends on the degree of separation of the latent roots, and the following deeper analysis reveals this more clearly. It is convenient to take the perturbation $\delta A = \epsilon B$, where every element of B is less than unity in modulus, and we assume also that A is scaled so that every $|a_{rs}| < 1$.

In the analysis we use two classical theorems of Gershgorin about the distribution of latent roots. The first states that every root of the matrix A lies in at least one of the circles with centres a_{rr} and radii $\sum_{s \neq r} |a_{rs}|$. The second states that if p of these circles form a connected region which is isolated from the other circles, then there are p latent roots within this region. In particular if any such circle does not intersect any other it contains just one latent root.

Consider now the matrix A, with its normalized vectors $x^{(r)}$ forming the modal matrix X, and take Y' to be the matrix whose rows are the vectors $y^{(r)'}/q_r$, where

$$q_r = y^{(r)'}x^{(r)}. \tag{6}$$

Then $Y' = X^{-1}$, and we have the similarity transformation

$$Y'AX = \Lambda, \tag{7}$$

where Λ is diagonal with elements $\lambda_1, \lambda_2, ..., \lambda_n$.

We then examine the latent roots of the perturbation $A + \epsilon B$ by considering the corresponding similarity transformation

$$Y'(A + \epsilon B)X = \Lambda + \epsilon C, \tag{8}$$

where the element c_{rs} of C is given by

$$c_{rs} = y^{(r)'}Bx^{(s)}/q_r. \tag{9}$$

4. The first Gershgorin theorem shows that the latent roots $\lambda_r(\epsilon)$ of $A + \epsilon B$, which are those of $\Lambda + \epsilon C$, lie in circles with centres $\lambda_r + \epsilon c_{rr}$ and radii $\epsilon \sum_{s \neq r} |c_{rs}|$. But

$$|q_r c_{rs}| = |y^{(r)'}Bx^{(s)}| < \|y^{(r)}\| \, \|B\| \, \|x^{(s)}\| < n, \tag{10}$$

by the normalisation of the vectors and the restriction on the elements of B. The radii of the circles are then bounded by the quantities $n(n-1)\epsilon q_r^{-1}$.

It follows, as before, that the effect of the perturbation on the latent roots depends critically on q_r, which is the cosine of the angle between the right and left latent vectors of A. If any such vectors are nearly orthogonal the corresponding latent root is very badly determined, and q_r can be taken as a *condition number* for the root λ_r. For a symmetric matrix $q_r = 1$, and the latent root problem is reasonably well-conditioned.

5. Turning to the latent vectors, we suppose the vectors of $A + \epsilon B$ to be denoted by $x^{(r)}(\epsilon)$, and those of $Y'(A + \epsilon B)X$ by $z^{(r)}(\epsilon)$, so that $x^{(r)}(\epsilon) = Xz^{(r)}(\epsilon)$. We now consider a particular vector $z^{(r)}(\epsilon)$. It is clear that the rth component of $z^{(r)}(0)$ is unity, and all the rest zero, since $x^{(r)}(0) = Xz^{(r)}(0)$ and $x^{(r)}(0)$ is the rth column of X. We assume that for sufficiently small ϵ the rth component of $z^{(r)}(\epsilon)$ is the largest, and we normalize so that this component is unity. Then the equation

$$\lambda_r(\epsilon)z^{(r)}(\epsilon) = Y'(A + \epsilon B)Xz^{(r)}(\epsilon) \quad (\Lambda + \epsilon C)z^{(r)}(\epsilon) \tag{11}$$

gives for the sth component of $z^{(r)}(\epsilon)$, with $s \neq r$, the result

$$\lambda_r(\epsilon) z_s^{(r)}(\epsilon) = \lambda_s z_s^{(r)}(\epsilon) + \epsilon \sum_{t=1}^{n} c_{st} z_t^{(r)}(\epsilon), \qquad (12)$$

so that

$$|\lambda_r(\epsilon) - \lambda_s| \, |z_s^{(r)}(\epsilon)| < \epsilon \, |q_s^{-1}| \sum_{t=1}^{n} |y^{(s)'} B x^{(t)}|, \qquad (13)$$

which gives

$$|z_s^{(r)}(\epsilon)| < \frac{\epsilon \, |q_s^{-1}| \sum_{t=1}^{n} |y^{(s)'} B x^{(t)}|}{|\lambda_r(\epsilon) - \lambda_s|}. \qquad (14)$$

We see that those components $z_s^{(r)}(\epsilon)$ may be large for which a latent root λ_s is near to the λ_r corresponding to $z^{(r)}$, and all the components of $x^{(r)}(\epsilon) = X z^{(r)}(\epsilon)$ may be affected adversely.

It is clear from these results that the ill-conditioning of the latent root and vector problem is a somewhat complicated phenomenon. For a symmetric matrix the latent roots are well determined, but if two roots are close together the corresponding latent vectors are poorly determined. In the unsymmetric case any particular root may be poorly determined whereas the corresponding vector can be quite accurate.

Corrections to approximate roots and vectors

†6. We turn now to methods of improving approximate solutions, particularly for mathematical problems when this is worth while. Suppose that we have determined approximations to all the roots and vectors, so that in particular we have a first estimate X_0 of the modal matrix X which should satisfy $AX = X\Lambda$.

Some measure of the error of our solution is given by the difference from a diagonal of the matrix $X_0^{-1} A X_0$, and we write

$$X_0^{-1} A X_0 = D + E, \qquad (15)$$

where D is diagonal and close to Λ, and the elements e_{rs} of E are small.

Consider now the matrix B, defined by

$$BD - DB = E, \qquad (16)$$

so that if d_r is the rth element of D the element b_{rs} of B is given by

$$b_{rs} = e_{rs}/(d_s - d_r), \qquad (17)$$

and is also small unless d_s is near to d_r. An improved modal matrix is then given by

$$X_1 = X_0(I + B). \qquad (18)$$

For
$$AX_1 = AX_0(I+B) = X_0(D+E)(I+B)$$
$$= X_0\{(I+B)D - DB\}(I+B), \qquad (19)$$

giving
$$AX_1 = X_1 D + X_0 EB. \qquad (20)$$

The last term in (20) has components which are of second order of smallness, so that the elements of D are better approximations to the latent roots and the columns of X_1 are better approximations to the latent vectors. Further corrections can obviously be made by repetition of the process.

†7. A slightly different analysis gives bounds for the errors in the first approximation and also in the improved estimates. We consider first the symmetric case, where we can find bounds for the error in a single vector, and an improvement to the latent root, without very much information about the other roots and vectors.

Consider approximations λ and x to a true latent root λ_1 and latent vector $x^{(1)}$. If x is normalized so that $x'x = 1$, and we compute the residual vector
$$r = Ax - \lambda x, \qquad (21)$$

we show that if $r'r = \epsilon^2$ there is a root λ_1 of A such that
$$|\lambda - \lambda_1| < \epsilon. \qquad (22)$$

For we can expand our x in terms of the latent vectors in the form
$$x = \sum_1^n \alpha_r x^{(r)}, \qquad \sum_1^n \alpha_r^2 = 1, \qquad (23)$$

and then
$$r = \sum_1^n (\lambda_r - \lambda)\alpha_r x^{(r)}, \qquad r'r = \epsilon^2 = \sum_1^n (\lambda_r - \lambda)^2 \alpha_r^2. \qquad (24)$$

It follows that not all values of $|\lambda - \lambda_r|$ can be greater than ϵ, and for small enough ϵ we would expect to have $|\lambda - \lambda_1| < \epsilon$.

†8. To correct the approximate latent root λ, and to find a bound for the error in the vector x, we need only some knowledge of the distribution of the latent roots. We have seen that the Rayleigh quotient $\lambda_R = x'Ax$ gives the residual vector for which $r'r$ is a minimum, so that certainly
$$(Ax - \lambda_R x)'(Ax - \lambda_R x) < \epsilon^2, \qquad (25)$$

and the distance between λ_1 and λ_R cannot exceed ϵ. We can show that $|\lambda_1 - \lambda_R|$ is considerably less than ϵ provided that the other roots are at a distance from λ_R much greater than ϵ, so that
$$|\lambda_R - \lambda_r| > a, \qquad r \neq 1, a \gg \epsilon. \qquad (26)$$

For we have

$$\lambda_R = \sum_1^n \lambda_r \alpha_r^2, \tag{27}$$

and from (25) we find

$$\sum_1^n \alpha_r^2 (\lambda_r - \lambda_R)^2 < \epsilon^2, \tag{28}$$

so that certainly

$$\epsilon^2 \geqslant a^2 \sum_2^n \alpha_r^2 = a^2(1 - \alpha_1^2), \quad \alpha_1^2 \geqslant 1 - \epsilon^2/a^2. \tag{29}$$

The Rayleigh quotient is our corrected estimate for the latent root, and to find a bound for its error we have

$$\lambda_R = \lambda_R \sum_1^n \alpha_r^2 = \sum_1^n \lambda_r \alpha_r^2, \tag{30}$$

so that

$$|\lambda_R - \lambda_1|\, \alpha_1^2 < \sum_2^n |\lambda_r - \lambda_R|\, \alpha_r^2 = \sum_2^n |(\lambda_r - \lambda_R)^{-1}|\, (\lambda_r - \lambda_R)^2 \alpha_r^2$$
$$\leqslant a^{-1} \sum_2^n (\lambda_r - \lambda_R)^2 \alpha_r^2$$
$$\leqslant a^{-1}\epsilon^2 \tag{31}$$

from (28). Then from (29) we have, for the error in the Rayleigh quotient, the result

$$|\lambda_R - \lambda_1| < a^{-1}\epsilon^2/(1 - \epsilon^2 a^{-2}), \tag{32}$$

which is of order ϵ^2.

9. For an error bound for the approximate latent vector x we have

$$x - x^{(1)} = (\alpha_1 - 1)x^{(1)} + \sum_2^n \alpha_r x^{(r)}, \tag{33}$$

so that

$$(x - x^{(1)})'(x - x^{(1)}) = (\alpha_1 - 1)^2 + \sum_2^n \alpha_r^2, \tag{34}$$

and from (29) we find, in terms of a vector norm, the result

$$\|x - x^{(1)}\|_2^2 < \frac{\epsilon^4}{a^4(1 + \alpha_1)^2} + \frac{\epsilon^2}{a^2} < \frac{\epsilon^4}{a^4} + \frac{\epsilon^2}{a^2}, \tag{35}$$

so that, with neglect of ϵ^4/a^4, we have the error bound

$$\|x - x^{(1)}\|_2 < \epsilon/a. \tag{36}$$

10. In the unsymmetric case we need some approximation to the whole solution even to find a bound for the error in any latent root. With the assumptions and notations of § 6 we see that the roots of A are those of $D + E$, and as in that section we take the elements d_r of D as our better approximations to the latent roots.

To find bounds for the errors we note, from Gershgorin's first theorem, that there is a root in a circle with centre d_r and radius

$p_r = \sum\limits_{s=1}^{n} |e_{rs}|$, in which $e_{rr} = 0$. If the elements of E are small enough, so that the circles are isolated, our estimate d_r of this root has an error of upper bound p_r. This we can reduce by a simple device. Multiplication of the rth row of $D+E$ by a number k, and of the rth column by k^{-1}, gives a very simple similarity transformation which does not change the latent roots. The Gershgorin theorem, applied to the matrix so formed, gives one circle with centre d_r and radius kp_r, and $(n-1)$ circles with centres d_s and radii $\sum\limits_{t \neq r} |e_{st}| + k^{-1} |e_{sr}|$. For $k = 1$ the circles are isolated, and we can decrease k until one of the other circles just touches that with centre d_r and radius kp_r.

If the circle with centre d_s is the first to touch, then the critical value of k is given by

$$|d_r - d_s| = kp_r + \sum_{t \neq r} |e_{st}| + k^{-1} |e_{sr}|, \tag{37}$$

and a good approximation to k is

$$k = \max \frac{|e_{sr}|}{|d_r - d_s|}. \tag{38}$$

If $|d_r - d_s|$ is not small this value of k is of the first order of small quantities, so that the new error kp_r of our approximate latent root is of the second order, and we have an error bound comparable to that of the symmetric case.

There is no simple estimate, such as that of (36) in the symmetric case, for an error bound for the vectors. But we can improve them, in both symmetric and unsymmetric cases, either simultaneously by the methods of § 6, or individually by computing a vector $y^{(r)}$ of $D+E$, and computing $x^{(r)} = X_0 y^{(r)}$ to produce a vector of A. The matrix $D+E$ is almost diagonal, and we can omit the rth equation of the set $(D+E)x^{(r)} = d_r x^{(r)}$ and solve the rest easily by iteration.

General perturbation analysis

†11. Here we seek the perturbed matrix $A + \delta A$ for which our computed solution is exact, and in order to find reasonable bounds for the elements of δA our computing process must be 'stable', so that the incidence of rounding or other errors does not cause catastrophic results. We have already discussed possibilities of instability in various methods of numerical linear algebra. For example the solution of linear equations with Gauss elimination and with pivots taken down

the diagonal can give very poor results if the multipliers are large, and this we avoid by partial or complete pivoting for size.

Again, in the latent root and vector problem, we have seen the precautions which must be taken in the determination of the vectors of triple-diagonal matrices, particularly those which arise from transformation methods of Givens, Householder and Lanczos types. Row interchanges, we noted, were a necessary precaution in the L-R method of Rutishauser and in the elementary similarity transformations in the reduction to Hessenberg form. The stability of the further elementary transformation to triple-diagonal form cannot be guaranteed, since row interchanges would destroy the necessary zeros and we are left with the possibility of large multipliers. Transformations even of orthogonal or unitary type, in which the elements of the matrices of the transformation have modulus less than unity, do not necessarily give stable processes, though the reasons are more subtle.

12. Some analysis has been published for processes using orthogonal transformations. The details are somewhat lengthy, involving the précise nature of the type of arithmetic used at all stages, including for example the determination of square roots, and here we give an indication of the methods, leaving the reader to consult the literature for detail.

We consider transformation methods for symmetric matrices involving the sequence

$$A_r = Y'_r A_{r-1} Y_r, \tag{39}$$

with $A_0 = A$. If all arithmetic were exact all matrices A_r would have the same latent roots. There are, however, various sources of error. For example the non-zero elements of Y_r are computed from some of those of A_{r-1}, so that Y_r may not be exactly orthogonal. Again the matrix multiplications in (39) cannot be performed exactly in finite-length arithmetic so that (39) is not satisfied.

In the sequel we denote by A_r the matrix actually computed and stored, and denote its latent roots by $\lambda_s^{(r)}$, $s = 1, 2, ..., n$, so that $\lambda_s^{(0)}$ is a true root of A. If we can find bounds for $|\lambda_s^{(r-1)} - \lambda_s^{(r)}| = \delta_s^{(r)}$, then those for $|\lambda_s^{(0)} - \lambda_s^{(r)}|$ are given by

$$|\lambda_s^{(0)} - \lambda_s^{(r)}| < |\lambda_s^{(0)} - \lambda_s^{(1)}| + |\lambda_s^{(1)} - \lambda_s^{(2)}| + ... + |\lambda_s^{(r-1)} - \lambda_s^{(r)}|$$
$$= \delta_s^{(1)} + \delta_s^{(2)} + ... + \delta_s^{(r)}. \tag{40}$$

We can find $\delta_s^{(r)}$ by considering the matrix

$$X = \text{Calculated } (Y'_r A_{r-1} Y_r) - \text{Exact } (Y'_r A_{r-1} Y_r), \tag{41}$$

where here 'Exact $(Y_r'A_{r-1}Y_r)$' means the exact use of the transformation calculated exactly from the matrix A_{r-1} accepted at this stage, and *not* the matrix which would have been obtained by exact computation all the way from A_0. The roots of Calculated $(Y_r'A_{r-1}Y_r)$ are $\lambda_s^{(r)}$ by definition, and those of Exact $(Y_r'A_{r-1}Y_r)$ are those of A_{r-1} which are $\lambda_s^{(r-1)}$. A classical theorem states that

$$|\lambda_s^{(r)}-\lambda_s^{(r-1)}| < |\mu|, \tag{42}$$

where μ is the arithmetically largest latent root of the matrix X.

The advantage of this type of analysis is that we are not concerned with the difference between A_r and the corresponding matrix which would have been obtained by exact computation throughout. This difference, indeed, might be quite large without affecting significantly the accuracy of our results, and this is an important point which led to previous suspicions that accumulation of rounding errors could be catastrophic.

†13. In more detail, we suppose that in the transformation we use a $\overline{Y}_r$ which is as good an approximation to the true Y_r (that is the true Y_r relevant to the accepted A_{r-1}) as we can manage. But $\overline{Y}_r$ may not be orthogonal, and has an error S_r, so that

$$\overline{Y}_r = Y_r+S_r, \tag{43}$$

and our process and analysis must give a bound for S_r. Second, we compute A_r using $\overline{Y}_r$, and make further rounding errors. Our analysis tries to show that the computed A_r is near to the true $Y_r'A_{r-1}Y_r$, an exact orthogonal transformation of A_{r-1}. We define

$$A_r = Y_r'A_{r-1}Y_r+F_r, \tag{44}$$

and our process and analysis must give a bound for F_r, the matrix X in (41).

Consider now the computation of A_r from $\overline{Y}_r$ and A_{r-1}. In all our orthogonal transformations part of A_r is the same as that of A_{r-1}. In the Givens process, for example, this is the triple-diagonal form already produced in the top left-hand corner. Another part of A_r will come from $\overline{Y}_r'A_{r-1}\overline{Y}_r$, but we make errors in the computation and produce

$$(A_r)_1 = (\overline{Y}_r'A_{r-1}\overline{Y}_r)_1+(E_r)_1, \tag{45}$$

where the suffix 1 refers to the part of A_r affected. In the Givens process this consists of the sub-matrix in the bottom right-hand corner just below the row in which a new zero is being produced.

Finally there is $(A_r)_2$, the rest of A_r, consisting partly of those elements which, in most of our transformations, are designed to be zero and which are taken to be zero, so that here the error is nil, and some specially computed elements for which

$$(A_r)_2 = (\overline{Y}'_r A_{r-1} \overline{Y}_r)_2 + (E_r)_2. \tag{46}$$

In the Givens process these special elements are the other elements in the row and column in which a zero is being produced.

From (43) we have

$$(\overline{Y}'_r A_{r-1} \overline{Y}_r) = (Y'_r A_{r-1} Y_r + S'_r A_{r-1} Y_r + Y'_r A_{r-1} S_r + S'_r A_{r-1} S_r), \tag{47}$$

so that

$$(F_r)_1 = (A_r - Y'_r A_{r-1} Y_r)_1 = (S'_r A_{r-1} Y_r + Y'_r A_{r-1} S_r + S'_r A_{r-1} S_r)_1 + (E_r)_1. \tag{48}$$

Then

$$\|(F_r)_1\| < \|(S'_r A_{r-1} Y_r)_1\| + \|(Y'_r A_{r-1} S_r)_1\| + \|(S'_r A_{r-1} S_r)_1\| + \|(E_r)_1\|, \tag{49}$$

and if we take a matrix norm for which $\|Y_r\| = 1$ this becomes

$$\|(F_r)_1\| < \|A_{r-1}\| \{2 \|S_r\| + \|S_r\|^2\} + \|(E_r)_1\|. \tag{50}$$

We can also write (46) in the form

$$(F_r)_2 = (A_r)_2 - (Y'_r A_{r-1} Y_r)_2, \tag{51}$$

and we expect to be able to compute this matrix. Then

$$\begin{aligned}
\|F_r\| &= \|(F_r)_1 + (F_r)_2\| \\
&< \|(F_r)_1\| + \|(F_r)_2\| \\
&< \|A_{r-1}\| \{2 \|S_r\| + \|S_r\|^2\} + \|(E_r)_1\| + \|(F_r)_2\|.
\end{aligned} \tag{52}$$

14. Using (44) for all the transformations (39) as far as the computation of A_r, we find

$$\begin{aligned}
A_r &= Y'_r Y'_{r-1} \dots Y'_1 A_0 Y_1 Y_2 \dots Y_r + Y'_r Y'_{r-1} \dots Y'_2 F_1 Y_2 \dots Y_r + \dots + F_r \\
&= P'_{r1} A_0 P_{r1} + P'_{r2} F_1 P_{r2} + \dots + P'_{rr} F_{r-1} P_{rr} + F_r, \tag{53}
\end{aligned}$$

and every P_{rs} is exactly orthogonal. We write this in the form

$$A_r = P'_{r1} A_0 P_{r1} + G_r, \tag{54}$$

where

$$\|G_r\| < \|F_1\| + \|F_2\| + \dots + \|F_r\|. \tag{55}$$

Finally we write

$$A_r = P'_{r1}(A_0 + H_r) P_{r1}, \qquad H_r = P_{r1} G_r P'_{r1}, \tag{56}$$

and $\|H_r\| = \|G_r\|$. Our transformation is such that the roots of A_r are exactly those of A_0 perturbed by H_r. The remaining task is the determination of a bound for $\|H_r\| = \|G_r\|$, and this is accomplished by use of statements like (52).

Again we see the point that A_r may differ substantially from the true transformed matrix, since P_{r1} is not the true orthogonal matrix of the complete transformation. All we require for a good solution to the latent roots, however, is that H_r should have small elements.

15. The effect on the latent vectors needs a little more considera tion. Suppose that we compute exactly a vector of our accepted A_r. The corresponding vector of A_0+H_r is equal to this vector pre- multiplied by P_{r1}. But in this multiplication we use the $\overline{Y}$ matrices instead of the exact Y, and also make some rounding errors in the multiplications. In fact if $y^{(k)}$ is the computed vector of A_k, so that $y^{(0)}$ is the computed vector of A_0, we have

$$y^{(k)} = \overline{Y}_{k+1}y^{(k+1)}+e^{(k+1)}, \tag{57}$$

where $e^{(k+1)}$ is the vector of errors in the multiplication. Then we can write
$$y^{(k)} = (Y_{k+1}+S_{k+1})y^{(k+1)}+e^{(k+1)} = Y_{k+1}y^{(k+1)}+f^{(k+1)}, \tag{58}$$
and

$$y^{(0)} = Y_1Y_2...Y_ry^{(r)}+Y_1...Y_{r-1}f^{(r)}+Y_1...Y_{r-2}f^{(r-1)}+...+f^{(1)}. \tag{59}$$

Then
$$\|y^{(0)} - P_{r1}y^{(r)}\| < \|f^{(r)}\| + \|f^{(r-1)}\| +...+ \|f^{(1)}\|, \tag{60}$$

and bounds for the norms on the right of (60) can be obtained by methods similar to those resulting in equation (52).

Equation (60) gives a bound for the error in the determination of a vector of A_0+H_r, and the difference between this vector and that of A_0 depends on the degree of ill-conditioning of the latent vector problem, already discussed in § 5.

16. The analysis can be used also for orthogonal transformations for reducing an unsymmetric matrix, say to Hessenberg form, but there is nothing comparable for elementary similarity transformations. Here we can show, by methods similar to those used in Gauss elimina- tion for linear equations, that the transformation actually obtained is given by
$$MA = HM+E, \tag{61}$$

where the elements of E cannot exceed 0.5×2^{-t} in a t-binary machine provided that all elements of M are less than unity in absolute value. Then
$$H = M(A-M^{-1}E)M^{-1}, \tag{62}$$

so that we are interested in bounds for the perturbation $M^{-1}E$. There is, unfortunately, no guarantee that M^{-1} is small even if no element of M exceeds unity in absolute value.

Deflation perturbation

17. We end with a brief look at a similar perturbation analysis for the deflation process of § 21 of Chapter 9, which has especially been viewed with suspicion because of the possibility of accumulation of rounding error.

The analysis is rather similar to that for Gauss elimination. We consider a matrix A for which we have computed, presumably by iteration, a latent root λ and vector x, normalized so that its largest component, here assumed to be the first, is unity. The computed solution is not of course exact, and we have the residual vector

$$Ax - \lambda x = r. \qquad (63)$$

We can, however, make this residual vector null by subtracting from the first column of A the components of r, so that for this perturbed matrix our solution is exact.

In the deflation process we take multiples of the first row from all the other rows. The resulting matrix is exact if we further perturb the elements of A, other than those in the first row and column, by the rounding errors of the multiplication. The doubly-perturbed matrix no longer satisfies (63) except in the first row, but we can correct this by making further changes in the other elements of the first column.

If the elements of r are of order ϵ, a single rounding error, then it is easily seen that the perturbation of A which would make exact both the latent solution and the deflation has elements with upper bounds in the array

$$\delta A = \epsilon \begin{bmatrix} 1 & 0 & 0 & 0 \\ n & 1 & 1 & 1 \\ n & 1 & 1 & 1 \\ . & . & . & . \end{bmatrix}, \qquad (64)$$

and we can perform a full analysis which gives credibility to the deflation process.

†18. Wilkinson has given the following example which illustrates this process and also the fact that good results are obtained even when the computed deflated matrix has little resemblance to the true deflated matrix.

The matrix
$$A = \begin{bmatrix} -0{\cdot}456 & -0{\cdot}253 & -0{\cdot}509 \\ 0{\cdot}052 & -0{\cdot}149 & 0{\cdot}103 \\ 0{\cdot}152 & 0{\cdot}151 & 0{\cdot}103 \end{bmatrix} \tag{65}$$

has the latent solutions
$$\left. \begin{aligned} \lambda_1 &= -0{\cdot}200 \\ \lambda_2 &= -0{\cdot}202 \\ \lambda_3 &= -0{\cdot}100 \end{aligned} \right\}, \qquad X = \begin{bmatrix} 1 & 3 & 5 \\ 1 & -1 & -1 \\ -1 & -1 & -3 \end{bmatrix}, \tag{66}$$

and we note two nearly equal roots which means that the matrix is rather ill-conditioned for the determination of latent vectors.

In fact we might suppose that an iteration, working to three decimal places, could produce a latent root $\lambda_1 = -0.200$, which happens to be correct, and a vector
$$x^{(1)'} = (1{\cdot}000,\ 0{\cdot}692,\ -0{\cdot}846), \tag{67}$$

which happens to be an exact linear combination of the true latent vectors corresponding to the nearly equal roots. The residual vector has small components; we find
$$Ax^{(1)} - \lambda_1 x^{(1)} = \begin{bmatrix} -0{\cdot}000462 \\ 0{\cdot}000154 \\ 0{\cdot}000154 \end{bmatrix}, \tag{68}$$

and to our working precision we should accept our $x^{(1)}$, which in fact has no accurate figures, as a good latent vector.

Deflation produces the matrix
$$\begin{aligned} A_1 &= \begin{bmatrix} -0{\cdot}149 & 0{\cdot}103 \\ 0{\cdot}151 & 0{\cdot}103 \end{bmatrix} - \begin{bmatrix} 0{\cdot}692 \\ -0{\cdot}846 \end{bmatrix} [-0{\cdot}253 \quad -0{\cdot}509] \\ &= \begin{bmatrix} 0{\cdot}026 & 0{\cdot}455 \\ -0{\cdot}063 & -0{\cdot}328 \end{bmatrix}, \end{aligned} \tag{69}$$

correctly rounded to three decimals. With the true vector corresponding to λ_1 we would produce the exact deflated matrix
$$\bar{A}_1 = \begin{bmatrix} 0{\cdot}104 & 0{\cdot}612 \\ -0{\cdot}102 & -0{\cdot}406 \end{bmatrix}, \tag{70}$$

and (69) and (70) have no figures in common.

The computed matrix A_1, however, has roots $-0\cdot203$, $-0\cdot099$, correct to three decimals, and these are very close approximations to the other roots of A. To this precision, also, the corresponding vectors of A_1 are

$$y^{(2)'} = (0\cdot455, -0\cdot229), \qquad y^{(3)'} = (0\cdot455, -0\cdot125). \qquad (71)$$

Building up to vectors of A, by the process described in Chapter 9, we find
$$x^{(3)'} = (1, -0\cdot201, -0\cdot601), \qquad (72)$$

which is a very accurate approximation to the true vector. For $x^{(2)}$ we get a result, as we should expect, no more accurate than our first vector $x^{(1)}$.

19. The conclusion is that our deflation, in spite of the very poor accuracy of our first vector, has given very accurate results for all the roots and for those vectors which are not affected by nearly equal roots. The reason for this, of course, is that our solution is an exact solution for a perturbation $(A + \delta A)$, in which δA has small elements, and the results are good except for those affected by ill-conditioning.

We can easily find the perturbed matrix, and here it is simplest first to change the elements of the relevant part of A so that the deflation is exact. We find

$$A + \delta_1 A = \begin{bmatrix} -0\cdot456 & -0\cdot253 & -0\cdot509 \\ 0\cdot052 & -0\cdot149076 & 0\cdot102772 \\ 0\cdot152 & 0\cdot151038 & 0\cdot102614 \end{bmatrix}, \qquad (73)$$

and to make $\lambda_1 = -0\cdot200$ and the vector in (67) exact solutions of this we must perturb further to produce

$$A + \delta A = \begin{bmatrix} -0\cdot455538 & -0\cdot253 & -0\cdot509 \\ 0\cdot051705704 & -0\cdot149076 & 0\cdot102772 \\ 0\cdot151493148 & 0\cdot151038 & 0\cdot102614 \end{bmatrix}, \qquad (74)$$

the maximum element of δA being little more than half a unit in the third decimal.

Additional notes and bibliography

§ 3. The first Gershgorin theorem gives an indication of when to stop in the theoretically infinite sequence of transformations in the Jacobi method for reducing a symmetric matrix A to diagonal form. If we make no rounding errors in the process, so that at any stage the roots of our current matrix are exactly those of A, then we should continue until the sum of the moduli of the off-diagonal terms in each row is less than the error we can tolerate in the latent roots. A deeper analysis, of course, is needed to deal with the various errors which inevitably appear.

§ 6. This analysis is given in detail in

JAHN, H. A., 1948, Improvement of an approximate set of latent roots and modal columns of a matrix by methods akin to those of classical perturbation theory. *Quart. J. Mech.* **1**, 131–44, and

COLLAR, A. R., 1948, Some notes on Jahn's method for the improvement of approximate latent roots and vectors of a square matrix. *Quart. J. Mech.* **1**, 145–8.

§§ 7–10. This analysis is given, with extensions and computational detail, in

WILKINSON, J. H., 1961, Rigorous error bounds for computed eigensystems. *Computer J.* **4**, 230–41.

§ 8. It is worth noting that we can compute the Rayleigh quotient in the form

$$\lambda_R = x'Ax = x'r + \lambda,$$

and since r is likely to have small components we can use single-length arithmetic to find a correction $x'r$ to the approximate λ. Double-length arithmetic, or at least the accurate accumulation of double-length results of products of single-length numbers, is of course necessary in the determination of the residual vector.

§ 11. The instability of elementary similarity transformations in the reduction to triple-diagonal form is analysed in detail in

WILKINSON, J. H., 1962, Instability of the elimination method of reducing a matrix to tri-diagonal form. *Computer J.* **5**, 61–70.

He shows that instabilities can be negotiated by the use in certain parts of the computation of multiple-length arithmetic, and suggests some modifications of the method which removes some instability, though not all.

He also shows, in

WILKINSON, J. H., 1959, Stability of the reduction of a matrix to almost triangular and triangular forms by elementary similarity transformations. *J. Ass. comp. Mach.* **6**, 336–59,

that the process of Givens (1958) *loc. cit.*, p. 274 described at the end of Chapter 10, for reducing the Hessenberg matrix to triangular form by unitary transformations, has instabilities similar to those encountered in the determination of vectors of triple-diagonal matrices. He finds that this can be avoided in general by combining the technique of finding a transformation $H^{-1}(A - I\lambda)H$, such that one column of the resulting matrix is null, with another which operates on the latent vector x corresponding to λ. Here we find an H such that $Hx = ke^{(n)}$, for then

$$HAx = \lambda Hx, \qquad HA(H^{-1}H)x = \lambda Hx, \qquad HAH^{-1}e^{(n)} = \lambda e^{(n)}.$$

The matrix HAH^{-1} therefore has its (n, n) element equal to λ and the other elements of the last column equal to zero. Hence $H(A - \lambda I)H^{-1}$ has its last column null, and H is of course the same matrix as that of the previous method.

Its computation, however, is different, and one method might have stability at any stage while the other may not. One or other method is therefore applied at each stage according to the size of the relevant numbers governing stability.

The instability of the orthogonal reduction from Hessenberg form to triangular form is also analysed by Wilkinson (1962), *loc. cit.*, this page. He shows, however, that the orthogonal deflation of a full symmetric matrix, essentially the method of Feller and Forsythe (1951) *loc. cit.*, p. 236, is stable.

§ 13. This analysis, with detailed applications to the processes of Jacobi, Givens, Householder, and others, is described in

WILKINSON, J. H., 1962, Error analysis of eigenvalue techniques based on orthogonal transformations. *J. Soc. industr. appl. Math.* **10,** 162–95.

This paper considers fixed-point computation, and similar results for floating-point arithmetic are given in

WILKINSON, J. H., 1963, Plane rotations in floating-point arithmetic. *Proceedings, A.M.S. Symposium in Applied Mathematics* Vol. 15.

In all cases the norm of the perturbation is expressed as a small power of n multiplying 2^{-t}, with possibly some norm of A as an additional factor.

§ 18. This example, with some correction here for unimportant arithmetic errors in the original, appears in

WILKINSON, J. H., 1958, Matrix Computations. Notes on lectures presented in the Advanced Numerical Analysis course at the Engineering Summer Conference, University of Michigan.

Index

Exercises

(In order to understand the various techniques the student should work through the examples given in the text. The following exercises are of two kinds, one to confirm the knowledge gained from reading this book and the other to introduce some extension of the theory and to remove ambiguities in the text. The arithmetical exercises are simple, in the sense that only mental arithmetic is required, except for one or two in Chapters 6, 9, 10, 11. Students with access to a computing machine would benefit by solving more formidable examples, both for an appreciation of the error analysis which is difficult to introduce in easy exercises, and also for practice in coding the various techniques.)

Chapter 1

1. (a) Construct a flow diagram for evaluating the roots of the quadratic equation $x^2 + 2b_r x + c_r = 0$, for 10 pairs of real values of b_r and c_r, taking into account the possibilities of both real and complex roots, and that in each case the modulus of the two roots should have the same relative error.

(b) It is known in advance that $b_r^2 - c_r$ changes sign from positive to negative at some unknown value q of r. Modify the flow diagram so that, in addition to the requirements of (a), the programme also computes the value of q.

2. Code the flow diagram of question 1 for any machine you know.

3. (a) In the formula

$$f(x) = \frac{5 \cdot 0x - 0 \cdot 19}{0 \cdot 40x + 0 \cdot 01}$$

the coefficients have possible errors of half a unit in their last figures. Find the possible error in $f(x)$ if $x = 0 \cdot 5 \pm 0 \cdot 01$.

(b) Show that $x^{1/64}$ is correct to five significant figures for $x = 2 \cdot 46 \pm 0 \cdot 005$.

4. Convert to binary the decimal numbers 1561 and 0·1561, rounding in the second case to 10 binary digits.

Convert to decimal the binary numbers 1011101011 and 0·1011101011.

5. The standardized floating-point number x in the range $\frac{1}{2} \leqslant x < 1$ is stored in two machines, one operating in the decimal system with p digits, and one operating in the binary scale with q digits. Find the relation between p and q such that the maximum absolute error is the same in both machines.

6. In a floating-point machine the fractional part of the standardized number $10^a \times b$ can be stored with p digits after the decimal point. The addition $(10^{a_1} \times b_1) + (10^{a_2} \times b_2)$, with $a_1 > a_2$, is obtained in the form

$$10^{a_1}(b_1 + 10^{a_2-a_1}b_2),$$

in which the second term in brackets is rounded to p figures after the decimal point before the addition, and the result is then standardized. Find the maximum absolute error of the addition when (a) the result exceeds 10^{a_1}, and (b) the result does not exceed 10^{a_1}.

What are the maximum absolute errors in the two cases if we have a 'double-length accumulator' so that the rounding and standardization to a single-length number are performed after the addition? When in this case is the absolute error zero?

(It is assumed throughout that b_1 and b_2 have no rounding errors and can be represented exactly in p digits.)

7. The expression $\mathrm{fl}(x_1 * x_2)$ represents the number stored in the machine after a floating-point operation $*$ on the floating-point numbers x_1 and x_2. If the arithmetic is performed by the second method of exercise 6, show that there is a number ϵ such that

$\mathrm{fl}(x_1 * x_2) \equiv (x_1 * x_2)(1+\epsilon)$, where $|\epsilon| \leqslant 2^{-p}$ in a binary machine, and $|\epsilon| \leqslant \frac{1}{2} . 10^{1-p}$ in a decimal machine, for $* = +, -, \times, \div$.

If the arithmetic is performed by the first method of exercise 6 show that

$\mathrm{fl}(x_1 + x_2) \equiv x_1(1+\epsilon_1) + x_2(1+\epsilon_2)$, where $|\epsilon_1|, |\epsilon_2| \leqslant 3.2^{-p-1}$ in a binary machine, and $|\epsilon_1|, |\epsilon_2| \leqslant 55.10^{-p-1}$ in a decimal machine.

Deduce that rounding is more satisfactory in binary than in decimal.

8. We try to compute e^x from the series

$$e^x = 1 + \frac{x}{1!} + \frac{x^2}{2!} + \ldots + \frac{x^{n-1}}{(n-1)!} + \ldots,$$

with a four-digit floating-point decimal machine. If the series is truncated after the term $x^{(n-1)}/(n-1)!$ show that for $x > 0$ the *truncation error* cannot exceed

$$x^n \left(1 - \frac{x}{n}\right)^{-1} \bigg/ n!$$

if $n > x$.

For $x = 4$ we assume that we can compute each term correctly rounded to four digits, and produce the series

$10^0(1 \cdot 000) + 10^1(0 \cdot 4000 + 0 \cdot 8000) + 10^2(0 \cdot 1067 + 0 \cdot 1067) +$

$+ 10^1(0 \cdot 8534 + 0 \cdot 5689 + 0 \cdot 3251 + 0 \cdot 1625) + 10^0(0 \cdot 7224 + 0 \cdot 2890 + 0 \cdot 1051) +$

$+ 10^{-1}(0 \cdot 3503 + 0 \cdot 1078),$

the last term representing $4^{13}/13!$

Show (i) that the truncation error does not exceed 0·0044,

(ii) that if the terms are added in order (the first three being exact) we can certainly assert that
$$|e^4 - 54\cdot60| < 0\cdot07,$$

(iii) that if the terms are added in reverse order we can find
$$|e^4 - 54\cdot60| < 0\cdot03.$$

9. Show that, in the computation of e^{-x} from the series of exercise 8, we cannot guarantee a rounding error of less than 0·012 at $x = 4$, so that we can guarantee no correct significant figures in our result. How can we compute e^{-4}, without using 'double-length' arithmetic, correct to nearly four significant figures?

10. The value of the polynomial
$$p_n(x) = a_0x^n + a_1x^{n-1} + \ldots + a_{n-1}x + a_n$$

can be calculated for given x by 'nested multiplication' from the recurrence relation
$$p_r = xp_{r-1} + a_r, \quad p_0 = a_0,$$

and $p_n = p_n(x)$. For the case
$$p_n(x) = 0\cdot1296x^3 + 0\cdot3424x^2 - 0\cdot2134x + 0\cdot7892$$

we compute each p_r by four-digit floating-point decimal arithmetic, producing for $x = 0\cdot1$ the approximations
$$\bar{p}_1 = 0\cdot3554, \quad \bar{p}_2 = -0\cdot1779, \quad \bar{p}_3 = 0\cdot7714.$$

Show that $\bar{p}_3$ is the *exact* value for $x = 0\cdot1$ of the 'perturbed polynomial'
$$\bar{p}_3(x) = 0\cdot1296x^3 + 0\cdot34244x^2 - 0\cdot21344x + 0\cdot78919.$$

(Hint: modify successive coefficients so that each $\bar{p}_r$ is exact. This is an example of 'perturbation' or 'backward error' analysis.)

Chapter 2

1. If $A^{-1} = A'$ the matrix A is *orthogonal*. Show (i) that the product of orthogonal matrices is an orthogonal matrix, (ii) that $|A| = \pm1$.

2. The real vectors x and y each have n components. Show that the matrix $A = xy' - yx'$ is *skew-symmetric* ($a_{rs} = -a_{sr}$). Verify that for $n = 2$ the determinant of A is a perfect square, and that for $n = 3$ the matrix is singular.

3. The rth column of the unit matrix is called $e^{(r)}$, and the column of row sums of the unit matrix is called e. Express the element a_{rs} of the square matrix A of order n as a product of three matrices of shapes $(1, n)$, (n, n) and $(n, 1)$. Express the sum of the elements of A in similar fashion.

4. The (n, n) unit lower triangular matrix L, with zero diagonal elements, is called *strictly* lower triangular. Show that L^n is the null matrix.

5. The operations (i) multiplying the rth row of the (n, m) matrix A by a constant k, (ii) interchanging rows r and s of A, (iii) adding a multiple k_r of the rth row to the sth row to form a new sth row, are called *elementary operations*. Express the resulting matrices as products of two matrices one of which is A, and repeat the exercise when the word 'row' is changed to 'column' throughout.

If $m = n$ find the relation between the determinants of the resulting matrices and that of A, using the fact that if $C = AB$ then $|C| = |A|\,|B|$.

6. The non-singular matrix A is partitioned in the form

$$A = \left[\begin{array}{c|c} P & Q \\ \hline R & S \end{array}\right],$$

where P and S are non-singular square matrices, not necessarily of the same order. The inverse is partitioned in the similar form

$$A^{-1} = \left[\begin{array}{c|c} C & D \\ \hline E & F \end{array}\right].$$

Show that $C = (P - QS^{-1}R)^{-1}$, $D = (-P^{-1}QF)$, $E = (-S^{-1}RC)$, $F = (S - RP^{-1}Q)^{-1}$.

When can we write $D = (R - SQ^{-1}P)^{-1}$, $E = (Q - PR^{-1}S)^{-1}$?

7. The matrix A is non-singular and is partitioned in the form

$$A = \left[\begin{array}{c|c} B & c \\ \hline r' & a_{nn} \end{array}\right],$$

where c is a column, r' a row, a_{nn} the (n, n) element of A, and B is non-singular. Show that

$$|A| = |B|\,(a_{nn} - r'B^{-1}c).$$

(Hint: multiply A on the left by a suitable lower triangular matrix so that the resulting partitioned matrix has a null row in place of r'.)

Hence show that if x is the solution of $Bx + c = o$, then $|A|$ is proportional to $(a_{nn} + r'x)$. Use this fact to find the value of λ for which the matrix

$$\begin{bmatrix} 1 & 2 & -1 \\ 2 & -8 & 1 \\ 4 & -4 & \lambda \end{bmatrix}$$

is singular.

8. Find the polynomial whose zeros are the latent roots of the matrix

$$A = \begin{bmatrix} 1 & 2 & 1 \\ 1 & 0 & 0 \\ -1 & 1 & 2 \end{bmatrix}.$$

If this is $\lambda^3 + a\lambda^2 + b\lambda + c$, verify that $A^3 + aA^2 + bA + cI$ is the null matrix. (The matrix satisfies its own characteristic equation, the *Cayley-Hamilton* theorem.)

Verify, and prove in general, that the sum of the latent roots is the sum of the diagonal elements of A (the *trace* of A).

9. The latent vectors $x^{(1)}, x^{(2)}, \ldots, x^{(n)}$ of a symmetric matrix A are normalized so that $x^{(r)'}x^{(r)} = 1$. The modal matrix X has columns $x^{(1)}, x^{(2)}, \ldots, x^{(n)}$. Show that

$$XX' = \sum_{r=1}^{n} x^{(r)}x^{(r)'}.$$

We want to solve the linear equations $Ax = b$, with a symmetric A, and we know the latent vectors $x^{(1)}, x^{(2)}, \ldots, x^{(n)}$ and corresponding non-zero latent roots $\lambda_1, \lambda_2, \ldots, \lambda_n$, the elements of a diagonal matrix D.

If $b = \sum \alpha_r x^{(r)}$, $x = \sum \beta_r x^{(r)}$,
  (i) express b in terms of the matrix X and the vector α with components $\alpha_1, \alpha_2, \ldots, \alpha_n$;
  (ii) find an expression for α_r in terms of the vector b and the latent vectors;
  (iii) show that $\beta_r = \alpha_r/\lambda_r$,
  (iv) show that $x = (\sum \lambda_r^{-1} x^{(r)}x^{(r)'})b = XD^{-1}X'b = XD^{-1}\alpha$.
  (v) show that $A = \sum \lambda_r x^{(r)}x^{(r)'}$ (the *spectral resolution* of A).

Find the corresponding results if A is not symmetric, but we know also the latent vectors $y^{(1)}, y^{(2)}, \ldots, y^{(n)}$ of A', with normalization $y^{(r)'}x^{(r)} = 1$.

10. Exercise 9 connects the linear-equation problem with the latent roots and vectors. Note that if A is singular, so that one $\lambda_k = 0$, there is in general no solution. But a solution is possible if $\alpha_k = 0$, so that b contains no component of the latent vector $x^{(k)}$, that is if b is orthogonal to $x^{(k)}$ in the symmetric case and to $y^{(k)}$ in the unsymmetric case. The solution is not unique, however, since if x satisfies $Ax = b$ then $x + ax^{(k)}$ also satisfies this equation for any value of a, since $Ax^{(k)} = o$. Similarly if there are p zero latent roots there is a solution only if b contains no component of any of the corresponding latent vectors, assumed to be independent, and there are p arbitrary constants in the solution $x + \sum_{k=1}^{p} a_k x^{(k)}$. Notice also that in this case the spectral resolution of A has no contribution from the latent vectors corresponding to the zero latent roots. Show that the rank of the matrix is $n - p$.

11. The matrix A has n rows and k independent columns, and B has k independent rows and n columns. The matrix AB has rank k, with $n-k$ zero latent roots. Show that the non-zero latent roots of AB are the latent roots of the (k, k) matrix BA, and that if y is a latent vector of BA the corresponding latent vector of AB is Ay. Verify this for (3×1) and (1×3) matrices on page 56.

Verify that

$$
\begin{array}{ccc}
C & A & B \\
\end{array}
$$

$$
\begin{bmatrix} 0 & 1 & 1 & 1 \\ 4 & 3 & -1 & 1 \\ 8 & 7 & -1 & 3 \\ 8 & 5 & -3 & 1 \end{bmatrix}
=
\begin{bmatrix} 1 & 0 \\ 0 & 1 \\ 1 & 2 \\ -1 & 2 \end{bmatrix}
\begin{bmatrix} 0 & 1 & 1 & 1 \\ 4 & 3 & -1 & 1 \end{bmatrix}.
$$

Show that the non-zero latent roots of C, and the corresponding unnormalized latent vectors, are

$$
\lambda_1 = 5, \quad x^{(1)\prime} = (1,\ 1,\ 3,\ 1),
$$
$$
\lambda_2 = -2, \quad x^{(2)\prime} = (-5,\ 2,\ -1,\ 9).
$$

12. The matrix A is orthogonal. Show that $\|A\|_2 = 1$.

13. The notation for norms comes from the definition $\|x\|_p = (|x_1|^p + |x_2|^p + \ldots + |x_n|^p)^{\frac{1}{p}}$, $p = \infty$, 1, 2. Verify (137) for these values of p.

14. The number $\lambda_1(P)$ is the latent root of largest modulus of the matrix P. Using the third matrix norm show that

$$
\lambda_1^{\frac{1}{2}}\{(A'+B')(A+B)\} < \lambda_1^{\frac{1}{2}}(A'A) + \lambda_1^{\frac{1}{2}}(B'B),
$$
$$
\lambda_1\{(AB)'(AB)\} < \lambda_1(A'A)\lambda_1(B'B).
$$

15. The unnormalized latent vector corresponding to the largest latent root of the matrix

$$
A = \begin{bmatrix} 1\cdot 5 & -0\cdot 5 \\ -0\cdot 5 & 1\cdot 5 \end{bmatrix}
$$

is $x^{(1)\prime} = (1,\ -1)$. Without computing any latent roots show that $\|A\|_2 = 2$.

16. Show that $\|A\|_\infty$, $\|A\|_1$ and $\|A\|_2$ all exceed unity for the matrix

$$
A = \begin{bmatrix} 2 & -1 \\ 3 & -1\cdot 7 \end{bmatrix},
$$

but that the series $(I-A)^{-1} = I + A + A^2 + \ldots$ nevertheless converges.

17. In a *centro-symmetric* matrix the elements are symmetrical about both diagonals of the matrix. Show that, for such a matrix, $A = I_n A I_n$, where the matrix I_n is the permuted unit matrix in which the unit elements appear in the diagonal from top right to bottom left. Given the linear equations $Ax = b$, with centro-symmetric A, show that for even n we can reduce the problem to the solution of two independent sets of $\frac{1}{2}n$ equations. (Hint: partition in the form

$$\left[\begin{array}{c|c} O & \bar{I}_{n/2} \\ \hline \bar{I}_{n/2} & O \end{array}\right]\left[\begin{array}{c|c} P & Q \\ \hline R & S \end{array}\right]\left[\begin{array}{c|c} O & \bar{I}_{n/2} \\ \hline \bar{I}_{n/2} & O \end{array}\right]\left[\begin{array}{c} x^{(1)} \\ \hline x^{(2)} \end{array}\right] = \left[\begin{array}{c} b^{(1)} \\ \hline b^{(2)} \end{array}\right],$$

and deduce the equations

$$(\bar{I}_{n/2}P \pm R)(x^{(1)} \pm \bar{I}_{n/2}x^{(2)}) = (\bar{I}_{n/2}b_1 \pm b^{(2)}).$$

Find the similar result for an odd number of centro-symmetric equations.

18. If A is a symmetric matrix, and X its modal matrix, show that the transformation $x = Xy$ reduces the quadratic form $x'Ax$ to a 'sum of squares.'

Consider the problem $(A - \lambda B)x = 0$, in which A and B are symmetric. If B is positive definite show that all the roots are real, and if A is also positive definite show that all the roots are positive.

If X is the modal matrix of $(A - \lambda B)$, and D the diagonal matrix of latent roots, show that

(i) $AX = BXD$,

(ii) $X'BX$ is a diagonal matrix, and therefore also $X'AX$,

(iii) The transformation $x = Xy$ reduces both $x'Ax$ and $x'Bx$ to sums of squares.

Chapter 3

1. By partitioning the matrix J_1 of p. 68 and its inverse J_1^{-1} in the forms

$$J_1 = \left[\begin{array}{c|c} 1 & o' \\ \hline c & I \end{array}\right], \quad J_1^{-1} = \left[\begin{array}{c|c} 1 & r' \\ \hline \gamma & B \end{array}\right],$$

where c and γ are columns, r' a row, o' the null row, I the unit matrix and B a matrix of order $n-1$, show that $\gamma = -c$.

2. The use of the notation I_r for a row-permuting matrix has an ambiguity with the unit matrix, and does not indicate which rows are interchanged. Suppose we use the notation $I_{r,s}$ for the unit matrix with rows r and s interchanged. Show that $I_{r,s}A$ is the matrix obtained from A by interchanging its rows r and s. What is $AI_{r,s}$? Show that $I_{r,s}$ is symmetric and orthogonal.

Using the matrices of type $I_{r,s}$, and the method of exercise 1, prove that the inverse of J_2 on p. 69 is J_2 with the signs changed of the m_{r2}. (Hint: the matrix J_2 can be transformed to the form of J_1 by pre- and post-multiplication with a certain permuting matrix.)

3. With the notation of exercise 2 the Gauss elimination process with interchanges is performed by the computing machine in the following way. We start with matrix $A_1 = A$, and form $\bar{A}_1 = I_{1,r_1}A_1$ by interchanging the pivotal row r_1 with the first row (if necessary). We then do the first elimination to produce A_2. Then $\bar{A}_2 = I_{2,r_2}A_2$, where $r_2 \neq 1$ is the pivotal row of A_2, and so on. Show that the operation can be expressed by the matrix equations

$$\bar{A}_1 = I_{1,r_1}A_1, \quad A_2 = J_1\bar{A}_1 = J_1 I_{1,r_1}A_1, \quad \bar{A}_2 = I_{2,r_2}A_2 = \ldots,$$
$$A_4 = J_3 I_{3,r_3}J_2 I_{2,r_2}J_1 I_{1,r_1}A_1, \ldots,$$

where each J_r is a lower triangular matrix corresponding to elimination with the rth diagonal element of $\bar{A}_r$ as pivot. Show also that A_n is an upper triangle U.

For the example of pp. 72–75 we can then relate our new matrices to those labelled I_r, U, and J_1, J_2, J_3, with $J = J_3 J_2 J_1$, as in § 15. The U matrices are obviously the same, so that

$$U = I_r J A = I_r J_3 J_2 J_1 A = A_4 = J_3 I_{3,r_3}J_2 I_{2,r_2}J_1 I_{1,r_1}A.$$

Using the symmetry and orthogonality properties of the matrices $I_{r,s}$ show that this equation can be written in the form

$$I_r J_3 J_2 J_1 A = (J_3 I_{3,r_3}I_{2,r_2}I_{1,r_1})(I_{1,r_1}I_{2,r_2}J_2 I_{2,r_2}I_{1,r_1})(I_{1,r_1}J_1 I_{1,r_1})A.$$

Deduce that $I_r = I_{3,r_3}I_{2,r_2}I_{1,r_1}$, and find the relations between the J_r and J_r. Show finally that

$$I_r A = I_{3,r_3}I_{2,r_2}J_1^{-1}I_{2,r_2}J_2^{-1}I_{3,r_3}J_3^{-1}U = LU,$$

and prove that the matrix L is lower triangular. (Hint: show that the form of J_1^{-1} ensures that $I_{2,r_2}J_1^{-1}I_{2,r_2}$ is lower triangular of the same form, with the elements in the first column interchanged in rows 2 and r_2.)

Verify all these conclusions for the example of § 15, recording all the relevant matrices.

4. With the same notation, show that the Jordan process for inversion illustrated in Table 6 and §§ 16–17 can be represented by the equation

$$\bar{J}_3 I_{3,r_3}\bar{J}_2 I_{2,r_2}\bar{J}_1 I_{1,r_1}A_1 = \bar{D},$$

where $\bar{D}$ is a diagonal matrix and each $\bar{J}_r$ is lower triangular except in its rth column. Find the relations between the matrices in this equation and the D, I_r and J_r of § 17.

5. Without evaluating x_1, x_2 or x_3 in the equations

$$\left.\begin{array}{r} 0{\cdot}6x_1 - 0{\cdot}3x_2 + 0{\cdot}5x_3 = -0{\cdot}3 \\ 1{\cdot}2x_1 + 2{\cdot}6x_2 - 0{\cdot}4x_3 = 1{\cdot}4 \\ -0{\cdot}6x_1 + 1{\cdot}9x_2 + 0{\cdot}8x_3 = 1{\cdot}3 \end{array}\right\},$$

evaluate the quantity $\quad 2{\cdot}4x_1 + 3{\cdot}6x_2 + 0{\cdot}3x_3.$

Find also the determinant of the matrix of the linear equations. (Use 'partial pivoting' in the elimination.)

6. In the equations

$$0 \cdot 900 x_1 + 0 \cdot 120 x_2 + 0 \cdot 936 x_3 = 1 \cdot 200$$
$$0 \cdot 450 x_1 - 0 \cdot 780 x_2 + 0 \cdot 708 x_3 = 0 \cdot 000$$
$$0 \cdot 150 x_1 - 0 \cdot 820 x_2 + (0 \cdot 376 + \epsilon_1) x_3 = -0 \cdot 410 + \epsilon_2$$

we know that $|\epsilon_1| \leqslant 0 \cdot 001$, $|\epsilon_2| \leqslant 0 \cdot 001$.

Find the possible range of values of x_3, and show, without evaluating x_1, x_2 or x_3, that

$$0 \cdot 300 x_1 + 0 \cdot 460 x_2 + 0 \cdot 992 x_3 = 1 \cdot 10 \pm 0 \cdot 06.$$

7. Find the 'least-squares' solution of the equations at the top of page 57. Find also the values of x_1 and x_2 which minimise the sum of squares of residuals subject to the 'constraint' $x_1 - 3x_2 = 1$. (Hint: minimise

$$S - \mu(x_1 - 3x_2 - 1)$$

with respect to x_1 and x_2, hence eliminating μ. This is Lagrange's 'method of undetermined coefficients'.)

8. Show that all the diagonal elements of a symmetric positive-definite matrix A must be positive. (Hint: use exercise 3, Chapter 2, and the fact that $x'Ax > 0$.)

Show also that the largest element of A lies on the diagonal. (Hint: it is sufficient to prove that $a_{pp}a_{qq} > a_{pq}^2$. Write $a_{pp} = e_p'Ae_p$, express e_p as $\sum p_r x^{(r)}$, where $x^{(r)}$ is a latent vector of A corresponding to latent root λ_r, and reduce the problem to the proof of Schwarz's inequality

$$\left(\sum \lambda_r p_r^2\right)\left(\sum \lambda_r q_r^2\right) > \left(\sum \lambda_r p_r q_r\right)^2, \text{ for } \lambda \text{ real and positive.})$$

9. If the matrix A_1 is symmetric the elimination with pivot a_{11} produces a new matrix A_2 expressed in the form $A_2 = J_1 A_1$, that is

$$A_2 = \left[\begin{array}{c|c} 1 & o' \\ \hline -a_{11}^{-1}c & I_{n-1} \end{array}\right]\left[\begin{array}{c|c} a_{11} & c' \\ \hline c & A_1^{(1)} \end{array}\right] = \left[\begin{array}{c|c} a_{11} & c' \\ \hline o & A_1^{(1)} - a_{11}^{-1}cc' \end{array}\right].$$

Show that if A_1 is positive definite then the 'reduced' symmetric matrix

$$B_2 = A_1^{(1)} - a_{11}^{-1}cc'$$

is also positive definite.

(Hint: Consider the partitioned vector $x' = [x_1, \xi']$, where ξ' has $n-1$ components. Show that $x'A_1x = \xi'B_2\xi + a_{11}^{-1}(a_{11}x_1 + \xi'c)^2$, and hence that unless $\xi'B_2\xi > 0$ for all ξ there is some ξ, and an x_1 equal to $-a_{11}^{-1}\xi'c$, for which $x'A_1x \not> 0$.)

10. Solve by the method of §§ 30 and 33 the equations

$$\left.\begin{array}{r} 2x_1 + x_2 + x_3 + x_4 = 1 \\ x_1 - 2x_2 + 3x_3 - 4x_4 = 0 \\ x_1 - x_2 + x_3 - x_4 = -1 \\ x_1 + x_2 + x_3 + x_4 = 2 \end{array}\right\},$$

(i) using successive diagonal elements as pivots, and (ii) using smallest element in successive columns as pivots, and in each case evaluate $|A|$.

11. In complete 'pivoting' we might interchange both rows and columns so that the pivot is brought at each stage into the relevant diagonal position. If the pivot is in row r_1 and column c_1 show, with the notation of exercises 2 and 3, that the first elimination process for the equations $Ax = b$ is expressible in matrix form

$$J_1 I_{1,r_1} A_1 I_{1.c_1} y = J_1 I_{1,r_1} b, \qquad y = I_{1.c_1} x.$$

Show also that, for order four,

$$J_3 I_{3.r_3} J_2 I_{2.r_2} J_1 I_{1.r_1} A_1 I_{1.c_1} I_{2.c_2} I_{3.c_3} = U,$$

an upper triangle, which in terms of § 15 and exercise 3 is expressible in the form

$$I_r J_3 J_2 J_1 A = U I_c, \qquad I_c = I_{3.c_3} I_{2.c_2} I_{1.c_1}.$$

The results of exercise 3 still apply. Hence show that $I_r A = L U I_c$, so that $I_r A I'_c = LU$, the triangular decomposition of a row and column permutation of A.

Perform the arithmetic of complete pivoting for the matrix of exercise 10 (using exact fractions), recording all relevant matrices and achieving the decomposition

$$\begin{bmatrix} -4 & 1 & 3 & -2 \\ 1 & 2 & 1 & 1 \\ 1 & 1 & 1 & 1 \\ -1 & 1 & 1 & -1 \end{bmatrix} = \begin{bmatrix} 1 & & & \\ -\frac{1}{4} & 1 & & \\ -\frac{1}{4} & \frac{5}{9} & 1 & \\ \frac{1}{4} & \frac{1}{3} & -\frac{3}{7} & 1 \end{bmatrix} \begin{bmatrix} -4 & 1 & 3 & -2 \\ & \frac{9}{4} & \frac{7}{4} & \frac{1}{2} \\ & & \frac{7}{9} & \frac{2}{9} \\ & & & -\frac{4}{7} \end{bmatrix}.$$

12. Perform similarly the arithmetic for the rectangular resolution with complete pivoting of the matrix C in Table 13, recording all relevant matrices and making all relevant interchanges, showing that

$$I_r C I'_c = \begin{bmatrix} 5 & 0 & 3 & 2 & 1 \\ 3 & -2 & 1 & 0 & -1 \\ 3 & 1 & 1 & 1\cdot8 & 1\cdot4 \\ 4 & -1 & 1\cdot7 & 1 & 0 \\ 1 & 1 & 0\cdot4 & 1 & 1 \end{bmatrix} =$$

$$= \begin{bmatrix} 1 & & \\ 0\cdot6 & 1 & \\ 0\cdot6 & -0\cdot5 & 1 \\ 0\cdot8 & 0\cdot5 & 0\cdot25 \\ 0\cdot2 & -0\cdot5 & 0\cdot5 \end{bmatrix} \begin{bmatrix} 5 & 0 & 3 & 2 & 1 \\ 0 & -2 & -0\cdot8 & -1\cdot2 & -1\cdot6 \\ 0 & 0 & -1\cdot2 & 0 & 0 \end{bmatrix}.$$

This I_r is that of (94) except for the order of its last two rows, corresponding to those labelled (1) and (3) in Table 13 which were reduced to null rows, and which therefore in that method had no particular preference for order. There is a corresponding reordering in our first matrix on the right in relation to the $\left[\dfrac{L_1}{Q_1}\right]$ of equation (96). Show that our I_c is

$$I_c = \begin{bmatrix} 0 & 0 & 0 & 0 & 1 \\ 1 & 0 & 0 & 0 & 0 \\ 0 & 0 & 0 & 1 & 0 \\ 0 & 0 & 1 & 0 & 0 \\ 0 & 1 & 0 & 0 & 0 \end{bmatrix},$$

and that post-multiplication of $I_r C I_c'$ with I_c gives on the right a new (3×5) matrix identical with that labelled $I_r B$ in equation (95).

13. This use of column as well as row interchanges is important in the analysis of § 40 for the determination of compatibility of linear equations. For example in (101) we cannot guarantee that U_1 is upper triangular and non-singular unless pivots can be taken and are taken from successive *columns*, which is assumed implicitly in § 40. In the decomposition of (96), on the other hand, $I_r B$ is not in the form $[U_1 | Q_2]$, and indeed the matrix of its first three columns is singular.

The final post-multiplication with I_c' is necessary to order the columns appropriately. Note that in

$$I_r C = (I_r A I_r')(I_r B) = \left[\dfrac{L_1}{Q_1}\right][I_r B]$$

the insertion of I_r' ensures that L_1 is a unit lower triangle, and therefore non-singular. To guarantee success we therefore write in place of (101) the equation

$$I_r C I_c' = \left[\dfrac{L_1}{Q_1}\right][U_1 \mid Q_2],$$

with non-singular L_1 and U_1, so that if we replace x throughout by $y = I_c x$ the rest of the analysis is unchanged.

Perform the arithmetic for the linear equations whose matrix is that of Table 13 and exercise (12), and find all possible solutions for (i) right-hand sides of (4·8, 2·0, 5·4, 10·0, 6·0), and (ii) right-hand sides of (1, 1, 1, 1, 1).

14. Solve the linear equations with the matrix of Table 10, with right-hand sides (1, 1, 1, 1), by each of the methods relevant to Table 10.

Chapter 4

1. Solve by the methods of Doolittle, Crout and Banachiewicz the equations of exercise 10 of Chapter 3.

2. Find the inverse of the matrix of these equations by each method, and with each of the variations of Tables 12, 13, 14 and 15 for the Banachiewicz method.

3. Show that all these methods fail for the system with matrix

$$
A = \begin{bmatrix} 2 & 1 & 1 & 1 \\ 1 & -2 & 3 & -4 \\ 1 & -1 & 2 & -1 \\ 1 & 1 & 1 & 1 \end{bmatrix},
$$

even though A is non-singular.

4. Solve by the methods of Doolittle, Crout and Cholesky the linear equations

$$
\left.\begin{aligned}
4x_1 + 6x_2 + 10x_3 + 4x_4 &= 16 \\
6x_1 + 13x_2 + 13x_3 + 6x_4 &= 18 \\
10x_1 + 13x_2 + 27x_3 + 2x_4 &= 28 \\
4x_1 + 6x_2 + 2x_3 + 72x_4 &= 144
\end{aligned}\right\}.
$$

5. Find the inverse of the matrix of exercise 4 by each of the methods of Tables 16 and 17.

6. Show that each method of exercise 4 fails if the $(3, 3)$ element is changed from 27 to 26.

Repeat the computation with the $(3, 3)$ element replaced by 25 and the $(4, 4)$ element by -64, and find also the inverse of the matrix.

7. Find by the method of Aitken of equations (47)–(50) the inverse of the matrix of Table 18, comparing the various computed numbers with those of the Banachiewicz method of Table 14.

8. In exercise 3 of Chapter 3 we observed that the machine use of partial pivoting was expressible in matrix notation $I_r A = LU$, where, for order four,

$$
I_r = I_{3,r_3} I_{2,r_2} I_{1,r_1}, \qquad L = I_{3,r_3} I_{2,r_2} \bar{J}_1^{-1} I_{2,r_2} \bar{J}_2^{-1} I_{3,r_3} \bar{J}_3^{-1}.
$$

This is the process used in § 30 with the addition that the row-permuting matrices are determined during the process of the computation.

Record, for this example, the matrices of type $I_{r,s}$, $\bar{J}_r$, and the successive steps in the transformation of A to U. Show that, in the computation of L, we perform the following steps.

(i) $\bar{J}_1^{-1}$ is a unit lower triangle with off-diagonal non-zero elements only in the first column.

(ii) The matrix $P = (I_{2,r_2} \bar{J}_1^{-1} I_{2,r_2})$ interchanges the elements in rows 2 and r_2 of the first column of $\bar{J}_1^{-1}$.

(iii) $Q = P \bar{J}_2^{-1}$ inserts into the second column of P the elements of $\bar{J}_2^{-1}$. In fact $P \bar{J}_2^{-1} = P + \bar{J}_2^{-1} - I$.

(iv) $R = I_{3.r_3} Q I_{3.r_3}$ interchanges rows 3 and r_3 in the first two columns of Q.

(v) $R J_3^{-1}$ adds to R the elements of J_3^{-1} in the appropriate positions in the third column. In fact $R J_3^{-1} = R + J_3^{-1} - I$.

9. Perform the computation of exercise 8 for the problem of exercise 4. Notice that symmetry is not maintained. (Use fractions and exact arithmetic.)

10. The main reason for performing the 'compact' elimination process, with interchanges, rather than the theoretically identical Gauss process with largest pivots in successive columns, is that when rounding errors are committed the full effect is smaller with the compact methods. Complete pivoting is useful because it tends to limit the possible growth of numbers in the elimination. The compact use of 'complete pivoting', in which the pivot is taken as the largest element in the rows and columns not already used as pivotal rows and columns, cannot, however, be performed economically. We might on the other hand find by the Gauss method of complete pivoting which row and column changes are necessary and then repeat the computation in the compact way. Try this for the matrix of exercise 3. (If exact arithmetic is used the results should be identical.)

11. We can, however, reasonably use complete pivoting for the Cholesky decomposition in the symmetric case, with largest diagonal element selected as pivot. Here we choose permutations so that $I_r A I_r = L L'$, where L is lower triangular and I_r is the product of row-permuting matrices. Consider an example of order 4. Starting with the symmetric matrix A_1, we choose the largest diagonal term. Suppose this is in the third row and column. Then I_{1,r_1} is $I_{1,3}$ and we have

$$A_1 = \begin{bmatrix} a_{11} & a_{12} & a_{13} & a_{14} \\ a_{12} & a_{22} & a_{23} & a_{24} \\ a_{13} & a_{23} & a_{33} & a_{34} \\ a_{14} & a_{24} & a_{34} & a_{44} \end{bmatrix}, \quad A_2 = I_{1,r_1} A_1 I_{1,r_1} = \begin{bmatrix} a_{33} & a_{23} & a_{13} & a_{34} \\ a_{23} & a_{22} & a_{12} & a_{24} \\ a_{13} & a_{12} & a_{11} & a_{14} \\ a_{34} & a_{24} & a_{14} & a_{44} \end{bmatrix}.$$

Then the first row of L' has elements $(l_{33}, l_{23}, l_{13}, l_{43})$, where

$$l_{33}^2 = a_{33}, \quad l_{23} = a_{23}/l_{33}, \dots,$$

but the position of the last three elements may change. The possible candidates for the next diagonal term of L' are

$$(a_{22} - l_{23}^2)^{\frac{1}{2}}, \quad (a_{11} - l_{13}^2)^{\frac{1}{2}}, \quad (a_{44} - l_{43}^2)^{\frac{1}{2}},$$

and we take the largest. Suppose it is the second of these. Then we take $I_{2.r_2} = I_{2,3}$ and find

$$A_3 = I_{2,r_2} I_{1,r_1} A_1 I_{1,r_1} I_{2,r_2} = \begin{bmatrix} a_{33} & a_{13} & a_{23} & a_{34} \\ a_{13} & a_{11} & a_{12} & a_{14} \\ a_{23} & a_{12} & a_{22} & a_{24} \\ a_{34} & a_{14} & a_{24} & a_{44} \end{bmatrix},$$

and the computed part of L' now looks like

$$\begin{bmatrix} l_{33} & l_{13} & l_{23} & l_{43} \\ & l_{11} & l_{12} & l_{14} \end{bmatrix},$$

where l_{12}, l_{14} come from $l_{13}l_{11}+l_{23}l_{12} = a_{12}$, etc., but the order of the last two columns is not yet fixed. The continuation should be obvious.

Try this method for solving the linear equations

$$\left.\begin{aligned} 16x_1+8x_2+4x_3+ 2x_4 &= 14 \\ 8x_1+9x_2+5x_3+ 6x_4 &= 3 \\ 4x_1+5x_2+4x_3+ x_4 &= 2 \\ 2x_1+6x_2+ x_3+25x_4 &= -27 \end{aligned}\right\},$$

using exact arithmetic throughout.

12. Consider the solution of the system of difference equations of 'boundary-value type'

$$a_r y_{r-1}+b_r y_r+c_r y_{r+1} = d_r, \quad r = 1, 2,..., n-1, \quad y_0 = p, \quad y_n = q.$$

Show that the solution can be obtained by recurrence from the equations

$$a_r y^{(1)}_{r-1}+b_r y^{(1)}_r+c_r y^{(1)}_{r+1} = d_r, \quad y^{(1)}_0 = p, \quad y^{(1)}_1 = 0,$$
$$a_r y^{(2)}_{r-1}+b_r y^{(2)}_r+c_r y^{(2)}_{r+1} = 0, \quad y^{(2)}_0 = 0, \quad y^{(2)}_1 = 1,$$

with $y_r = y^{(1)}_r+k y^{(2)}_r$, the constant k coming from $y^{(1)}_n+k y^{(2)}_n = q$.

The solution can also be obtained directly from the linear equations

$$\left.\begin{aligned} b_1 y_1+c_1 y_2 &= d_1-a_1 p \\ a_2 y_1+b_2 y_2+c_2 y_3 &= d_2 \\ \cdots\cdots\cdots\cdots\cdots \\ a_{n-2} y_{n-3}+b_{n-2} y_{n-2}+c_{n-2} y_{n-1} &= d_{n-2} \\ a_{n-1} y_{n-2}+b_{n-1} y_{n-1} &= d_{n-1}-c_{n-1}q \end{aligned}\right\}.$$

Show that $y^{(2)}_n = (-1)^{n+1}(c_1 c_2 ... c_{n-1})^{-1}|A|$, where A is the matrix of the linear equations.

Show also that in the triangular decomposition $A = LU$, with L a unit lower triangle, the triangular matrices have elements only on the diagonal and adjacent sloping line (assuming that no leading submatrix of A is singular). Show finally that

$$u_{r,r+1}/u_{r,r} = -y^{(2)}_r/y^{(2)}_{r+1}.$$

13. Express in matrix notation the operation defined at the top of page 124.

14. Discuss the use of Gauss elimination with (a) partial pivoting (b) complete pivoting, for the method of § 36.

Chapter 5

1. If we choose the $e^{(r)}$ in § 3 to be the latent vectors of the symmetric matrix A, show that the analysis reduces to that of exercise 9 in Chapter 2. Discuss in similar terms the analysis of §§ 6 and 7 for the unsymmetric case.

2. With the notation of §§ 3–4, show that $d_3 = 0$ if $a_{33} = \alpha_1 d_{13}^2 + \alpha_2 d_{23}^2$. Hence deduce an amended value of a_{33} in the matrix of Table 1 for which the matrix is singular.

3. Find the corresponding result for the unsymmetric case of § 6, and find a value of a_{33} in the matrix A of Table 2 which would give a singular matrix.

4. Solve the problem of Table 3 by writing the equations $Ax = b$ in the form $AA'y = b$, $y = (A^{-1})'x$, and using the method of §§ 3–5. Note the relations between the numbers obtained in this computation and those of Table 3.

5. The rth row of B is $e^{(r)'}B$, and the sth column is $Be^{(s)}$, where $e^{(r)}$ is the rth row of the unit matrix. Express the analysis of § 12 in corresponding matrix algebra, verifying all the conclusions reached in § 12.

6. Perform the computation of § 12 for the example of Table 3, starting with $B_0 = I$, confirming the similarity with the Jordan method.

7. Solve exercise 6, starting with $B_0 = A'$, as in § 14.

Chapter 6

1. In the solution of the equations $Ax = b$, with A given in (4), show that if the elements of b have uncertainties with equal upper bound ϵ the maximum uncertainties in the elements of x are approximately 235ϵ, 44ϵ, 118ϵ and 265ϵ, but that the maximum uncertainty in $x_1 - x_2 + x_4$ cannot exceed about 14ϵ.

2. For the example of Table 3, in which the elements of A are exact, show that, if the right-hand sides have uncertainties δb_r, then the problem is well-conditioned if $\delta b_3 - \delta b_4 = 0$. Verify this by solving the equations with the two sets of right-hand sides $b^{(1)'} = (1, 1, 0\cdot4, -0\cdot8)$, $b^{(2)'} = (1, 1, 0\cdot45, -0\cdot75)$, and compare with the rather different results for $b^{(3)'} = (1, 1, 0\cdot45, -0\cdot85)$.

3. If $A^{-1}E = F$ in § 8, show that

$$\|A^{-1} - (A+E)^{-1}\| \leqslant \|A^{-1}\| \, \|F\|/(1 - \|F\|).$$

The result (11) comes from replacing $\|F\|$ by $\|A^{-1}\| \, \|E\|$, and gives for the inverse an absolute error proportional to $\|A^{-1}\|^2$. Show that this is a considerable overestimate for particular perturbations of the form $E = PA$ or $E = AQ$, where the elements of P and Q are of the same order of magnitude as those of A, and $\|A^{-1}\|$ is large.

4. Show that if $Ax = b$, $A(x+\delta x) = b+\delta b$, then

$$\frac{\|\delta x\|}{\|x\|} \leqslant k \frac{\|\delta b\|}{\|b\|},$$

where k is the condition number $\|A\| \, \|A^{-1}\|$.

Show also that if $Ax = b$, $(A+E)(x+\delta x) = b$, then, with the notation of § 8 and exercise 3,

$$\frac{\|\delta x\|}{\|x\|} \leqslant \frac{\|F\|}{1- \|F\|},$$

and finally that if $Ax = b$, $(A+E)(x+\delta x) = b+\delta b$, then

$$\frac{\|\delta x\|}{\|x\|} \leqslant \frac{1}{1- \|F\|} \left\{ \|F\| + k \frac{\|\delta b\|}{\|b\|} \right\}.$$

5. Prove formula (28) for the Aitken acceleration method in the case of complex roots.

6. In the process of §§ 12–13 a certain problem had the approximations 1·000, 0·910, 0·864 and 0·840 for one component of successive vectors $x^{(r)}$, $r = 1, 2, 3, 4$. If the Aitken extrapolation process is valid estimate the largest latent root of the 'iteration matrix' C, and a better approximation to this particular component of the required solution.

7. Complete the inversion of the matrix A in Table 4, and show that if the right-hand sides of the equations $Ax = b$ have uncertainties of upper bound ϵ then those of x_1, x_2, x_3 and x_4 have approximate upper bounds of 5205ϵ, 3004ϵ, 2604ϵ, and 3204ϵ respectively. Show, however, that the quantity

$$0 \cdot 776x_1 - 0 \cdot 667x_2 + 0 \cdot 449x_3 + 1 \cdot 000x_4$$

has an uncertainty of upper bound of about $1 \cdot 6\epsilon$, and verify that the vector $c' = (0 \cdot 776, -0 \cdot 667, 0 \cdot 449, 1 \cdot 000)$ is a close approximation to a latent vector of the transpose of A, corresponding to a latent root of about 6·960.

8. The following computation shows the results of the 'lengthy' Gauss elimination process for transforming a matrix A into an upper triangle U, so that if the arithmetic were exact we would have $|A| = |U|$. In the arithmetic shown all entries, including the multipliers, are correctly rounded to one decimal place. The multipliers and the pivots are in italics.

	A					*U*				*U*	
0·9	0·8	−0·4		0·9	0·8	−0·4	0·9	0·8	−0·4		
−0·2	0·2	−0·5	0·3		*−0·7*	0·4		−0·7	0·4		
−0·4	0·4	−0·1	0·4	−0·6	−0·4	0·6			0·4		

Find the perturbed matrix $A+\delta_1 A$ which, with exact arithmetic, gives rise to the recorded multipliers and elements of U.

Find also by another process a U whose determinant is that of another perturbation $A+\delta_2 A$, in which the largest element of $\delta_2 A$ does not exceed 0·05 in modulus, and in which no computations are recorded to more than one decimal place.

9. The analysis of § 32 can be performed by considering the 'history' of the element occupying the (r, s) position in successive 'reduced' matrices $A^{(r)}$. For $r < s$ the element a_{rs} is changed at each step until we obtain $A^{(r)}$, after which it remains constant. Show that

$$a_{rs}^{(2)} = a_{rs}^{(1)} + m_{r1}a_{1s}^{(1)} + \epsilon_{rs}^{(1)}$$

$$a_{rs}^{(3)} = a_{rs}^{(2)} + m_{r2}a_{2s}^{(2)} + \epsilon_{rs}^{(2)}$$

$$\cdots\cdots\cdots\cdots\cdots\cdots\cdots$$

$$a_{rs}^{(r)} = a_{rs}^{(r-1)} + m_{r,r-1}a_{r-1,s}^{(r-1)} + \epsilon_{rs}^{(r-1)},$$

and deduce the result

$$a_{rs}^{(r)} = a_{rs}^{(1)} + m_{r1}a_{rs}^{(1)} + m_{r2}a_{rs}^{(2)} + \ldots + m_{r,r-1}a_{r-1,s}^{(r-1)} + \epsilon_{rs},$$

where $\epsilon_{rs} = \sum\limits_{p=1}^{r-1} \epsilon_{rs}^{(p)}$.

For $r > s$ the element a_{rs} is changed as before until we obtain $A^{(s)}$. It is then used to compute multipliers, and $a_{rs}^{(s+1)}$ to $a_{rs}^{(n)}$ are taken to be exactly zero. Show that

$$a_{rs}^{(2)} = a_{rs}^{(1)} + m_{r1}a_{1s}^{(1)} + \epsilon_{rs}^{(1)}$$

$$a_{rs}^{(3)} = a_{rs}^{(2)} + m_{r2}a_{2s}^{(2)} + \epsilon_{rs}^{(2)}$$

$$\cdots\cdots\cdots\cdots\cdots\cdots\cdots$$

$$a_{rs}^{(s)} = a_{rs}^{(s-1)} + m_{r,s-1}a_{s-1,s}^{(s-1)} + \epsilon_{rs}^{(s-1)}$$

$$0 = a_{r,s}^{(s)} + m_{rs}a_{ss}^{(s)} + \epsilon_{rs}^{(s)}.$$

Hence deduce the result

$$0 = a_{rs}^{(1)} + m_{r1}a_{1s}^{(1)} + m_{r2}a_{2s}^{(2)} + \ldots + m_{rs}a_{ss}^{(s)} + \epsilon_{rs},$$

where $\epsilon_{rs} = \sum\limits_{p=1}^{s} \epsilon_{rs}^{(p)}$.

Finally show that the two sets of equations can be combined to give

$$LU = A_1 + E,$$

where L is the matrix of (negative) multipliers, U the final upper triangle, and E a 'perturbation' whose (r, s) element is the sum of $r-1$ rounding errors for $r < s$ and of s rounding errors for $r > s$.

10. Consider the application of the analysis of p. 173 to the example of p. 151. We find

$$LL' = A + 10^{-4}\begin{bmatrix} 0\cdot3184 & 0\cdot2800 & 0\cdot2992 & 0\cdot3184 \\ 0\cdot2800 & 0\cdot0921 & -0\cdot0196 & -0\cdot0022 \\ 0\cdot2992 & -0\cdot0196 & 0\cdot0248 & -0\cdot2146 \\ 0\cdot3184 & -0\cdot0022 & -0\cdot2146 & 0\cdot1782 \end{bmatrix},$$

$LL'x^{(0)} = b + 10^{-4}(-0.1917, \ -0.0009, \ -0.2170, \ -0.1894)$ approximately, and, to sufficient accuracy,

$$(LU)^{-1} = \begin{bmatrix} 698 & -421 & -175 & 103 \\ -421 & 257 & 103 & -62 \\ -175 & 103 & 51 & -31 \\ 103 & -62 & -31 & 20 \end{bmatrix}.$$

Using the first vector and matrix norms, show that the error in $x^{(0)}$, compared with the true solution of $Ax = b$ in (34), is approximately

$$\|\delta x^{(0)}\|_\infty < 0.06,$$

so that the second decimal of the approximate solution is certainly doubtful. (Note that in the analysis of p. 173 the quantities δA and δb are here null.)

Show also that if the equations (34) are not exact, but are correctly rounded from equations with coefficients containing more figures, the difference between the approximate $x^{(0)}$ and the solution of the true equations will have a norm not exceeding 0.3, so that even the first figure after the decimal point is doubtful.

11. In matrix algebra, with fixed-point arithmetic, one often associates a single power of 10 (or 2 in a binary machine) with all the elements of a vector (or matrix). In a decimal machine the power is chosen so that the element of largest modulus is $10^r y_k$, where $0.1 \leqslant y_k < 1$. Show that $0.1 \leqslant \|y\|_\infty < 1$.

Suppose that we solve linear equations by compact Gauss elimination (triangular decomposition), and that all elements of A and U are less than unity in modulus. In the back-substitution we associate the scaling factor 10^{-r} with every element of the computed x, and write $10^r \bar{y} = \bar{x}$, where $0.1 \leqslant \|\bar{y}\|_\infty < 1$. Show that we have solved exactly the system

$$(A + \delta_1 A) 10^r \bar{y} = b + \delta b,$$

where $|(\delta_1 A)_{ij}| < 0.5 \times 10^{-p}$, $|\delta b_i| < 0.5 \times 10^{-p} + i0.5.10^{r-p}$.

Show that we can write this in the form

$$(A + \delta_1 A + \delta_2 A) 10^r \bar{y} = b,$$

and that we can select a $\delta_2 A$ for which $\|\delta_2 A\|_\infty < 0.5.10^{1-p}(n + 10^{-r})$. (Hint: $\delta_2 A$ has non-zero elements in only one column.)

Show finally that we have solved exactly the system $(A + E)10^r \bar{y} = b$, where $\|E\|_\infty < 0.5.10^{-p}(11n + 10^{1-r})$, and that a bound for the residual vector $R = b - A.10^r \bar{y}$ is given by $\|R\|_\infty < 5.5n10^{-p} \|10^r \bar{y}\|_\infty$, with neglect of a very small term.

(Note that the factor 5.5 has appeared, greater than that of (98) in the text, due to closer attention to the precise nature of the arithmetic. In a binary machine the factor 5.5 becomes 1.5. Similarly in computing the residual of the true solution $x = 10^r y$ rounded to p figures, we can write only $\bar{x} = x(1 + \epsilon)$, where $\epsilon < 5.10^{-p}$, so that $\|b - A\bar{x}\|_\infty < 5n.10^{-p} \|10^r y\|_\infty$. In binary the factor 5 becomes 1.)

12. For correcting an approximate solution we might perform the compact elimination, and use the method of § 12 with $B = LU = A + \delta_1 A$ as in exercise 11. The iterative scheme is given by the equations

$$x^{(0)} = (LU)^{-1}b, \quad R^{(r)} = b - Ax^{(r)}, \quad x^{(r+1)} = x^{(r)} + (LU)^{-1}R^{(r)},$$

in which we actually use L and U to solve $LU\Delta^{(r)} = R^{(r)}$, that is we do not evaluate B^{-1}.

Show that
$$x - x^{(r+1)} = \{I - (LU)^{-1}A\}(x - x^{(r)}),$$
$$R^{(r+1)} = \{I - A(LU)^{-1}\}R^{(r)},$$

and that the process converges if $\|\delta_1 A\| \, \|A^{-1}\| < \tfrac{1}{2}$.

If the computed $\bar{x}^{(r+1)} = x^{(r)} + \Delta^{(r)} + \epsilon^{(r)}$, where $\epsilon^{(r)}$ is the effect of rounding to single-length numbers, show that the residual $R^{(r+1)}$ has from this source an error $A\epsilon^{(r)}$ with upper bound $5n \, . \, 10^{-p} \, \|x^{(r)}\|_\infty$.

(Note that it is essential to compute the residual $R^{(r)}$ to full accuracy for the accurate computation of $\Delta^{(r)}$. The approximation $x^{(r)}$ will then tend to x, at least to single-length accuracy, but the residual will not tend to zero, and will generally have much the same size at all stages. Moreover provided the first approximation is of the same order as the true solution the residuals are 'misleadingly small', as in the example of § 21.)

13. The method of § 12, and its examples of §§ 15–18, in which we actually compute an approximation B^{-1} to A^{-1} and use this in the iteration, in practice gives different behaviour for the residual vector. Suppose that in a decimal machine B^{-1} is computed by solving linear equations with matrix A and columns $\epsilon^{(k)}$ of the unit matrix on the right-hand side. If the corresponding columns of B^{-1} are denoted by $x^{(k)}$, then according to exercise 11 we have exactly

$$\{A + (\delta_1 A) + (\delta_2^{(k)} A)\}x^{(k)} = e^{(k)},$$

$$\text{where } \|\delta_1 A\|_\infty < 0{\cdot}5n \, . \, 10^{-p}, \ \|\delta_2^{(k)} A\|_\infty < 0{\cdot}5n \, . \, 10^{1-p}.$$

(with neglect of a much smaller term).

Show that B^{-1} satisfies exactly the equation

$$I - AB^{-1} = (\delta_1 A)B^{-1} + P, \qquad C_k(P) = (\delta_2^{(k)}A)x^{(k)}.$$

Hence show that $\|I - AB^{-1}\| < f(n) \, \|B^{-1}\|$, $f(n) = 10^{-p}(0{\cdot}5n + 5n^2)$, and that

$$\|B^{-1}\| < \frac{\|A^{-1}\|}{1 - f(n) \, \|A^{-1}\|}, \qquad \text{if} \qquad f(n) \, \|A^{-1}\| < 1.$$

If we use this method to find the inverse $(B^{-1})'$ of the transpose A' prove that we have similar results, and can write

$$\|I - B^{-1}A\| < f(n) \, \|B^{-1}\|.$$

This is the matrix called C in our iterative process of § 12. For the first approximation $x^{(0)} = B^{-1}b$ show that

$$\|x - x^{(0)}\| < \|C\| \, \|x\| < \frac{f(n) \, \|A^{-1}\| \, \|x\|}{1 - f(n) \, \|A^{-1}\|},$$

and that $R^{(0)} = b - Ax^{(0)} = ACx$, so that

$$\|R^{(0)}\| < \|A\|\,\|C\|\,\|x\| < \frac{nf(n)\,\|A^{-1}\|\,\|x\|}{1-f(n)\|A^{-1}\|},$$

if no element of A exceeds unity in modulus. We can also write

$$R^{(0)} = (I - AB^{-1})b, \qquad \|R^{(0)}\| < \|I - AB^{-1}\|\,\|b\|,$$

and if the norm of $I - AB^{-1}$ is very near to that of $I - B^{-1}A = C$, which is usual in practice, we have

$$\|R^{(0)}\| < \frac{f(n)\,\|A^{-1}\|\,\|b\|}{1-f(n)\,\|A^{-1}\|}.$$

(Note the presence of the factor $\|A^{-1}\|\,\|b\|$, compared with the factor $\|x\| = 10^r\,\|\bar{y}\|$ in the method of exercises 11 and 12. If $\|A^{-1}\|\,\|b\|$ is of the same order as $\|A^{-1}b\| = \|x\|$, a situation representing the true ill-conditioning of the problem, then the residuals are again 'misleadingly small', but if b is special so that $\|A^{-1}b\|$ is of the same order of magnitude as $\|b\|$, then the residuals give some measure of accuracy of the current approximation to the solution. This is true, for example, in the problem of § 15, in which $\|R^{(0)}\|_\infty = 0{\cdot}535$, $\|R^{(1)}\|_\infty = 0{\cdot}031$, and further computation gives $\|R^{(2)}\|_\infty = 0{\cdot}002$. Detailed analysis of the material of exercises 11, 12 and 13 is given by Wilkinson in his *Rounding errors in algebraic processes*, HMSO, 1963.)

Chapter 7

1. Prove the results of § 9.

2. The results of § 11 are misleading. Show that the computation of $CA^{-1}B$, where C is (p, n), A is (n, n), B is (n, m) and C, A^{-1} and B are known, depends on the order of the operations, and if $p < m$ it is quicker to compute in order $C(A^{-1}B)$ than in order $(CA^{-1})B$. In particular if B is a vector x the best method of computing $CA^{-1}x$ has the ordering $C(A^{-1}x)$.

Show similarly that we can evaluate $(CA^{-1})'$, by the Gauss method applied to linear equations with matrix A' and p right-hand sides, the columns of C', and that premultiplication with B' produces $(CA^{-1}B)'$ in a total of $n(\tfrac{1}{3}n^2 - \tfrac{1}{3} + np + pm)$ multiplications. Hence show that the 'better' Gauss method is always faster than the 'better' Aitken method by the number $\tfrac{1}{2}n(n-1)\,|m-p|$ of multiplications.

3. Prove the results of §§ 12 and 13.

4. Find a method for calculating the elements of a square matrix of order n, given its latent root and vectors, in approximately $\tfrac{4}{3}n^3$ multiplications.

5. Given an approximation to the latent vector corresponding to the largest latent root of a symmetric matrix A, how many multiplications are needed to compute a close approximation to $\|A\|_2$, the third matrix norm?

6. Compute the number of multiplications and the amount of storage needed for the method of pp. 186–7.

7. Express all the computations of pp. 186–7 in the form $PAx = Pb$, where P is the product of certain matrices corresponding to the row interchanges and the elimination. Write down the individual matrices and evaluate the matrix P. Note that P is not a lower triangle, permuted or otherwise, even though PA is an upper triangle U.

8. Evaluate the number of multiplications for the solution of linear equations and the inversion of the matrix when the elements are complex, using the method of § 36, in Chapter 4, pages 122–3.

Chapter 8

1. In the solution of certain partial differential equations an approximation is found by solving linear algebraic equations. In the diagram the value at

each numbered point is the average of the values at the four surrounding points, and the function is given on the boundary $PQRS$. If $x^{(1)}$ is the vector of values along the line 123, and $x^{(2)}$ and $x^{(3)}$ have similar respective meanings for the next two lines, show that the linear equations can be expressed in partitioned form

$$\begin{bmatrix} X & I & O \\ I & X & I \\ O & I & X \end{bmatrix} \begin{bmatrix} x^{(1)} \\ x^{(2)} \\ x^{(3)} \end{bmatrix} = \begin{bmatrix} b^{(1)} \\ b^{(2)} \\ b^{(3)} \end{bmatrix},$$

where I is the unit matrix and O the null matrix of order 3, and $b^{(1)}$, $b^{(2)}$ and $b^{(3)}$ are vectors depending on the boundary values. Record the elements of the matrix X.

2. If in exercise 1 the solution is symmetrical to the extent that

$$x_1 = x_3 = x_7 = x_9, \qquad x_2 = x_4 = x_6 = x_8,$$

show that the equations can be written in the form $Cx = d$, where x has components x_2, x_3 and x_5, and

$$C = \begin{bmatrix} 1 & -\tfrac{1}{2} & -\tfrac{1}{4} \\ -\tfrac{1}{2} & 1 & 0 \\ -1 & 0 & 1 \end{bmatrix}.$$

Find the largest latent roots of the Jacobi and Gauss-Seidel iteration matrices for the solution of $Cx = d$, hence demonstrating convergence of both methods. (Find the actual iteration matrices and their characteristic polynomials.)

Can we guarantee convergence by inspecting any norm of the iteration matrices?

3. Suppose that in the iteration $x^{(r+1)} = Cx^{(r)} + c$ we know the latent roots λ_k and latent vectors $y^{(k)}$ of C. If

$$x^{(0)} = \sum \alpha_k y^{(k)}, \qquad c = \sum \beta_k y^{(k)},$$

show that

$$x^{(r)} = \sum \{\alpha_k \lambda_k^r + \beta_k(1 + \lambda_k + \lambda_k^2 + \ldots + \lambda_k^{r-1})\} y^{(k)}.$$

Hence verify that the process converges if all $|\lambda_k| < 1$, that in this case the solution is

$$x = \sum \left(\frac{\beta_k}{1 - \lambda_k}\right) y^{(k)},$$

and that at any stage the error is

$$x - x^{(r)} = \sum \left(\frac{\beta_k}{1 - \lambda_k} - \alpha_k\right) \lambda_k^r y^{(k)}.$$

4. Use the formula of exercise 3 for the Gauss-Seidel solution of the equations

$$\left. \begin{array}{l} 2x_1 + 3x_2 = 1 \\ x_1 + 4x_2 = 3 \end{array} \right\},$$

by finding the latent roots and vectors of the relevant iteration matrix and its transpose, and computing the β_k using the biorthogonal properties of the latent vectors.

5. Find the latent root of largest modulus of the Gauss-Seidel iteration matrix for equations (31). Find the iteration matrix for the Aitken double-sweep method, and show that its largest latent root is $(9 + \sqrt{17})/32$, so that the Aitken double-sweep here converges slightly more slowly than the Gauss-Seidel method.

Estimate three values of this root from the last three approximations to the solution given in equation (32).

6. Why cannot we say that $\bar{y}' L D^{-1} L' y$ in equation (43) is positive rather than just non-negative? For the matrix of equations (31) find a non-null y for which this quantity is zero.

7. The equations $Cx = d$ are transformed to $Ax = b$ by premultiplication with a suitable diagonal matrix E, so that $A = EC$, $b = Ed$. Show that the respective rates of convergence for the Jacobi method and the Gauss-Seidel method are the same for both sets of equations.

If the matrix A is now symmetric and positive definite, and we carry out the transformation of equation (48) to produce the equations

$$By = c, \qquad y = D^{\frac{1}{2}}x, \qquad c = D^{-\frac{1}{2}}b,$$

show that there is still no change in the rates of convergence of the Jacobi and Gauss-Seidel iterative methods. (Hint: Note that the D in (48) is in fact the D of the expression $A = L + D + U$. Then if $A = (L_A + D_A + U_A)$, $B = (L_B + D_B + U_B)$, and

$$B = D^{-\frac{1}{2}} A D^{-\frac{1}{2}},$$

show that the Jacobi and Gauss-Seidel iteration matrices for B are similarity transformations of the corresponding matrices for A. Symmetry and positive definiteness of A are unnecessary here, except that D must not have a zero element. If it has, of course, the original methods also fail.)

8. Carry out the operations of exercise 7, and verify its conclusions, for the matrix C of exercise 2.

9. Observe that the matrix C in exercise 2 has the desirable form of equation (56), so that the optimum value of w for the extrapolated Gauss-Seidel method is given by (55). Find this value and show that the iteration can be written in the form

$$\begin{bmatrix} \frac{1}{4}(2+\sqrt{2}) & & \\ -\frac{1}{2} & \frac{1}{4}(2+\sqrt{2}) & \\ -1 & 0 & \frac{1}{4}(2+\sqrt{2}) \end{bmatrix} \begin{bmatrix} x_1^{(r+1)} \\ x_2^{(r+1)} \\ x_3^{(r+1)} \end{bmatrix} =$$

$$= d + \begin{bmatrix} \frac{1}{4}(\sqrt{2}-2) & \frac{1}{2} & \frac{1}{4} \\ & \frac{1}{4}(\sqrt{2}-2) & 0 \\ & & \frac{1}{4}(\sqrt{2}-2) \end{bmatrix} \begin{bmatrix} x_1^{(r)} \\ x_2^{(r)} \\ x_3^{(r)} \end{bmatrix}.$$

Show, on the other hand, that the best extrapolated Jacobi method for the matrix B, obtained by the operation of exercise 7 on the matrix C of exercise 2, is the same as the ordinary Jacobi method, that is $w = 1$ in equation (50).

10. Show that in exercise 1 the writing of the equations in order of the points 1, 3, 5, 7, 9; 2, 4, 6, 8, gives rise to a matrix equation of the form

$$\left[\begin{array}{c|c} D_1 & P \\ \hline Q & D_2 \end{array}\right] \left[\begin{array}{c} x^{(1)} \\ \hline x^{(2)} \end{array}\right] = \left[\begin{array}{c} b^{(1)} \\ \hline b^{(2)} \end{array}\right]$$

where here the vector $x^{(1)}$ has components x_1, x_3, x_5, x_7, x_9, where $x^{(2)}$, $b^{(1)}$ and $b^{(2)}$ have similar obvious definitions, and where D_1 and D_2 (giving a special case of (56)) are diagonal matrices. Here all the odd points are first corrected using the 'old' even values, and the ordering of this operation is immaterial. All the new 'even' values are then obtained using the new 'odd' values, and again the ordering of the equations in this sequence is unimportant. Show that the ordering 1; 4, 2; 7, 5, 3; 8, 6; 9 gives another matrix of favourable type (56), and that the natural ordering of exercise 1, whose matrix is not in favourable form since the matrix X is not diagonal, is consistent with this.

11. Show that in exercise 1 the ordering 1, 2, 3; 6, 5, 4; 7, 8, 9 is not consistent with either of the two favourable orderings of exercise 10. (The rate of convergence cannot then be optimised by the use of equations (55).)

Show, however, that for the special case of three equations of the form (57) and exercise 2, all orderings, whether consistent or not, have the representation (56) and therefore the same rate of convergence for the Gauss-Seidel iteration.

12. The exact solution of the equations

$$
\left.\begin{aligned}
5x_1 + 7x_2 + 6x_3 + 5x_4 &= 2\cdot 3 \\
7x_1 + 10x_2 + 8x_3 + 7x_4 &= 3\cdot 2 \\
6x_1 + 8x_2 + 10x_3 + 9x_4 &= 3\cdot 3 \\
5x_1 + 7x_2 + 9x_3 + 10x_4 &= 3\cdot 1
\end{aligned}\right\}
$$

is $x_1 = x_2 = x_3 = x_4 = 0\cdot 1$.

In the Gauss-Seidel method we produced after some steps the approximation

$$
x^{(r)\prime} = (0\cdot 060,\ 0\cdot 12,\ 0\cdot 11,\ 0\cdot 10).
$$

We work in two-digit floating-point arithmetic, with method specified in the determination of $x_1^{(r+1)}$ given here by

$$
5x_1^{(r+1)} + 0\cdot 84 + 0\cdot 66 + 0\cdot 50 = 2\cdot 3.
$$

The addition $0\cdot 84 + 0\cdot 66$ gives $1\cdot 5$ (rounded if necessary), the addition of $0\cdot 50$ (rounded if necessary) gives $2\cdot 0$, and subtraction from $2\cdot 3$ gives $0\cdot 30$. So $x_1^{(r+1)} = 0\cdot 060$ (rounded if necessary).

Show that the approximation $x^{(r+1)}$ is identical with $x^{(r)}$, and that this is the exact solution of the given equations with right-hand sides changed to $2\cdot 3$, $3\cdot 2$, $3\cdot 32$ and $3\cdot 13$.

13. In the method of §§ 23–24 we take the cyclic order, changing the components of $x^{(r)}$ in turn, and at the end of the first complete step, in which all the components have been changed once, the resulting approximation is now called $x^{(r+1)}$. Show that

$$
r^{(r+1)} = \left(I - \frac{C_n C_n'}{C_n' C_n}\right)\left(I - \frac{C_{n-1}C_{n-1}'}{C_{n-1}'C_{n-1}}\right) \cdots \left(I - \frac{C_1 C_1'}{C_1' C_1}\right) r^{(r)} = Br^{(r)},
$$

where C_k is the kth column of A. Hence show that the process will generally converge. (Hint: show that each matrix $(C_k C_k')/(C_k' C_k)$ has rank 1, with just one non-zero latent root of value unity, and deduce that $\| B \|_2 < 1$.)

14. Repeat the calculation of the problem of equation (102), working with rounded arithmetic, proceeding as far as the evaluation of $x^{(4)}$.

15. For a connexion between the Gauss-Seidel iteration and the determination of latent roots and vectors by direct iteration see exercise 7 of Chapter 9.

16. For a connexion between the method of conjugate gradients and the Lanczos method for determining latent roots and vectors see exercise 8 of Chapter 10.

Chapter 9

1. Using the first Gershgorin theorem (see page 276) show that the latent roots of the matrix A of equation (8) lie in the range $-4 < \lambda < 7$. Evaluate the determinant of $A - Ip$ for four equidistant values of p covering this range, and by inverse interpolation in the table of $|A - Ip|$ find the value of λ of largest modulus.

2. If $f(\lambda) = \lambda^n + a_1\lambda^{n-1} + a_2\lambda^{n-2} + \ldots + a_{n-1}\lambda + a_n$ is the characteristic polynomial of a matrix A, so that $f(\lambda_r) = 0$ for each latent root λ_r, show that

$$A^n + a_1 A^{n-1} + a_2 A^{n-2} + \ldots + a_{n-1}A + a_n I = 0.$$

(Hint: show that, for an arbitrary vector y defined by equation (3),

$$f(A)y = \sum f(\lambda_r)\alpha_r x^{(r)}.)$$

3. The matrix A has real roots, the two of largest modulus being equal and opposite in sign. Show how to find these roots, and the corresponding vectors, by direct iteration. Apply the method to the matrix

$$A_1 = \begin{bmatrix} 5 & 5 & -7 \\ 5 & -7 & 5 \\ -7 & 5 & 5 \end{bmatrix},$$

starting with the vector $y^{(0)\prime} = (1, 1, -2)$, and deduce the root of smallest modulus.

Find, by direct iteration, the nature of the roots of the matrix

$$A_2 = \begin{bmatrix} 3 & -1 & -1 \\ -1 & 3 & -1 \\ -1 & -1 & 3 \end{bmatrix}.$$

4. Note that in the computation of b and c in §8 the consistency of the results obtained using any two components and those obtained from the least squares solution is no guarantee of accuracy. Verify this remark by computing another vector in the sequence of equation (18) and finding new values of b and c.

5. Find an appropriate value of q, and perform direct iteration with $(A_1 - Iq)$ in exercise 3 to find directly the algebraically largest root and corresponding vector.

6. If the columns of the matrix A are denoted by $c^{(1)}, c^{(2)}, \ldots, c^{(n)}$, show that A^2 has columns $(Ac^{(1)}, Ac^{(2)}, \ldots, Ac^{(n)})$, and in general A^{2^p} has columns $(A^{2^p-1}c^{(1)}, A^{2^p-1}c^{(2)}, \ldots, A^{2^p-1}c^{(n)})$. Show that if $c^{(i)} = \sum_{k=1}^{n} \alpha_{ik}x^{(k)}$, where $x^{(k)}$ is a latent vector of A, and the latent root of largest modulus is real, then

$$A^{2^p} \to \lambda_1^{2^p-1}(\alpha_{11}x^{(1)}, \alpha_{21}x^{(1)}, \ldots, \alpha_{n1}x^{(1)}), \quad |\lambda_1| > |\lambda_2| > \ldots > |\lambda_n|,$$

so that each column of A^{2^p} tends to a multiple of the latent vector corresponding to λ_1, and λ_1 can be estimated from the ratio of corresponding terms for different values of p.

Show that the rate of convergence depends on the ratios $(\lambda_r/\lambda_1)^{2^p-1}$, compared with $(\lambda_r/\lambda_1)^p$ for the method of § 3. Find by this process of matrix squaring a two-figure approximation to the largest latent root and corresponding vector for the matrix of equation (8).

7. Show that the Gauss-Seidel iteration can be expressed in the form

$$\left[\begin{array}{c} 1 \\ \hline x^{(r+1)} \end{array}\right] = \left[\begin{array}{c|c} 1 & o' \\ \hline (L+D)^{-1}b & -(L+D)^{-1}U \end{array}\right]\left[\begin{array}{c} 1 \\ \hline x^{(r)} \end{array}\right],$$

or $y^{(r+1)} = Cy^{(r)}$. Then $y^{(2^p)} = C^{2^p}y^{(0)}$, and the standard iterative method can be replaced by a process of matrix squaring.

Show that one latent root of C is unity and the others are those of

$$-(L+D)^{-1}U;$$

that the latent vector corresponding to the unit root is the vector $[1|x']'$, where x is the solution of $Ax = b$; that the other latent vectors all have a zero in their first component; and that only the first column of C has a non-zero component of the vector corresponding to the unit root in its expression in terms of latent vectors. Hence show that the first column of C^{2^p} tends to the required solution $[1|x']'$, and that all the other columns tend to null columns.

Perform a few steps of the computation for this method applied to equations (31), page 195. (Hint: the terms $(L+D)^{-1}b$, $-(L+D)^{-1}U$ in the matrix C are obtained most easily by recording $[(L+D), b, -U]$ and reducing $(L+D)$ to a diagonal by elimination, with corresponding operations on b and $-U$.)

8. Show that the general solution of the *recurrence relation* (or *difference equation*)

$$y_r + p_1 y_{r-1} + p_2 y_{r-2} + \ldots + p_n y_{r-n} = 0$$

is

$$y_r = A_1 \lambda_1^r + A_2 \lambda_2^r + \ldots + A_n \lambda_n^r,$$

where $\lambda_1, \lambda_2, \ldots, \lambda_n$ are the roots of the polynomial equation

$$\lambda^n + p_1 \lambda^{n-1} + p_2 \lambda^{n-2} + \ldots + p_{n-1}\lambda + p_n = 0,$$

and $A_1, \ldots, A_n$ are arbitrary constants whose values could, if required, be computed from a knowledge of the n initial values $y_0, y_1, \ldots, y_{n-1}$.

If the root λ_1 of largest modulus is real, show that, for any starting conditions, successive values $y_n, y_{n+1}, \ldots$ obtained by the recurrence will ultimately have the behaviour $y_r \to \alpha_1 \lambda_1^r$, so that $y_{r+1}/y_r \to \lambda_1$.

Consider the matrix decomposition

$$\begin{bmatrix} b & c & & & \\ a & b & c & & \\ & a & b & c & \\ & & a & b & c \\ & & & & \cdots \end{bmatrix} = \begin{bmatrix} 1 & & & \\ l_2 & 1 & & \\ & l_3 & 1 & \\ & & l_4 & 1 \\ & & & \cdots \end{bmatrix}\begin{bmatrix} u_1 & v_1 & & \\ & u_2 & v_2 & \\ & & u_3 & v_3 \\ & & & u_4 & v_4 \\ & & & \cdots \end{bmatrix}.$$

If $b^2 > 4ac$, show that $l_r \to -\lambda_2$, $u_r/v_r \to -\lambda_1$, where $|\lambda_1| > |\lambda_2|$ and λ_1 and λ_2 are roots of the quadratic equation $a + b\lambda + c\lambda^2 = 0$. (Hint: use the ideas of this exercise and the results of exercise 12 of Chapter 4.)

9. For the matrix A_2 in exercise 3 perform one step of inverse iteration with the matrix $A_2 - 3I$ for each of the starting vectors

$$y^{(0)\prime} = (1, -1, 0), \quad y^{(0)\prime} = (0, 1, -1).$$

What can you deduce about the latent roots and vectors of A_2?

10. Prove (as distinct from 'verify') the result of equations (50).

11. Using exact arithmetic, show that inverse iteration fails to find the latent vector corresponding to the latent root of smallest modulus for the matrix A_1 of exercise 3, with starting vector $(1, -1, 0)$. What can we deduce from the inverse iteration? Note that we shall usually succeed ultimately with rounded arithmetic.

12. By using suitable row and column permuting matrices, show that the analysis of the deflation method of §§ 21–25 can be expressed in such a way that, in any relevant deflation, the 'pivotal' row and column is always the first of the corresponding "reduced" matrix. (Hint: if $I_{r,s}$ is the permuting matrix, the matrix $I_{r,s} A I_{r,s}$ is a similarity transformation of A.) Record all the relevant computations, in this form, for the example of § 23.

13. In the deflation process of § 21 the computed $x^{(1)}$ has small errors, represented by a small contamination with $x^{(2)}$, so that instead of the true $x^{(1)}$ we use the vector $y = \alpha_1 x^{(1)} + \alpha_2 x^{(2)}$. If y is normalised to have its first component unity, so that for its remaining $(n-1)$ vector we take

$$\eta = \alpha_1 \xi^{(1)} + \alpha_2 \xi^{(2)},$$

with $\alpha_1 + \alpha_2 = 1$, show that

$$(B - \eta R_1')(\xi^{(r)} - \eta) = \lambda_r(\xi^{(r)} - \eta) - \alpha_1\alpha_2(\lambda_1 - \lambda_2)(\xi^{(1)} - \xi^{(2)}),$$

for $r = 1, 2, ..., n$.

14. Suppose that, for an unsymmetric matrix A, we have computed one latent root λ_1 and both the latent vectors $x^{(1)}$ of A and $y^{(1)}$ of A'. Both $x^{(1)}$ and $y^{(1)}$ are normalised to have unit first components, with 'remainders' $\xi^{(1)}$ and $\eta^{(1)}$. Show that, with the partitioning of equation (57),

$$(B - C_1\eta^{(1)\prime})\xi^{(r)} = \lambda_r\xi^{(r)}, \quad \text{for } r \neq 1,$$

and find the latent vectors of the matrix $(B - C_1\eta^{(1)\prime})'$.

15. Prove all the results given on page 236 for the Feller-Forsythe transformation.

16. Prove the results of Wilkinson at the bottom of page 235. (Hint: we have A with root $\mu + i\nu$ and vector $\alpha + i\beta$ with maximum element in rth position, A_1 with root $\mu - i\nu$ and vector β with maximum element in sth position, and A_2 with root λ and vector $z^{(3)}$. Building up the vector $z^{(3)}$ to $x^{(3)}$ we find easily the relation $x^{(3)} = \alpha + p\beta + qz^{(3)}$, where p and q are constants and $z^{(3)}$ has zeros in positions r and s. Premultiply this equation by

R' to find the first of the linear equations for the constants, and by $(R^*)'$ to find the second.)

17. Perform two deflation processes of § 21 for the matrix

$$C = \begin{bmatrix} 0 & 1 & 1 & 1 \\ 4 & 3 & -1 & 1 \\ 8 & 7 & -1 & 3 \\ 8 & 5 & -3 & 1 \end{bmatrix},$$

given that two of its latent roots are $\lambda_1 = 5$ and $\lambda_2 = -2$ with corresponding vectors $x^{(1)'} = (1, 1, 3, 1)$ and $x^{(2)'} = (-5, 2, -1, 9)$. (See exercise 11 of Chapter 2.)

18. Show that the equation

$$\lambda^n + p_1\lambda^{n-1} + p_2\lambda^{n-2} + \dots + p_{n-1}\lambda + p_n = 0$$

is the characteristic equation of the matrix

$$A = \begin{bmatrix} 0 & 1 & 0 & \dots & & 0 \\ 0 & 0 & 1 & 0 & \dots & 0 \\ \multicolumn{6}{c}{\dots\dots\dots\dots\dots} \\ 0 & 0 & 0 & \dots & & 1 \\ -p_n & -p_{n-1} & \dots & & -p_2 & -p_1 \end{bmatrix}.$$

Use this idea to find a two-figure approximation to the root of largest modulus of the equation
$$\lambda^3 + 3\lambda^2 - 2 = 0.$$

(This is the method of Bernoulli.)

Chapter 10

1. Prove the remark at the bottom of page 240, that in the Jacobi method the sum $\sum_{s=1}^{n} (a_{ss}^{(r)})^2$ of A_r is greater than that of A_{r-1} by the amount $2\{a_{pq}^{(r-1)}\}^2$.

2. Show that in the Givens process the values of the sines and cosines used to produce zeros in the first row are obtained only from the original elements in that row. In fact if c_{2r} and s_{2r} are the values involved in the rotation in the $(2, r)$ plane, show that

$$s_{2r} = \frac{a_{1r}}{(a_{12}^2 + a_{13}^2 + \dots + a_{1r}^2)^{\frac{1}{2}}}, \qquad c_{2r} = \frac{(a_{12}^2 + a_{13}^2 + \dots + a_{1,r-1}^2)^{\frac{1}{2}}}{(a_{12}^2 + a_{13}^2 + \dots + a_{1r}^2)^{\frac{1}{2}}},$$

and that after the complete operation on the first row the new $(1, 2)$ element is $(\sum_{r=2}^{n} a_{1r}^2)^{\frac{1}{2}}$.

Show also, for a matrix of order four, that we can perform all the transformations which produce zeros in the first row in the following way, the element a_{rs} denoting the number currently in row r and column s.

(i) Subject the elements a_{22}, a_{23} and a_{33} to the row and column operations involving c_{23} and s_{23}.

(ii) Perform the row transformations involving c_{23} and s_{23} on a_{24} and a_{34}, and the column transformations involving c_{24} and s_{24} on a_{23} and a_{34}, using the symmetry of a_{23} and a_{32}. At this stage all the elements of row 3 have their final values together with a_{23} of row 2, but a_{22} and a_{24} have been affected only by the transformations involving c_{23} and s_{23}.

(iii) Perform the row and column operations involving c_{24} and s_{24} on elements a_{22}, a_{24} and a_{44}. With the computed new a_{12} this completes the transformation involving the $(2, r)$ rotations, $r = 3$, 4.

(This device saves time in minimising transfers from the backing store to the main store, and a similar process can be used with Householder's method.)

3. Perform the escalator method for the matrix of equation (32), as far as the computation of the final characteristic equation (29) and the expression (30) for the vectors. Use the Newton iterative method to compute a three-figure approximation to the smallest latent root and corresponding latent vector.

4. By drawing suitable curves of $f_r(\lambda)$ against λ, where $f_r(\lambda)$ is defined in § 7, verify the following distribution of signs for the case $n = 3$ of a symmetric matrix, where λ_1 is the latent root of the leading submatrix of order 1, $\mu_2 > \mu_1$ are the two roots of the leading submatrix of order 2, $\nu_3 > \nu_2 > \nu_1$ are the three roots of the leading submatrix of order 3, and all these numbers are distinct.

If $\nu_3 > \mu_2 > \lambda_1 > \nu_2 > \mu_1 > \nu_1$, then

	$f_0(p)$	$f_1(p)$	$f_2(p)$	$f_3(p)$
$p > \nu_3$	$+$	$-$	$+$	$-$
$\mu_2 < p < \nu_3$	$+$	$-$	$+$	$+$
$\lambda_1 < p < \mu_2$	$+$	$-$	$-$	$+$
$\nu_2 < p < \lambda_1$	$+$	$+$	$-$	$+$
$\mu_1 < p < \nu_2$	$+$	$+$	$-$	$-$
$\nu_1 < p < \mu_1$	$+$	$+$	$+$	$-$
$p < \nu_1$	$+$	$+$	$+$	$+$

Form a similar table for any other possible ordering of the λ, μ and ν.

5. For the matrix of equation (58) find the value of the first significant figure of the second smallest latent root, in the algebraic sense, using the Stürm sequence property.

6. Find the Lanczos similarity transformation for the matrix

$$A = \begin{bmatrix} 5 & -1 & -2 \\ -1 & 3 & -2 \\ -2 & -2 & 5 \end{bmatrix},$$

with starting vectors $y^{(0)} = o$, $y^{(1)\prime} = (1, -1, \frac{1}{2})$.

7. Find the Lanczos similarity transformation for the matrix

$$A = \begin{bmatrix} -2 & -1 & 4 \\ 2 & 1 & -2 \\ -1 & -1 & 3 \end{bmatrix},$$

with starting vectors $y^{(0)} = o$, $y^{(1)'} = (1, 0, 1)$, $z^{(0)} = o$, $z^{(1)'} = (1, 0, 0)$.

8. There is a close connexion between the Lanczos method for symmetric matrices and the method of conjugate gradients for solving symmetric linear equations. Consider the formulae of equation (90), page 208. By eliminating the vectors $w^{(r)}$, show that the residual vectors for the equation $Ax = b$ satisfy the three-term recurrence relation

$$r^{(r+1)} = \left(1 + \frac{\alpha_r \beta_{r-1}}{\alpha_{r-1}}\right) r^{(r)} - \alpha_r A r^{(r)} - \frac{\alpha_r \beta_{r-1}}{\alpha_{r-1}} r^{(r-1)},$$

where α_r and β_r are defined in equation (90). Since $r^{(r+1)}$ is orthogonal to all previous residual vectors (page 209) show that

$$p_r = \alpha_r^{-1} + \beta_{r-1}\alpha_{r-1}^{-1} = \frac{r^{(r)'}Ar^{(r)}}{r^{(r)'}r^{(r)}}, \qquad q_r = \beta_{r-1}\alpha_{r-1}^{-1} = -\frac{r^{(r-1)'}Ar^{(r)}}{r^{(r-1)'}r^{(r-1)}}.$$

Show also that the conjugate gradient method produces the similarity transformation $AR = RB$, where the columns of R are the residual vectors $r^{(0)}, r^{(1)}, \ldots, r^{(n-1)}$, and

$$B = \begin{bmatrix} p_0 & -q_1 & & & & \\ -\alpha_0^{-1} & p_1 & -q_2 & & & \\ & -\alpha_1^{-1} & p_2 & -q_3 & & \\ & & \cdots\cdots\cdots & -\alpha_{n-3}^{-1} & p_{n-2} & -q_{n-1} \\ & & & & -\alpha_{n-2}^{-1} & p_{n-1} \end{bmatrix}.$$

Confirm these results from the computations of pages 210, 211.

9. Perform three steps of the Cholesky-type decomposition for the L–R method for the matrix

$$A_1 = \begin{bmatrix} 2 & -1 & 0 \\ -1 & 2 & -1 \\ 0 & -1 & 2 \end{bmatrix}.$$

The matrix A_4 is

$$\begin{bmatrix} 3 & -\sqrt{\frac{3}{7}} & 0 \\ -\sqrt{\frac{3}{7}} & \frac{7}{3} & -\sqrt{\frac{8}{63}} \\ 0 & -\sqrt{\frac{8}{63}} & \frac{2}{3} \end{bmatrix},$$

and the diagonal elements are approximations to the latent roots $2+\sqrt{2}$, 2, $2-\sqrt{2}$. Repeat the process for the matrix $(A_1 - 0 \cdot 5I)$, noting the faster convergence to the smallest root.

10. Express the analysis of the method of § 32 in terms of the row-permuting matrices of exercise 2 of Chapter 3. Transform to upper Hessenberg form the matrix

$$\begin{bmatrix} 2 & 1 & 1 & 1 \\ 1 & -2 & 3 & -4 \\ 1 & -1 & 2 & -1 \\ 1 & 1 & 1 & 1 \end{bmatrix}$$

by the method of § 32, and by the method of § 34 find an approximation to the smallest latent root.

11. Using the results of exercise 10 perform the deflation transformation of page 273, and find an approximation to the latent vector corresponding to the approximate smallest latent root.

Chapter 11

1. Prove the two theorems of Gershgorin quoted in § 3. (Hint: for the first theorem, suppose a latent vector normalised to have its largest component unity, and suppose this is the rth, then consider the rth equation of $Ax = \lambda x$. For the second, write $A = D + (L + U)$, and study the continuous variation in the eigenvalues of the matrix $D + k(L + U)$ as k varies from 0 to 1.)

2. Show that the condition number for the real root of the matrix (17) on p. 219 is $q = \frac{1}{6}$.

3. Show that the reciprocals of the condition numbers for the latent roots 2, 1, -1 of the matrix

$$A = \begin{bmatrix} -2 & -1 & 4 \\ 2 & 1 & -2 \\ -1 & -1 & 3 \end{bmatrix}$$

are respectively $2\sqrt{3}$, 3 and 2. Verify that the perturbed matrix

$$A + \epsilon B = \begin{bmatrix} -2 \cdot 1 & -0 \cdot 9 & 3 \cdot 9 \\ 1 \cdot 9 & 1 \cdot 1 & -2 \cdot 1 \\ -0 \cdot 9 & -1 \cdot 1 & 3 \cdot 1 \end{bmatrix},$$

for which $|\epsilon| \leq 0 \cdot 1$, has largest latent root of about 2·62, considerably changed from that of A.

4. Starting with the approximate modal matrix

$$X_0 = \begin{bmatrix} 1 & 1 & 1 \\ 0 & 1 & -1 \\ 1 & 1 & 0 \end{bmatrix},$$

estimate approximations to the latent roots and a better modal matrix X_1 for the matrix $A + \epsilon B$ in exercise 3. Find also better approximations to the latent roots. (Note the *quadratic* convergence, the first approximations to the roots being 2·8, 0·7, −1·4 and the second 2·616, 0·837, −1·353 with almost three figures correct.)

5. The matrix

$$A = \begin{bmatrix} 1 & 1 & 3 \\ 1 & -2 & 1 \\ 3 & 1 & 3 \end{bmatrix}$$

has approximate latent roots and unnormalized vectors

$$\lambda_1 = 5\cdot4, \qquad x^{(1)\prime} = (0\cdot7,\ 0\cdot2,\ 1\cdot0)$$
$$\lambda_2 = -2\cdot3, \qquad x^{(2)\prime} = (-0\cdot3,\ 1\cdot0,\ 0\cdot0)$$
$$\lambda_3 = -1\cdot1, \qquad x^{(3)\prime} = (1,\ 0\cdot3,\ -0\cdot8).$$

Show from the analysis of § 7 that the upper bounds for the errors in the latent roots are respectively 0·22, 0·10, and 0·05 to two decimal places. Show also that the Rayleigh quotient gives for the largest latent root the value 5·418 with upper bound of 0·008 for its error, and that the norm of the error of our first approximate latent vector is little greater than 0·04.

6. For the matrix $A + \epsilon B = C$ of exercise 3, and with the approximate modal matrix X_0 of exercise 4, we find

$$X_0^{-1}CX_0 = \begin{bmatrix} 2\cdot8 & 0\cdot4 & 0\cdot8 \\ -0\cdot6 & 0\cdot7 & -0\cdot6 \\ -0\cdot4 & -0\cdot2 & -1\cdot4 \end{bmatrix},$$

and the diagonal elements are our approximations to the latent roots. Show that the Gershgorin circle for the smallest (algebraic) root does not intersect those for the two larger roots.

Find a similarity transformation which reduces to a minimum the Gershgorin estimate of the error of the smallest root, and give this error bound. (Note that the approximation (38) is not valid, since the off-diagonal terms are too large.)

Repeat this process for a modal matrix X_1 which gives

$$X_1^{-1}CX_1 = \begin{bmatrix} 2\cdot616 & -0\cdot034 & 0\cdot151 \\ -0\cdot007 & 0\cdot837 & 0\cdot144 \\ 0\cdot039 & 0\cdot086 & -1\cdot353 \end{bmatrix},$$

showing that the root λ_3 is in the range $-1\cdot353 \pm 0\cdot008$, and that the approximation (38) is here nearly the optimum.

7. If A and B are symmetric, with largest latent roots $\lambda_1(A)$ and $\lambda_1(B)$, prove that, for $X = A - B$, we have $\lambda_1(X) > |\lambda_1(A) - \lambda_1(B)|$. (Hint: use norms. This is a special case of equation (42).)